AF749685

Mae'r eitem yma ar gael i chi hyd y dyddiad neu'r amser a nodir isod, oni bai fod darllenydd arall yn ei alw'n ôl ynghynt.

This item is on loan to you until the date or time shown below, unless it is recalled sooner by another borrower.

16.5.06

19-11-07

27.3.09.

2/6/10

22/3

9/5/11

ENGINEERED COASTS

Coastal Systems and Continental Margins

VOLUME 6

Series Editor

Bilal U. Haq

Editorial Advisory Board

M. Collins, *Dept. of Oceanography, University of Southampton, U.K.*
D. Eisma, *Emeritus Professor, Utrecht University and Netherlands Institute for Sea Research, Texel, The Netherlands*
K.E. Louden, *Dept. of Oceanography, Dalhousie University, Halifax, NS, Canada*
J.D. Milliman, *School of Marine Science, The College of William & Mary, Gloucester Point, VA, U.S.A.*
H.W. Posamentier, *Anadarko Canada Corporation, Calgary, AB, Canada*
A. Watts, *Dept. of Earth Sciences, University of Oxford, U.K.*

Engineered Coasts

Edited by

Jiyu Chen

State Key Laboratory of Estuarine and Coastal Research,
East China Normal University, Shanghai, P.R. China

Doeke Eisma

Emeritus Professor, Utrecht University
and Netherlands Institute for Sea Research,
Texel, The Netherlands

Kenji Hotta

College of Science and Technology,
Nihon University, Chiba, Japan

and

H. Jesse Walker

Department of Geography and Anthropology,
Louisiana State University, Baton Rouge, U.S.A.

KLUWER ACADEMIC PUBLISHERS

DORDRECHT / BOSTON / LONDON

A C.I.P. Catalogue record for this book is available from the Library of Congress.

UWB Wolfson Library
TC330 .E53 2002
30110007360903

ISBN 1-4020-0521-0

Published by Kluwer Academic Publishers,
P.O. Box 17, 3300 AA Dordrecht, The Netherlands.

Sold and distributed in North, Central and South America
by Kluwer Academic Publishers,
101 Philip Drive, Norwell, MA 02061, U.S.A.

In all other countries, sold and distributed
by Kluwer Academic Publishers,
P.O. Box 322, 3300 AH Dordrecht, The Netherlands.

Printed on acid-free paper

All Rights Reserved
© 2002 Kluwer Academic Publishers
No part of this work may be reproduced, stored in a retrieval system, or transmitted
in any form or by any means, electronic, mechanical, photocopying, microfilming, recording
or otherwise, without written permission from the Publisher, with the exception
of any material supplied specifically for the purpose of being entered
and executed on a computer system, for exclusive use by the purchaser of the work.

Printed in the Netherlands.

TABLE OF CONTENTS

INTRODUCTION

This book is concerned with the types of structure that are found today along the sea coasts of the world: dikes, dams, artificial channels, harbours, storm barriers, artificial islands, reclamation works and artificially nourished beaches. The structures discussed, because their size and/or methods of construction, have had major impacts on their respective coastal zones. The modifications emphasized are scattered along the coasts of China, Japan, the United States and western Europe (England and The Netherlands).

The motive for writing this book was the realization that coastal areas will be more intensively exploited during this century than they have been in the past because of economic and demographic demands which involve an increasing need for new land, better marine facilities for trade and other communications, and better protection against the sea. This will not only lead to increased development of the coastal area and reclamation of shallow sea areas but also to a greater expansion of activities into deeper water including the construction of artificial platforms and islands on the continental shelf. This book presents a survey of existing structures as well as a general introduction into their history, planning and construction. Technical details are not presented here – with some minor exceptions – as it was not our intention to write a technical manual; i.e. one that deals with advanced techniques and advanced design including physical needs and conditions in particular areas. In addition, this book does not go into the details of the economic, legal, social and political factors that must be considered before artificial engineering of coastal areas is undertaken. It does, however, provide a series of studies that describes the history, scope and ramifications of a number of major coastal engineering projects completed or under construction to date.

Each chapter summarizes the experience gained in specific areas, which in some cases (e.g. China and The Netherlands) covers several centuries. Emphasis is on recent large construction efforts in China, Japan, the United States and The Netherlands. The discussion of various structures in these countries is divided into sections covering coastal protection, damming of bays or straits, artificial islands and beaches, artificial channels, and large projects in deltas or bays. The information presented will provide a basis for those individuals who are engaged in coastal studies, for those involved in planning and constructing coastal engineering projects, and for all those who have a vested interest in coastal environments.

LIST OF AUTHORS

Jiyu Chen
State Key Laboratory of Estuarine and Coastal Research
East China Normal University
3663 Zhongshan North Road
Shanghai 200062
P.R. China

Zeheng Dai
Zhejiang Institute of Estuary and Coast
185 Qingjian Road
Hangzhou 310016
P.R. China

Donald W. Davis
Energy Center/Oil Spill Coordinator's Office
Louisiana State University
Baton Rouge, Louisiana 70803-4105
U.S.A.

Doeke Eisma
Paulineweg 17
1865 AD Bergen aan Zee
The Netherlands

C.W. Finkl
Coastal Education and Research
1656 Cypress Row Drive
West Palm Beach, Florida 33411
U.S.A.

Zengcui Han
Qiantang River Administration
185 Qingjian Road

Hangzhou 310016
P.R. China

Kenji Hotta
College of Science and Technology
Nihon University
7-24-1 Narashinodai Funabashi-shi
Chiba 274-8501
Japan

Wei Jiang
Zhejiang Institute of Estuary and Coast
185 Qianjiang Road
Hangzhou 310016
P.R. China

Yaping Lei
Institute of Coastal and Estuarine Studies
Zhongshan University
Guangzhou 510275
P.R. China

Daoji Li
State Key Laboratory of Estuarine and Coastal Research
East China Normal University
3663 Zhongshan North Road
Shanghai 200062
P.R. China

G.J. Schiereck
Ministerie van Verkeer en Waterstaat
Afd. Q
Postbus 20901
2500 EX Den Haag
The Netherlands

Marcel Stive
Netherlands Centre for Coastal Research
Delft University of Technology
Faculty of Civil Engineering
P.O. Box 5048
2600 GA Delft
The Netherlands

Cunhuan Tao
Qiantang River Administration
185 Qingjiang Road
Hangzhou 310016
P.R. China

Jan van de Graaff
Delft University of Technology
Stevinweg 1
2628 CN Delft
The Netherlands

Charles J. Vos†
Delft University of Technology
Faculty of Civil Engineering
P.O. Box 5048
2600 GA Delft
The Netherlands

H. Jesse Walker
Department of Geography and Anthropology
Louisiana State University
Baton Rouge, Louisiana 70803-4105
U.S.A.

R.E. Waterman
Netherlands Centre for Coastal Research
Delft University of Technology
Faculty of Civil Engineering
P.O. Box 5048
2600 GA Delft
The Netherlands

Chao-yu Wu
Institute of Coastal and Estuarine Studies
Zhongshan University
Guangzhou 510275
P.R. China

BEACH NOURISHMENT

C. W. FINKL
Coastal Education and Research
1656 Cypress Row Drive
West Palm Beach, Florida 33411, U.S.A.

H. J. WALKER
Department of Geography and Anthropology
Louisiana State University
Baton Rouge, LA 70803—4105

1. Introduction

The coastal zone has long been one of the most intensely utilized segments of the landscape. Its juxtaposition between land and sea and its varied and abundant resources have been so attractive to humans that today within the industrialized countries of the world some 50% of the population lives near a coast[1]. With such a concentration of people within such a finite area, coastal problems including resource destruction, flooding, pollution, and shoreline erosion are becoming increasingly important. Although these problems are very serious along many coasts today, there is concern that they will be amplified in the near future because of global warming induced acceleration in the rate of sea-level rise[2].

Because coastal erosion impinges directly on many of the industrial, commercial, transportational, residential, and recreational (Fig. 1) constructs of society, it has been receiving increasing amounts of attention. It has been customary to combat coastal erosion by the use of "hard structures" such as sea walls, breakwaters, groins, and jetties. These structures are usually expensive and unsightly, and frequently have aggravated the erosion problem they were designed to solve.

In recent years a new technique for mitigating erosion, especially along sandy shores, has been developed. This technique stems from the fact that beaches themselves serve to protect against shoreline erosion and follows logically from the presently advocated notion that "...the best protection is afforded by using methods as similar as possible to natural ones"[3]. Often labeled "soft engineering" it involves the replenishment of beach materials along eroding shorelines. It has become the favored procedure by many coastal engineers, scientists, and coastal managers and in many countries. However, it is still evolving as a subfield of coastal engineering.

2. The World's Beaches

Beaches are composed of unconsolidated materials that border water bodies. Although the textural range in beaches may be large, it is sand that is most commonly meant when beach

J. Chen et al. (eds.), Engineered Coasts, 1–22.
© 2002 *Kluwer Academic Publishers. Printed in the Netherlands.*

Fig. 1. Beach erosion at Holly Beach, Louisiana, U.S.A.*

nourishment is being considered (Fig. 2). There are exceptions, for example, the shingle (gravel) that exists on some beaches such as those in Massachusetts, U.S.A. (Fig. 3) and Chesil Beach in southern England and boulders as in northern Baja California, Brazil (Fig. 4), and Wales, U.K.

2.1. SEDIMENT SOURCES

Unconsolidated sediments along shorelines have been extensive since sea level reached its nearly still-stand position some 4000-6000 years ago. During much of that period (especially the early part) the supply of sediment insured that beaches increased in size and distribution. Sand was sufficiently abundant so that along many coastal sectors progradation resulted in the development of coastal plains, beach ridges, offshore barriers, and dune systems[4]. These sediments come from a number of sources, including especially: inland deposits which are carried shoreward by rivers and winds, updrift cliffs from which eroded material is transported by longshore currents, and offshore from where deposits are carried onshore by waves and currents.

These sources are still important although, along most coastal sectors, contributions from them have decreased drastically. Indeed, Russell maintained that contributions from offshore began decreasing as soon as sea level became stable. He even went so far as to

*All photographs by H. J. Walker unless otherwise noted.

Fig. 2. Sandy beach with sand fencing, Florida, USA

Fig. 3. Gravel Beach, Massachusetts, U.S.A.

Fig. 4. Boulder beach, Brazil

Fig. 5. Pocket beach with gravel and driftwood, Washington State, U.S.A.

write: "The beach dune system reached greatest volume as the still stand was approached. But once that level was attained, new sediment supplies were no longer encountered and marine processes brought about a net loss to the system"[5]. Thus, beach erosion, on a world-wide basis, is very much a natural phenomenon! It has been calculated that today 75% of the world's sandy shorelines are being eroded[6]. Leatherman in 1988 stated that for the United States the percentage is even greater than the world average and may be as much as 90%[7].

2.2. MORPHOLOGY AND COMPOSITION

Sandy shores exist along about 13% of the world's coastline[8] and exhibit many different sizes and shapes. Some of them are long and straight, e.g. Ninety Mile Beach in southeastern Australia and Padre Island in Texas, U.S.A. which is more than 200 km long[9]. Frequently, long beaches are a major feature of offshore barriers that are separated from the mainland by lagoons such as Padre Island, U.S.A.. The continuity of such beaches is often broken by inlets. At the other extreme are those small beaches found in coves often along cliffy and even mountainous coastlines, known as pocket beaches. They tend to be isolated between rocky headlands (Fig. 5).

The beach environment in cross-section consists of a number of components (Fig. 6). The beach proper extends from the low-water line inland to a position established by sand dunes, permanent vegetation, storm wave action, or a change in material. The beach is often divided into the foreshore and backshore, a separation established by frequency of wave action. From the standpoint of beach nourishment the zones both seaward of the low-water line and landward of the backshore are extremely important.

Beaches are depositional features and exist only where material is available. Although quartz is the dominant mineral of most beaches, some have sizable fractions of other minerals including shell, coral, and algal fragments and other biotic matter[10].

3. Needs For Beach Nourishment

Beach erosion is a common problem along shorelines and yet it is often difficult to recognize in the field unless there are obvious indications of sediment removal. The development of beach scarps (Fig. 7), the presence of tree stumps or marsh muds on the beachface (Fig. 8), and the location or damage of buildings (Fig. 1) precariously close to uprush levels are all signs that beaches are moving landward due to sediment loss. The

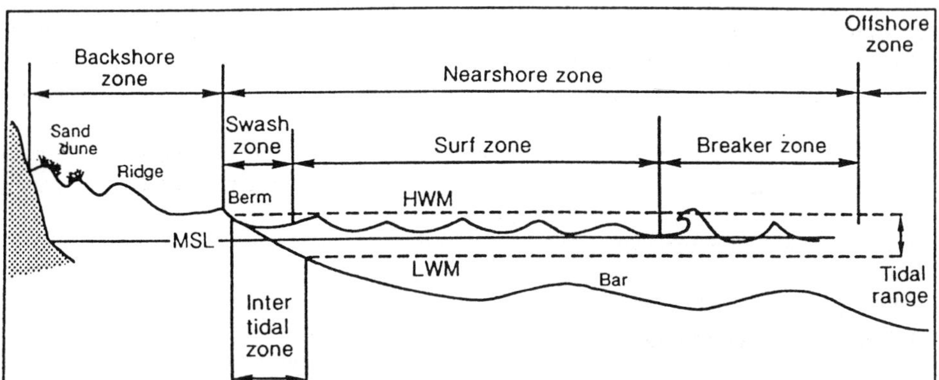

Fig. 6. Cross-section of a typical sandy shore zone. Modified from Viles and Spencer[2].

removal of beach materials is by wave action, tidal or littoral currents, or wind. A range of countermeasures provide protection from beach erosion, foremost among them during the last quarter of the 20th century, being artificial nourishment (i.e., the mechanical placement of sand on the beach).

Beach protection measures are necessary because beaches are important natural resources that support multipurpose activities. When well maintained, beaches provide storm surge protection, flood control, recreational activities, and habitat for numerous species of plants and animals[11]. Lack of proper coastal maintenance may allow beach erosion to reduce natural upland protection, increase loss of natural habitats, degrade a major source of revenue, and shrink the overall economy[12,13]. Beaches thus need to be protected because they reduce vulnerability to coastal development in high hazard areas.

Although beaches provide a measure of protection to the shore from damage by coastal storms and hurricanes (typhoons and tropical cyclones), their effectiveness as natural barriers against surge flooding depends on their size and shape and on the severity of storms. Beaches are also highly valued as recreational resources that contribute to the economic well being of many coastal regions in the world. The trend of increasing beachfront development since WW II has resulted in the replacement of dune systems with buildings. This practice has increased exposure of buildings to damage from natural forces. The presence of buildings close to an eroding shoreline reduces beach width which in turn adversely impacts both natural storm protection and recreational quality of affected beaches[14].

Because beaches provide natural protection from storms and have economic value, their deterioration or degradation is regarded as undesirable. The unwanted effects of beach erosion commonly place life and property at risk, usually from flooding, and decrease a community's ability to maintain a viable tourist-based economy. Commercial and residential development on upland areas behind beaches and in close proximity to eroding beachfronts are jeopardized by decreasing (eroding) beach widths. Increased potential for economic loss and safety concerns for human life thus drive desires to remediate beach erosion by artificially replacing sand that is lost to erosion.

In addition to the use of beach nourishment for combating coastal erosion, it has been advocated because: (1) it tends to be less expensive and easier to prepare than hard structures, (2) it is aesthetically more desirable (contrast Figs. 2 and 7 with 9) and it is "user friendly"[14,15], (3) it can provide a source of sand for wind-created or artificially created dunes which add to the protection of inshore areas, (4) it can often utilize the "waste products" from dredging or construction projects, (5) it adds to the sediment budget and may benefit down-drift locations unlike most hard structures, (6) it capitalizes on natural processes and thus is more acceptable to society, and (7) it restores habitat for biota, and as Healy et al. state, it may also "...appease local perception that popular beaches have been eroding within historical times…"[16].

4. Definitions, Terminology, And Models

The term beach nourishment came into general use after the first renourishment project in the United States at Coney Island, New York, in 1922[17]. In engineering parlance, beach nourishment (or replenishment) refers to the artificial placement of sand along an eroded stretch of coast where only a small beach, or no beach, previously existed. A beach nourishment project involves placement of sand on a beach to form a designed structure so that an appropriate level of protection from storms is achieved. The placement of sand is commonly by methods such as dredging sand from the seafloor, bypassing sand around obstructions along the coast, or delivery of sand from inland quarries to the coast. Although pumping of sand from offshore is the most widespread method of application, due to the large volumes of sediment that are required for most projects, other developments feature placement of sand by trucking or barging from quarries or construction sites, as well as removal of sand from dunes, or relocation of sediment on the berm via beach scraping. It is now known, however, that removal of sand from dunes is not an appropriate option for sand supply because dune and beach sediment budgets are interrelated[18]. Although the terms beach nourishment and replenishment are used interchangeably, it is implicit that the placement of new sand on the beach is by artificial (mechanical) means and not the result of natural (nourishment) processes.

Beach life spans are sometimes expressed in terms of beach lifetime categories. Beach lifetime is defined as the period between the times of initial emplacement of the sand and the earliest documented loss of at least 50% of the fill material[19]. Durability of replenished beaches can then be expressed in terms of beach half-life and percentage of fill remaining after one year. Both of these expressions are based on a linear extrapolation of beach loss but it is appreciated that under natural conditions, fill loss from artificial beaches occurs non-linearly due to the singularity of storms or other high energy crescendo events[20]. Even though reported rates of sediment loss are usually only rough estimates of volumetric attrition obtained during a limited monitoring period, the concept of lifetime categories offers a means for quantitatively representing durability of replenished beaches. By comparing projected life spans of nourishment projects to actual field experience, Leonard et al[19] found that in the United States post-replenishment loss rates are one and a half to twelve times greater than pre-replenishment loss rates, with the exception of Miami Beach, Florida, which lasted much longer than the project design life.

Several theoretical models have been developed to predict the durability of an artificial beach[21,22,23]. James[23], for example, introduced the nourishment factor (Rj), which is the ratio of the predicted erosion rate for the allochthonous beachfill relative to the existing erosion rate of natural beach material. In theory, the renourishment factor predicts the length of time, after initial replenishment, before subsequent nourishment will be required. There are, however, many variables that can affect the time span between nourishments. A single large storm occurring just after completion of a nourishment project can erode away much of the recently placed beachfill. Dean[21], examining the effect of variable fill length (i.e. the contiguous length of the beach nourishment project) on longevity (durability), derived a numerical relationship which predicts the length of time before a subsequent nourishment is required, based on the life spans of prior nourishments. Related in concept is another model[22] that predicts the time required for a specified percentage of beach fill material to be lost from a recently replenished beach. Like the nourishment factor, accuracy of these kinds of quantitative models has been questioned in relation to their value or relevance to observations of actual beach behavior on barrier islands[19].

5. Causes Of Beach Erosion

It should perhaps be emphasized that artificial beach nourishment became necessary only when beachfronts were developed for recreational, urban, industrial, and military uses. Although it is difficult to understand at once what are the reasons of the erosion, the causes can be natural or introduced by human activity along the shore. Beach erosion is a natural process that takes place in response to dynamic conditions along the coast. When induced or accelerated by engineering structures (Figs. 9 and 10), the process is sometimes referred to as structural erosion[24]. Beach erosion is influenced by such factors as uplift (e.g. neotectonism) or subsidence (e.g. groundwater withdrawal, compaction of sediments) of the land surface, change in climate patterns (especially storm frequency, deviation of prevailing wind direction), interruption of sediment supply to the coast, eustatic fluctuations of the sea surface, blockage of littoral drift, and construction on the coast. An increase in relative sea level (i.e. drowning of the coast and landward retreat of shorelines) is, however, often cited as the primary geophysical cause of beach erosion but many other factors are involved. Construction of dams on the nation's rivers withhold sediment delivery to the coast. In the case of the Mississippi River, for example, the sediment load today is about half what it was in pre-dam construction days. Further, the texture of the sediment that bypasses a dam is usually fine and therefore is more likely to be carried out to sea and lost to the coast. The dredging of deep inlets and navigational entrances and the construction of shore protection structures such as jetties are other interrelated factors that contribute to the degradation of natural beach systems. This list is by no means comprehensive and it must be concluded that the causes of beach erosion are manifestly complicated and often interrelated.

Even though a range of natural processes are at work, it is urban development along the shore that has necessitated placement of sand on eroding shores in an effort to protect infrastructure. The real problem, however, is not the natural adjustment of coastlines to fluctuating conditions and not just any kind of development but construction too close to the shore. Most shoreline development is deliberate for reasons of access, proximity, or aesthetics. Whatever the initial impetus for developing shorelines, the result has been an expensive exercise in what is usually nationally funded coastal protection. Retreat of the coast in Australia for example, is not as problematical as it is in the United States due to foresight of the Commonwealth government in establishing Crown lands along most of the coast, which kept development some distance inland. In other countries such as The Netherlands where land has been reclaimed from the sea, large dikes and other engineering structures[25] such as dunes and renourished beaches are part of the efforts to hold back the sea. Although these areas have multipurpose uses (e.g. storm protection, conservation, recreation, water catchment), they are not open to intensive urban uses[24].

Along with coastal development comes the dredging of ports and harbors and the navigational channels that serve them. Many entrances contain jetties that stabilize the cut or provide protection from waves in the channel. Jetties and deep inlets, as well as other shore protection structures such as groins (Fig. 11), interrupt the natural littoral drift by impounding sediment or causing it to be jetted offshore (by ebb-tidal flow) out of the longshore sediment transport system. It is now widely recognized that the interruption of sediment transport along the shore by artificial structures causes the downdrift shores to

Fig. 7. Beach scarp, Hawaii, U.S.A.

Fig. 8. Exposed tree roots due to beach erosion, Ravenna, Italy.

Fig. 9. Seawall and erosion, Japan.

Fig. 10. Groynes and shore protection using tetrapods, Japan.

become sediment starved which in turn results in shore erosion and retreat of the shoreline. Large littoral drift blockers can initiate downdrift erosion that propagates for several tens of kilometers[27]. In a study of 1238 km of Florida shoreline, distributed among 25 coastal counties and covering about 95% of the state's beaches, Finkl and Esteves[28] concluded that littoral drift blockers on Florida's Atlantic and Gulf coasts accounted for 72% of the state-wide beach erosion. Along the southeastern coast where there are numerous stabilized inlets, and the volume of sediment transported in the littoral drift is relatively small (e.g. < 50,000 to 100,000 $m^3\ a^{-1}$), littoral drift blockers appear to cause at least 90% of the beach erosion. Most of the beach erosion here is thus anthropogenically induced and is, at least theoretically, quite preventable from a technical point of view. In practice, however, remediation of the beach erosion problem is politically recalcitrant.

Although the causes of beach erosion are complicated interacting processes, it should be noted that beach nourishment only treats erosional symptoms and does not eliminate the causes. Beach fills are sacrificial in the sense that they are not meant to be permanent solutions to the beach erosion problem. Beach fills thus provide temporary protection and it is anticipated that replenishment will be repetitive.

6. Beach Nourishment

Beach nourishment is normally along open ocean coasts but there have been projects in enclosed or semi-enclosed lakes and seas as well. Technically speaking, beaches can be comprised by a variety of grain sizes ranging from clays (muds) to cobbles and boulders. In engineering practice it is commonly understood that the artificial nourishment process refers to beaches comprised by sand-sized grains although in many parts of the world it is gravel (shingle) that is used. In the United States, Florida has more beach erosion control projects (i.e. beach nourishments) than all other states combined, although some states (e.g. California and North Carolina) have beaches that have been renourished more times than any in Florida.

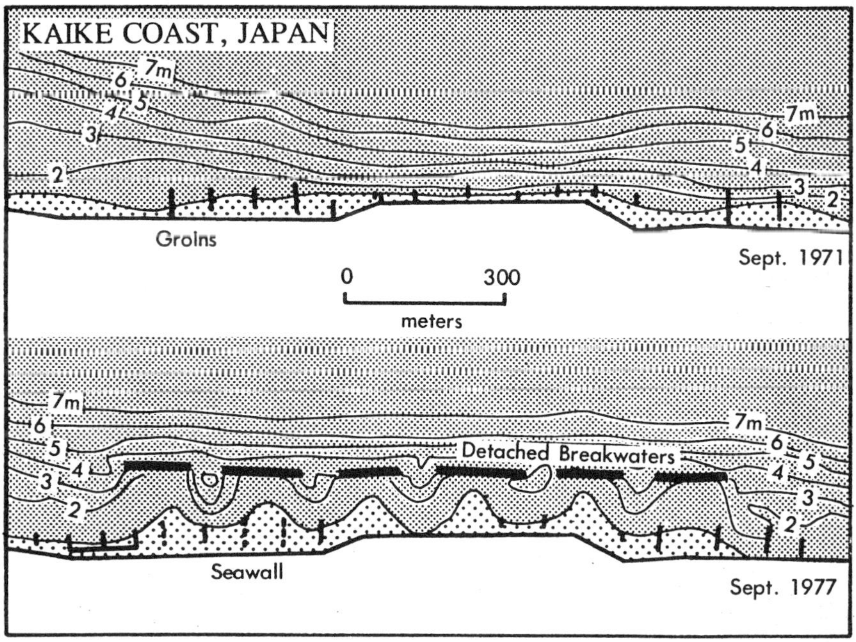

Fig. 11. Groynes, detached breakwaters, and beach development at Kaike, Japan illustrating rapidity of beach accretion. Modified from Toyoshima[26].

6.1. PROS AND CONS

Before discussing the various engineering techniques for renourishing a beach, it is perhaps worth while to consider the pros and cons of the process and its results. In the United States there are two schools of thought regarding beach nourishment; the larger group favors artificial placement of sand on eroding beaches as part of shore protection measures while the other smaller group discourages the practice on the grounds of environmental, economic, sociological, and political grounds. A recent study by the United States National Research Council (NRC)[14] examined the diversity of viewpoints about the success or failure of nourishment projects (Table 1). It was concluded that the factors involved include the large number of interested groups who have different "...viewpoints, objectives, needs, and ideas..."[14].

In the 1980s and early 1990s, there was much debate about the pros and cons of beach nourishment with many illuminating facts coming from both sides of the issue[29]. As persuasive as many arguments were, the end result was that federal support for new beach nourishment was largely withdrawn when the United States Congress removed the USACE from many projects by reducing or eliminating funding. If beach renourishment is to continue as a shore protection measure in the United States, local funds will have to support the practice along many stretches of the coast.

Proponents of beach nourishment favor continuance of the engineering practice for many reasons, the most important of which feature shore protection (mainly flood control against storm surge) and economic value in terms of income from recreational use. The arguments for beach nourishment are legion and include those factors already indicated as part of the needs for shore protection.

Antagonistic to views of beach nourishment are concepts that focus on environmental impacts of offshore dredging[15], especially near sensitive environments such as coral reefs and seagrass beds, and placement of sand along the shore which buries meiofauna and infauna, or which may adversely impact rare species such as sabellariid worm reefs. Other views focus on expenditures of public funds for coastal segments that do not provide public access to the beach or interpretations of coastal management practices that call for retreat from the coast. Other issues that are sometimes raised feature concerns about beachfill performance (i.e. durability, longevity, half-lives of replenishments) which is keyed into the design life of renourished beaches. The USACE, for example, often estimated and advertised life spans of a decade or so for many proposed projects. Studies subsequently found that, on average, the life spans of renourished beaches were usually less than anticipated. Durabilities of renourished Atlantic beaches, for example, were found to have a half life of about four to five years[30].

6.2. DESIGN OF BEACH NOURISHMENT PROJECTS

The purposes of beach nourishment projects are to: (1) increase dune and berm dimensions (i.e. height, length, and width), (2) advance the shoreline seaward, (3) reduce storm damage from flooding and wave action, and (4) widen the recreational beach area. Beach nourishment projects are complicated technical procedures that require careful preparation for successful execution of site-specific engineering design.

Table 1. Evaluation of Beach Nourishment Projects*

Objective	Criteria for Success	Measures of Performance
Provide, enhance, or maintain a recreational beach	A viable (acceptable width and carrying capacity) recreational asset during the beach-going season usually expressed as dry berm width.	Periodic survey of beach width using quantifiable observation techniques. Assessment of a number of beach visits, aerial photography useful.
Protect facilities from wave attack	Sufficient sand, gravel, or cobbles remaining in a configuration suitable to block or dissipate wave energy prior to its striking facilities, protection possibly including hard structures IN the solution.	Evaluation of structural flooding damage following storms that do not exceed the limit for which the project was designed.
Maintain an intact dune seawall system	No overtopping during a storm that does not exceed design water-level and wave-height limits.	Verification of stabilization of the shoreline position
Create, restore, or maintain beach habitat	Seasonal extremes in erosion not exceeding the design profile. Structures, if allowed, remaining intact. Postfill erosion rates comparable to historical values.	Profile surveys to establish that the amount and configuration of the sediment meet or exceed the design profile
Protect the environment	Sediment extent and condition and the vegetation of the backbeach or dune meeting environmental needs	Observations of habitat characteristics and condition.
Avoid long-term ecological changes in affected habitats	Return to prenourishment conditions within an acceptable time period	Periodic monitoring of faunal assemblages of great concern

*From National Research Council, 1995[14].

The design process specifies the quantity, configuration, and timing of sediment distribution along a specific coastal segment to provide natural storm protection or recreational area, or both. The design must consider rates of long-term erosion as well as the erosional impacts of storms and wave energy to adequately address variables associated with the quantity, quality, and placement of beachfill along the shore. As a general rule, sediment comprising the nourished beach is anticipated to erode at least as fast as the background rate of the prenourished shoreline. In practice, however, it is usually observed that sediment volume loss rates and shoreline retreat for artificial beaches are significantly greater than historical rates for the natural beach[22], even when differences in grain size and sorting are taken into account[31]. Although an allowance for continued erosion of beach fill is usually part of the design assessment, the purpose of beachfill design is to maximize the longevity of artificial beaches. The designs can only be optimized by changing the morphological configuration of the beach fill or by the choice of the fill material. The grain-size of borrow material was traditionally considered to be the most important factor for optimizing beachfill. Studies by Eitner[32], however, indicate that grain-size has little effect on beachfill longevity because grain-size influences the critical threshold stress to a lesser extent that does the grain density. Only a coarser material such as gravel, which also has a significantly higher critical threshold stress, may effectively extend beachfill longevity.

Various methods of beach nourishment design are complimentary in the overall process of optimizing project performance. Potential designs are initially evaluated at a preliminary level in which the anticipated project performance is predicted using simple and relatively inexpensive methods. After the performance characteristics are compared with project objectives, the design is refined until the performance predictions confirm an optimal design. For sites without complex boundaries (i.e. straight beaches without terminal groins, inlets, or headlands), prediction tools correctly estimate the time required for renourishment to within approximately 30% of actual project performance[14]. Subsequent to establishing the preliminary design, more sophisticated predictive methods are used to optimize the design. This bimodal approach checks preliminary and advanced methods of design, facilitates a rapid and efficient convergence to final design, and provides a clear perspective of how well the design parameters fit project requirements.

6.3. THE DESIGN BEACH

The design profile is the shore normal cross-section that the equilibrated beach is expected to take. The best estimate of this profile is obtained by the seaward transfer of the natural beach profile by the amount of beach widening that is required[33]. Estimates of beachfill volumes are generally increased if the borrow material is finer grained than the native sand. The construction profile is the cross section that the contractor is required to achieve. Because the constructed beach, which contains design fill and the advanced-fill volumes, is often steeper than the design cross section due to construction limitations, it is also usually significantly wider than the design profile. Wave action adjusts the construction cross-section to a flatter dynamically equilibrated slope within the first few months to a year after placement of the beachfill. Because the dynamically adjusted profile contains design and advanced fills, it is wider than the design width during the nourishment interval. At the time of renourishment, the design and equilibrium profile are theoretically equal[14].

Mechanical deposition of sediment along a beach nourishment site, during initial construction or renourishment, may not correspond to the natural profile of the beach at the time of placement. In the United States, use of a construction profile rather than a natural profile is the normal placement practice. It is customary for nourishment designs in the United States to establish uniform beach width along a project's length. It is also standard practice to provide sufficient sand to nourish the entire profile from the dune to the depth of significant sand removal*. Estimates of fill requirements are based on the geometric

*The depth of closure (DoC) is a term used by engineers to define the depth of active sediment movement on the seabed. Other terms that are related in concept include profile pinch-off depth, critical depth, depth of active profile, maximum depth of beach erosion, seaward limit of nearshore eroding wave processes, and seaward limit of constructive wave processes. The DoC in beachfill design is defined as "The depth of closure for a given or characteristic time interval is the most landward depth seaward of which there is no significant change in bottom elevation and no significant net sediment transport between the nearshore and the offshore[34]". This definition applies to the open coast where nearshore waves and wave-induced currents are the dominant sediment-transporting mechanisms. The definition of the DoC infers or stipulates that: (1) a DoC can be reliably identified, (2) the DoC is defined as the most landward depth at which no sediment change occurs, (3) there is an estimate of no significant change in bottom elevation and no significant net cross-shore sediment transport, (4) the DoC concept contains a time frame related to the renourishment interval or design life of the project, and (5) at the DoC cross-shore transport processes are effectively decoupled from transport processes occurring farther offshore.

transfer of the active cross-shore profile seaward by the design amount. If the beachfill grain size matches the native sand and there are no rock outcrops, seawalls, or groins, the design profile (shore-normal cross section) at each alongshore range marker (permanent locations of cross-shore survey sites are typically spaced every 330 m along Atlantic and Gulf coast beaches in Florida). Cross sections may be more closely spaced in beach nourishment project areas for better engineering control. Enough sediment is included in the design to nourish the entire profile[35].

The total sediment volume is independent of the cross-shore profile because the shape of the renourished profile is parallel and similar to the existing natural profile. Extra fill is required, however, in front of seawalls in order to achieve the proposed berm elevation. After these seawall volumes are calculated, estimates of nourishment fill volumes are based on seaward transfer of the entire profile. It is emphasized that sand is usually needed along the entire profile, both above and below the water because the beach, by definition, retains subaerial (berm) and submarine (beachface) sections. Placement of the required extra fill volume in front of seawalls typically moves the shoreline farther seaward than adjoining non-seawall areas. This design requirement, however, causes alongshore gradients in littoral drift that tend to become erosional hot spots (localized sites of increased erosion and rapid shoreline retreat[36]). An alternative to providing the additional seawall volumes is to build narrower berms in front of seawalls. Narrower berms can be used to advantage because they reduce the littoral drift gradients that are set up by overly wide sections of nourished beaches in front of seawalls. Similar levels of storm protection (for uplands) are provided by narrower berms when they are backed by seawalls compared to wider berms without them. In many instances, however, beach nourishment in front of seawalls can become problematic, especially where shoreline retreat extends landward along coastal segments adjacent to the seawall and where there is deep wave scour in front of the wall (see discussion in Kraus and Pilkey[34]).

Coastal engineers frequently use beach and dune recession models to predict beach washout and profile response seaward of seawalls. Commonly used approaches include EDUNE[37] and SBEACH[38]. These numerical models predict the evolution of the cross-shore beach profile toward the so-called equilibrium storm profile[14]. Both models are based on principles related to the disparity between actual and equilibrium (theoretical) wave energy dissipation per unit volume of water within the surf zone. For convenience of calculations, the models assume that sand eroded from the upper beach deposits offshore, with no loss or gain of material to the profile. It is well known, however, that beach sediment is often transported offshore and is lost from the littoral drift system[24]. Estimates of storm surge used in shoreline recession models, and for calculating runup, are based on USACE[39,40,41] engineering manuals. Storm-surge frequencies and extents of coastal flooding are also deployed by the Federal Emergency Management Agency (FEMA) and the National Oceanographic and Atmospheric Administration (NOAA). Storm hydrographs are thus obtained from FEMA, NOAA, and universities to generate probabilities of storm-induced shoreline recession. Wave statistics can be obtained from wave gauge records, published summaries of observations, or wave hindcast estimates such as the Wave Information Study[41].

6.4. PROTOCOLS FOR OVERFILL ON DESIGN BEACHES

Beachfill usually moves out of the nourishment area (i.e. the replenished or artificial beach) to adjacent shores or to deeper water. This process leading to a decrease in beachfill volume is simply referred to as "loss", although this sand temporarily contributes to the stability of the shore in general, but not at the original location. From the point of view of sediment transport, the sand is not lost because it is partly retained in the littoral system. From the perspective of the beach manager, migration of sediment away from the beachfill represents a loss of dry beach area.

The erosion of a nourished beach features two distinct components: (1) the linear regression of the volume of sand in the coastal profile and (2) extra erosion arising from the newly nourished shoreline which becomes more exposed (lying more seaward) than adjacent shore up- and downdrift[42]. Sediment losses alongshore as well as adaptation of the cross-shore profile are responsible for the so-called 'extra erosion.' In cases where the volume loss associated with the coastal erosion is large compared to the quantity of the beachfill and where the previous rate of erosion is known, a multiplier is used to compensate for all losses of sand. Verhagen[42], for example, suggests a value of 40% extra fill. The presence of structures such as seawalls, due to their interaction with coastal processes, may also require additional fill.

The term "advanced fill" refers to the eroded part of the beach profile before nourishment becomes necessary. The volume and areal distribution of advanced fill is estimated from analysis of the historical rate of erosion and shoreline change. The potential impact of project fill on coastal processes is an additional consideration that is taken in account. Procedures used to make these estimates include the historical shoreline change method[43] or numerical methods[44]. The historical shoreline change method assumes that the nourished beach will erode at the same rate as the prenourished beach. This method is commonly employed by beach designers (based on survey results) but can yield a significant underestimate of nourishment requirements[14].

Most long-term erosion (as opposed to episodic storm erosion or development of erosional hot spots) of a nourished beach is initiated by increased gradients of littoral drift along the project length. Major littoral drift gradients affecting the stability of nourished beaches are the preexisting background (regional) rates or historical erosion of the prenourished shore and stresses associated with the shoreline salient that was advanced seaward by the project fill. These perturbations of normal coastal processes are the cause of end losses and spreading of the fill. All of these littoral drift gradients interact with the nourished beach causing a progressive loss of fill. Exclusive consideration of the background erosion rate neglects end (and spreading) losses, which causes an underestimate of nourishment volume and overestimate of project life. Although losses from the project due to spreading cause accretion on adjacent beaches, they must be included in the advanced-fill design in order to achieve performance objectives[14].

6.5. COMPATIBILITY OF BEACHFILL VERSUS NATIVE BEACH SAND

Compatibility of beachfill materials refers to grain size, clast shape, and mineral composition. There has been much discussion of these topics, and general agreement that grain size and composition should match the native beach sand as closely as possible. An additional factor important to the compatibility of fills is the shape of clasts. This consideration becomes relevant in transitional areas such as southeastern Florida where native sands are mixtures of subangular siliciclastics and flat biogenic materials such as broken shells and Halimeda platelets. Grain size, composition, and shape are considerations that need to be included in beach nourishment projects.

The distribution of grain sizes (i.e. frequency distribution of grain-size classes) in the borrow material (nourishment sand) affects beach durability and how the nourished beach responds to high-energy conditions. Nourished sand that is finer-grained than the native beach sediment tends to form a flatter beach profile with a narrower dry subaerial beach width. Because finer-grained beach sediments have a lower angle of repose than coarser beaches, they require a greater volume of fill to provide the same amount of beach widening. Generally, with increased energy levels (i.e. higher storm surges or increased duration of storms), a greater volume of fine sand is required to protect upland property from flooding or undermining by wave action than would be required if coarser sand were used. Coastal models such as SBEACH and EDUNE take grain size into account.

The coarse sand/fine sand debate is not as clear-cut as initial deliberations might suggest. Finer sands, for example, are not always more prone to erosion during storms than coarser sands because fine sands lessen the slope of the beach which dissipates wave energy over a wider surf zone. Conversely, beaches with coarser sand tend to have steeper slopes and narrower surf zones which concentrate wave energy over a more restricted zone. When affected by the same wave conditions without storm surge, coarse-sand beach profiles show more pronounced changes during storms of limited duration and surge than fine-sand beach profiles[45]. The selection of coarser sand for nourishment may, in some circumstances, lead to greater erosion from storms of limited duration and smaller storm surges. The response of fill grain size to storm conditions is a complicated matter that requires careful consideration in design of nourishment profiles.

6.6. NOURISHMENT PROFILES

Various design schemes are used to place nourished sediment on the beach. Some of the more common approaches include: (1) placing all of the sand in a dune behind the active beach, (2) using the nourished sand to build a wider and higher berm above mean water level, (3) distributing fill material over the entire beach profile (above and below water), or (4) placing sand offshore in an artificial bar[14]. The approach taken partly depends on the location of the source material and the method of delivery to the beach (Figs. 12 and 13). If the borrow site is a quarry on land and the sand is transported by trucks to the beach, placement on the berm or in a dune is generally most economical. If the material is pumped shoreward from offshore ocean-going dredges, it is usually more practical to place the sand directly on the beach, in the nearshore zone, or to build an artificial bar. If pumped onshore in a sand-water slurry, the sand is subsequently redistributed by grader or bulldozer across the shore to form a more natural profile (Fig. 14).

The use of large dunes (i.e. man-made dikes) fronted by renourished beaches as a coastal protection measure has long been recognized in The Netherlands[46,47]. These constructed dune-beach systems are designed to withstand the 1-in-10,000 years condition

Fig. 12. Placing fill material directly onto the beach from a suction dredge. Photo courtesy of C. W. Finkl.

Fig. 13. Transport of sand by dump truck, Brazil.

of wave intensity and storm surge. This extreme level of protection is justified because entire cities lie behind the coastal defenses.

Bruun[48] advocates nourishing the entire beach profile, which he terms profile nourishment. The main advantage of this approach is that the sand is placed in approximately the same configuration as the existing profile, so that drastic initial adjustments are mostly avoided, especially the rapid erosion of the nourished berm. When wave action undermines the newly constructed berm, a beach scarp frequently forms along the length of the project fill. These scarps (Fig. 7) can pose hazards to beachgoers trying to gain access to the water from the berm. In some cases, foot traffic across the scarp tramples the steep slope to a flatter one that cuts into the beachfill. These cuts or beach tracks can provide ingressive pathways for surge and run up which in turn can accelerate erosion of the project fill volume.

Beach nourishment can be indirectly achieved by placing dredged sand in the offshore zone[49]. Dredged material is deposited in shallow water, typically using a split-hull barge, either as a mound or shaped as a long liner ridge that simulates a shore-parallel sand bar. It is anticipated that the sand deposited in the offshore mound or artificial bar will migrate onto the beach. Prior to welding onto the beachface, the bar causes waves to break farther offshore, a process that reduces the wave energy on the beach in the lee of the bar. The disposal depth of the offshore nourishment should be shallower than the seaward boundary of active sediment transport (as defined by normal to moderately elevated energy conditions) so that sediment quickly moves onto the subaerial beach.

6.7. MECHANICAL BYPASS SYSTEMS

Where there is longshore transport of sediment, shore protection structures or other coastal construction works can interrupt littoral drift flow patterns and trap sediment near structures, within navigational entrances to port and harbors, and in flood- or ebb-tidal deltas. Sediment trapping by littoral drift blockers causes downdrift beach erosion. In order to mitigate the downdrift effects of sand starvation along the coast, it is necessary to move sand around barriers in order to supply beaches with sediment. Due to losses of sediment offshore, the quantity needed for downdrift beach nourishment may be greater than the trapped sediment volume. Bypassing only the trapped sediment volume may not be sufficient to adequately maintain downdrift beaches. Some bypassing systems that are geared for normal use may be overwhelmed during large storms. Other bypassing systems function best during or immediately after storms when sediment is brought to the dredge pit area.

Fixed bypassing systems generally are less effective and more expensive to run than floating systems[48]. Most bypassing plants work at less than 50% efficiency, and some at 30%, which means that less than half of the drift is bypassed to the downdrift beaches. The combination of periodic beach replenishment and innovative bypassing techniques is an option that can restore longshore sediment transport and greatly reduce beach erosion[50]. Suggested new alternatives include mobilization of the bypass intakes on rails or cranes, implementation of jet pumps, or seabed fluidizers[51].

Several different kinds of mechanical bypassing systems are used effectively in a variety of coastal settings: (1) mobile dredges in the harbor and or entrance (e.g. Santa Cruz, California), (2) movable dredge in the lee of a detached breakwater that forms an updrift sand trap (e.g. Channel Islands and Port Hueneme, California), (3) floating dredge within an entrance using a weir jetty on the updrift side (e.g. Hillsboro Inlet, Florida; Boca Raton, Florida; Masonboro Inlet, North Carolina; Perdido Pass, Alabama), (4) fixed pump with dredge mounted on a movable boom (Lake Worth Entrance, Florida; South Lake Worth Inlet, Florida, (Fig. 15), (5) jet pumps (eductor) mounted on a movable crane, with main water supply and booster pumps in a fixed building (e.g. Indian River Inlet, Delaware)[14].

These, and other installations, and their operational performances are described in engineering and design manuals[43] which provide guidance for the design and evaluation of sand bypassing systems.

6.8. VENEER BEACH FILLS

Veneer fills are placed over a relatively large volume of material that is generally not suitable for beach nourishment. The unsatisfactory materials, which may be either grossly coarser or finer than normal beach sand remain as an underlayer beneath the beach-quality sand. Veneer beachfills are thus used in situations where beach-quality sand is not available in sufficient quantities to economically undertake a nourishment project. The usual reason for placing a veneer fill is based in economics because the cost is prohibitive if a cross section is totally built of beach-quality sand. Veneer fills are of two basic types: (1) fills where the underlying materials are coarser than typical beach sands (e.g. boulders, coral, rocks) and (2) fills where the underlying materials are finer than typical beach sands (e.g. silts or silty sands where the median grain size is much smaller than native sand). In the United States, veneer beach fills have been used in Corpus Christi, Texas; Key West, Florida; and Grand Isle, Louisiana[14]. A fundamental design problem associated with veneer fills involves selection of a veneer that is thick enough so that it will not erode away and expose the underlayer during storms or before scheduled replenishment. Although variable, depending on the local conditions, the thickness of the veneer must provide a sedimentary envelope that incorporates profile variations without compromise.

7. Conclusion

Although there are different approaches to beach nourishment the world over, the various techniques essentially involve methods of placing suitable sediment along the shore to: (1) maintain an existing but eroding beach, (2) create a new beach where none existed before, or (3) improve a degraded beach. No matter the actual beach nourishment design used, a soft engineering approach to erosion mitigation must be regarded as a temporary solution to a chronic problem. In spite of the fact that in the United States of America, for example, there has been more than a half century of experience, beach nourishment remains a procedure with unclear universal application. Experience has shown that there are no simple rules that work everywhere because it is

Fig. 14. Bulldozer and beach maintenance.

Fig. 15. South Lake Worth Inlet sand transfer plant, Florida, U.S.A. Photo courtesy of C. W. Finkl.

now widely appreciated that local site characteristics must be important criteria in successful design. Peculiarities of local conditions related to bathymetry, sediment grain size or shape and composition, exposure and orientation of the beach to prevailing and storm wind patterns, wave climate, and para- and diabathic sediment flux pathways can all affect shore erosion and beach stability. Intricacies of shore processes and their interactions with engineering works such as jetties, dredged channels, groins, and breakwaters, for example, can exacerbate natural shore erosion.

Fortunately, it is now realized that many shore protection structures are themselves the main causes of accelerated beach erosion. In southeast Florida, for example, stabilized navigational entrances (i.e. jettied tidal inlets) are responsible for about 90% of the beach erosion problem[28]. As formidable as this figure may seem, it is now evident that improved sediment bypassing at littoral drift blockers, as described by Bruun[27], can significantly enhance beach nourishment efforts by prolonging what are relatively short life spans of placed sediments.

Shore protection via beach nourishment is, however, an expensive undertaking but there often are few options that are practical. Retreat from the shore in highly developed coastal infrastructures is not possible nor is a passive approach where structures or facilities are threatened by beach erosion or coastal flooding. In the developed countries the Dutch are the only people who have taken an aggressive approach by actually reclaiming land from the seabed by diking and poldering. Elsewhere, most of the world's developed shores face the prospect of attempting to maintain present shorelines via beach nourishment.

Although beach nourishment is the shore protection measure of choice for many coastal managers, the future of the procedure in the short term (less than 50 years beyond today) may seem bright but in the long term (more than 100 years from today) the prognosis would be poor. If the natural rise in mean sea level continues to be exacerbated by human action to the point that relative sea level continues to increase, many coastal areas will experience inability to locate suitable beachfill materials in sufficient quantity and quality to support artificial nourishment. As beachfill materials become more scarce due to increased demands, project costs will escalate but cost/benefit ratios will probably be favorably maintained because of higher property values per length of coastal segment. Further, if the general rise in mean sea level accelerates as some researchers predict, renourished and constructed beaches will be no match for increased vulnerabilities from erosion and storm surges. The problem is, unfortunately, growing as more and more stretches of shore are developed.

Recommendations to improve performance of beach nourishment projects (e.g. their life span and esthetic quality) include mapping of the shore zone (both its subaerial and submarine portions) to better understand the topographic features and sediment transport pathways that are related to coastal stability. Post-project monitoring is another important step that can help assimilate factors that are related to the degradation of beachfill project life. For now, beach nourishment projects meet the needs of many coastal communities that require protection of beaches.

8. References

1. Goldberg, E.D. 1994. *Coastal zone space – Prelude to conflict?* Paris, UNESCO.
2. Viles, H. and Spencer T. 1995. *Coastal problems: geomorphology, ecology and society at the coast.* Edward Arnold, London.
3. California, State of. 1976. *Shore protection in California.* Department of Navigation and Ocean Development.
4. Bird, E.C.F. 1981. Recent changes on the world's sandy shorelines. In: Bird, E.C.F. and Koike, K. Eds. *Coastal Dynamics and Scientific Sites.* Komazawa University, Tokyo. 5-30.
5. Russell, R.J. 1967. Aspects of coastal morphology. *Geografiska Annaler.* 49A, 299-309.
6. Bird, E.C.F. 1985. Coastline changes - a global review. John Wiley - *Interscience*, Chichester, U.K.

7. Leatherman, S.P. 1988. Beach response strategies to accelerated sea-level rise. *Proceedings 2nd North American Conference on Preparing for Climate Change.* The Climate Institute, Washington, D.C. 353-358.
8. Coleman, J.M. and Murray, S.P. 1976. Coastal sciences – recent advances and future outlook. *Science, Technology, and the modern Navy.* Department of the Navy, Arlington, VA. 346-370.
9. Russell, R.J. 1958. Long, straight beaches. *Ecolgae Geologicae Helvetiae*, 51, 3, 591-598.
10. Walker, H.J. 1981. Man and shoreline modification. In: Bird, E.C.F. and Koike, K. Eds. *Coastal Dynamics and Scientific Sites.* Komazawa University, Tokyo. 55-90.
11. Wiegel, R. L. 1988. Keynote address: Some notes on beach nourishment, problems and advancement in beach nourishment. Tallahassee: Florida Shore and Beach Preservation Association, *Proceedings of Beach Preservation Technology '88*, 1-18.
12. Finkl, CW., Jnr., 1996. What might happen to America's shorelines if artificial beach replenishment is curtailed: A prognosis for southeastern Florida and other sandy regions along regressive coasts. *Journal of Coastal Research*, 12(l), iii-ix.
13. Strong, W.B., 1994. Beaches, tourism and economic development. Shore and Beach, 62(2), 6-8.
14. National Research Council (NRC), Committee on Beach Nourishment and Protection, 1995. *Beach Nourishment and Protection.* Washington, DC: National Academy Press, 334 p.
15. Nelson, W.G., 1993. *Beach restoration in the southeastern US: environmental effects and biological monitoring.* Ocean and Coastal Management, 19(2), 157-182.
16. Healy, T.R., Kirk, R.M., and deLange, W.P. 1990. Beach renourishment in New Zealand. *Journal of Coastal Research.* SI 6, 77-90.
17. Dornhelm, R.B., 1995. The Coney Island public beach and boardwalk improvement of 1923. *Shore & Beach*, 63(l), 7-11.
18. Psuty, N.P., 1988. Sediment budget and dune/beach interaction. In: Psuty, N.P., Dune/Beach Interaction. *Journal of Coastal Research*, Sl 3, 1-4.
19. Leonard, L.A., Dixon, K.L., and Pilkey, O.H. 1990. A comparison of beach replenishment on the U.S. Atlantic, Pacific, and Gulf coasts. *Journal of Coastal Research.* SI 6, 127-140.
20. Fairbridge, R.W., 1989. Crescendo events in sea-level changes. *Journal of Coastal Research*, 5(l), ii-vi.
21. Dean, R.G., 1983. Principles of beach nourishment. In: Komar, P.D. (ed.), *CRC Handbook of Coastal Processes and Erosion.* Boca Raton, Florida: CRC Press, 217-232.
22. Dean, R.G., 1988. *Engineering Design Principles. Short Course on Principles and Applications of Beach Nourishment.* Gainesville, Florida. FSBPA, 42p.
23. James, W.R. 1975. *Techniques in evaluating suitability of borrow material for beach nourishment. Vicksburg, Mississippi:* U.S. Army, Coastal Engineering Research Center, Technical Memorandum No. 60 (December).
24. Pilarczyk, K.W., 1990. *Coastal Protection. Rotterdam, The Netherlands*: Balkema, 500 p.
25. Walker, H.J., (ed.), 1988. *Artificial Structures and Shorelines.* Dordrecht, The Netherlands: Kluwer Academic Publishers, Dordrecht. 708 p.
26. Toyoshima, O. 1983. Variation of foreshore due to detached breakwaters. *Coastal Engineering, 18, 1873-1892.*
27. Bruun, P., 1995. The development of downdrift erosion. *Journal of Coastal Research*, 11(4), 1242-1257.
28. Finkl, C.W., Jnr. and Esteves, L.S., 1998. *The state of our shores: A critical evaluation of the distribution, extension, and characterization of beach erosion and protection in Florida.* Tallahassee, Florida: Florida Shore & Beach Association, 302-318.
29. Pilkey, O.H. 1990. A time to look back at beach replenishment (editorial). *Journal of Coastal Research,* 6(1), iii-vii.
30. Leonard, L.; Clayton, T.D.; Dixon, K., and Pilkey, O.H., 1989. U.S. beach nourishment experience: A comparison of the Atlantic, Pacific, and Gulf coasts. *Proceedings of Coastal Zone '89* (American Society of Civil Engineers), 1994-2006.
31. Ashley, G.M.; Halsey, S.D., and Farrell, S.C., 1987. A study of beachfill longevity: Long Beach Island, NJ. In: Kraus, N.C. (ed.), *Coastal Sediments '87.* New York: American Society of Civil Engineers, 1188-1202.
32. Eitner, V., 1996. The effect of sedimentary texture on beach fill longevity. Journal of Coastal Research, 12(2), 447-461.
33. U.S. Army Corps of Engineers (USACE), 1992. *Monitoring Coastal Projects.* Washington, DC: U.S. Government Printing Office, Engineer Regulation ER 1110-2-815 1.
34. Kraus, N.C. and Pilkey, O.H., (eds.), 1988. The Effects of Seawalls on the Beach. *Journal of Coastal Research,* Sl 4, 146 p.

35. Hanson, M.E. and Lillycrop, W.J., 1988. *Evaluation of closure depth and its role in estimating beach fill volume. Proceedings of Beach Preservation* Technology '88. Tallahassee: Florida Shore and Beach Preservation Association, 107-114.
36. Finkl, C.W., Jnr. and Kerwin, L., 1997. Emergency beach fill from glass cullet: An environmentally green management technique for mitigating erosional 'hot spots' in Florida. *Proceedings 10th National Conference on Beach Preservation Technology. Tallahassee, Florida:* Florida Shore & Beach Association, 304-319.
37. Kriebel, D.L., 1986. Verification study of a dune erosion model. *Shore and Beach*, 54(3).
38. Larson, M. and Kraus, N.C., 1990. SBEACH: *Numerical Model for Simulating Storm Induced Beach Change. Report 2*: Numerical Foundation and Model Tests. Vicksburg, Mississippi: Coastal Engineering Research Center, Technical Report CERC-89-9.
39. U.S. Army Corps of Engineers (USACE), 1984. *Shore Protection Manual.* Washington, DC: U.S. Government Printing Office, U.S. Army Corps of Engineers Publication No. 008-002-00218-9.
40. U.S. Army Corps of Engineers (USACE), 1986. *Storm Surge Analysis.* Washington, DC: U.S. Government Printing Office. Engineer Manual No. EM 1110-2-1412.
41. U.S. Army Corps of Engineers (USACE), 1989. *Water Level and Wave Heights for Coastal Engineering Design.* Washington, DC: U.S. Government Printing Office. Engineering Manual 1110-2-1414.
42. Verhagen, H.J., 1996. Analysis of beach nourishment schemes. *Journal of Coastal Research*, 12(l), 179-185.
43. U.S. Army Corps of Engineers (USACE), 1991. *Sand Bypassing System, Engineering and Design Manual.* Washington, DC: U.S. Government Printing Office. Engineer Manual No. EM 1110-2-1616.
44. Hanson, M.E. and Kraus, N.C., 1989. GENESIS: *Generalized Model for Simulating Shoreline Change. Report 1*: Reference Manual and Users Guide. Vicksburg, Mississippi: Coastal Engineering Research Center.
45. Shih, S.M. and Komar, P.D., 1994. Sediments, beach morphology and sea cliff erosion within an Oregon coastal littoral cell. *Journal of Coastal Research*, 10(l), 144-157.
46. Verhagen, H.J., 1990. Coastal protection and dune management in the Netherlands. *Journal of Coastal Research,* 6(l), 169-179.
47. Watson, I. and Finkl, C.W., 1990. State of the art in storm surge protection: The Netherlands delta project. *Journal of Coastal Research*, 6, 739-764.
48. Bruun, P. 1993. An update on sand bypassing procedures and prices. *Journal of Coastal Research*, SI 18, 277-284.
49. McLellan, T.N., 1990. Nearshore mound construction using dredged material. *Journal of Coastal Research,* SI 7, 99-107.
50. Bruun, P., 1996. Navigation and sand bypassing procedures at inlets: Technical management and cost aspects. *Journal of Coastal Research*, SI 23, 113-119.
51. Bruun, P. and Willekes, G., 1992. Bypassing and backpassing at harbors, navigation channels, and tidal entrances: Use of shallow-draft hopper dredges with pump-out capabilities. *Journal of Coastal Research*, 4(4), 687-701.

BEACH NOURISHMENT: CASE STUDIES

H.J. WALKER
Department of Geography and Anthropology
Louisiana State University
Baton Rouge, Louisiana 70803-4105, USA

C.W. FINKL
Coastal Education and Research
1656 Cypress Row Drive
West Palm Beach, Florida 33411, USA

1. Introduction

Just when humans first renourished a beach is unknown. However, the first example may well have occurred more than 2000 years ago when Cleopatra had sand from Egypt shipped to "...Turkey so that she would not have to step on foreign soil"[1]. Nevertheless, the artificial nourishment of beaches is relatively new as a method of coping with coastal erosion.

One of the earliest voices advocating beach nourishment in the United States was that of Elliot J. Dent who, in 1916, wrote: "I know of no means by which exposed sandy beaches for surf bathing may be preserved except by feeding fresh beach material to them as rapidly as the old material is carried away"[2]. Soon after that (1922) the beach at Coney Island, New York, was the first to benefit from a concerted effort at beach nourishment. More than 10^6 m^3 of material were dredged from New York Harbor and transported to Coney Island[3]. There soon followed a number of other projects along the New York and New Jersey coastlines with a few in southern California. As in the case of Coney Island, these early coastal renourishment projects utilized materials that were dredged from harbors and ship channels.

Some other early beach nourishment projects include: (1) Durban, South Africa. The building of harbor entrance structures in 1850 initiated erosion of an adjacent beach. Over the years groins were built but they did not stop erosion so, on the recommendation of a Belgian engineer, additional long, low groins, in combination with sand bypassing, were added. It was the first attempt at renourishment in South Africa[4], (2) Waikiki Beach, Hawaii, United States (Fig. 1). Another early project related mainly to recreation was the renourishment of Waikiki Beach in 1939. This beach has continued to need renourishment ever since, and (3) Norderney, Germany. Sea walls and groins have been used at Norderney (one of Germany's barrier islands) since about 1850. Although they prevented dune erosion they did not stop beach erosion. To rectify this problem, the first large scale beach renourishment project in Europe was initiated at Norderney in 1951. By 1989 the beach had been renourished an additional six times[5].

J. Chen et al. (eds.), Engineered Coasts, 23–59.
© 2002 *Kluwer Academic Publishers. Printed in the Netherlands.*

Beach nourishment projects have been carried out in many counties viz. Australia, Belgium, Brazil, Cuba, Denmark, France, Germany, Great Britain, Japan, New Zealand, Portugal, Russia, South Africa, and the United States. Even though the basic aim of beach nourishment is to elevate the beach and advance the shoreline in order to realize all of the consequent benefits such as increased storm protection, the techniques of sand transfer to the shore and design parameters differ among national approaches. Some of those differences are compared and contrasted in the 21 examples selected for comment (see also Tables 1 and 2).

2. Europe

Europe, the smallest of the continents, has a long, highly indented, and diverse coastline. Ranging in latitude from the Arctic to the subtropics, it has coasts that show evidence of glaciation and rebound, tectonic activity and volcanism, subsidence and sea-level rise, as well as the erosion and deposition that accompanies marine, fluvial, and aeolian activity.

Parts of Europe face the North Sea and the open Atlantic Ocean whereas much of it borders large inland waters such as the Baltic, Mediterranean, and Black Seas. Tides also are highly varied with some of the highest in the world along the coast of France and some of the lowest in the Mediterranean.

Along the shorelines of Europe, beaches are equally as varied. Some coastlines such as the rebounding areas of Finland and Sweden and the hard rock coasts of western Britain and parts of Portugal have limited beaches because of the lack of materials. In contrast, beaches, many of which are suffering erosion, are extensive along many shorelines such as those of Denmark and the northern Adriatic coast of Italy. Many of these countries have generally relied on the use of hard structures to halt erosion, but often without success. Recently, beach nourishment, often in combination with hard structures, has been gaining in importance.

2.1 DENMARK

Most of the sandy coasts in Denmark, particularly those on the west coast of Jutland (Jylland), are subject to erosion. In consequence to the long fetch, the west coast of Jutland is influenced by strong wind and wave conditions, while tides are less important with a tidal range of approximately 0.8 m. This barrier coast experiences the same kinds of erosional problems that occur in many other areas viz. inlet migration, loss of beach volume, shoreline retreat, and expansion of downdrift erosional fronts in the lee of stabilized inlets[6]. In the example of beach erosion downdrift (south) of the inlet at Hvide Sande, construction of jetties increased shore erosion to the point that beach nourishment efforts began in 1973 with sand dredged from the outlet and the harbor (Table 1). Dunes were vegetated as part of the same coastal protection effort, although there were problems with tourists trampling the planted vegetation and destroying the sand fences. Along the coastline near Årgab (south of Hvide Sande), the annual shoreline retreat averaged about 3.1 m a^{-1} (varying from 5.8 to 0.2 m a^{-1}) from 1963 to 1972. The average annual retreat of the dune foot was about 30 m and the erosion affected the sea floor to a depth of -4 m msl.

Table 1. Examples of beach nourishment procedures employed in different countries.

General Location	Cause of Erosion ①	Procedure	Sediment Source	Equipment	Durability (yrs)	Source
Hvide Sande, Denmark	1	Beach nourishment, feeder bar storage, revegetation of dunes	Harbor, inlet, seabed	Ocean dredgers, bulldozers, trucks	≅10	Møller (1990)[6]
Zuid Holland, Netherlands	1, 4	Profile nourishment	Seabed	Hopper dredges, trailing hopper dredges	5-8	Roelse (1986)[7], de Ruig(1998)[8]
Bournemouth & Christchurch, UK	2, 3	Placement of mixed sand & gravel on beach	Offshore banks	Ocean dredgers, bulldozers	5-10	May (1990)[9]
Algarve, Portugal	1, 4	Nourishment of pocket beaches	Seabed, harbor, inlet	Ocean dredgers, harbor dredgers	10-13	Psuty and Moreira(1990)[10]
Georgian Black Sea coast	5, 6, 7	Dumping onshore, placement in feeder bar (4-6 m depth)	River deltas, coastal plains	Coastal dredger & split-hill barges, hydraulic dredging of onland sources, bulldozers, trucks	≅8	Zenkovich and Schwartz(1988)[11], Kiknadze et al. (1990)[12]
Tokyo Bay, Japan	8	Beach nourishment, beach construction, gravel cap	Harbor, shipped from remote coastal sectors	Offshore dredgers	Annual	Koike (1990)[13]
Port Phillip Bay, South Australia	1, 2, 5	Beach nourishment, beach scraping, feeder bar	Seabed, harbor	Ocean dredgers, bulldozers	8-10	Jones and Schaffer(1986)[14], Bird (1990)[15]
Omaha Beach, New Zealand	1, 9	Beach nourishment, reconstruction	Tidal inlet	Maintenance dredgers, some trucking	8 Injections over 5 yrs	Healy et al. (1990)[16]
Zeebrugge and De Haan, Belgium	1, 4	Profile nourishment, beach nourishment, feeder berm	Offshore, harbor	Trailing-suction split-hopper dredges	?	Charlier and De Meyer (1995[17], 1999[18])

① (1) Construction or deepening of tidal inlet, construction or lengthening of jetties, (2) presence of seawalls, revetments or other upland protection works, (3) construction or maintenance of groins, (4) dredging of inlets or harbors, (5) harbor breakwaters, (6) quarrying of beach sediments or removal of coastal plain sediments, (7) impoundment of fluvial sediments by damming of rivers, (8) land reclamation and dumping of landfill in bay waters, and (9) general construction including oceanside housing subdivisions that destroy or degrade natural dune systems.

After careful field study and construction of physical models, it was ascertained that that the most important measure against coastal erosion was continuous and regular beach nourishment with sand dredged in deep water and dumped in shallow water (about 4-5 m depth)[6]. The rationale for this method of sand placement is that a stock of sand can be built up, ready to be transported towards the beach by waves during relatively calm weather in the summertime. During late summer and early autumn, sand is transported in longshore bars towards the beach waves. When the bars weld to the beachface, the additional volume is moved by bulldozers towards the dune cliff to protect the dune itself and buffer winter storm surge. Small coastal segments with high rates of shoreline retreat (e.g. erosional hot spots), such as occurred near Årgab, were nourished through the year by sand pumped from the seabed and trucked overland. It takes about 150,000 m^3 of sand to stabilize the shore annually and to maintain a dynamic equilibrium (i.e. relative stability of shoreline position).

Durability of beach nourishment is an important issue. In Denmark protection plans generally call for estimates of shortsighted protection (approximately 10 years), longsighted protection (about 25-40 years), and total protection (at least 100 years)[6]. The main idea there is to keep the entire length of an eroding coast in a state of dynamic equilibrium, through periodic or nearly continuous nourishment, so that no coastal sector forms a salient or promontory that will induce lee side erosion that in turn could lead to unpredictable erosional conditions. The strategic position of the shore and level of adjacent infrastructure determines which beach durability plan is selected.

2.2 THE NETHERLANDS

The coastal defense system in The Netherlands consists of dunes and dikes that protect polders (Fig. 2). The dunes, which account for about 75% of the defense line, vary in width from 100 m to 5 km[8]. Shoreline retreat occurs along the entire coast and more than 50% of the dune coast suffers long-term structural erosion. In 1990 the Dutch government decided to stop any further long-term coastal recession by preserving the shoreline in its 1990 position[19]. This policy of 'dynamic preservation'[20] of the shore relies on beach nourishment, the main coastal defense measure.

On the international scene, the Dutch have been leaders in the use of beach nourishment and also in scientific advances that have reduced the costs associated with this soft-engineering procedure. Research in The Netherlands leads to a national policy that supports an integrated approach to coastal protection that includes both beach and dune systems. Beach replenishment projects in The Netherlands thus rely on combined engineering efforts that supply additional sediment to the shoreface, beach (berm), and dunes. These combined efforts provide a major element in coastal defenses against storms. Application of this strategy in fifty projects along the Dutch coast between 1952 and 1989, with most projects occurring after 1970, resulted in more than $60x10^6$ m^3 of fill being placed along the shore[7]. Nourishment of the entire shore profile (i.e. shoreface, beach, dune systems) provides advantages over traditional strategies because it is flexible, does not result in intractable decisions, and is, in many cases, the only thorough solution that does not involve the possibility of periodic repetition.

Fig. 1. Waikiki Beach, Hawaii, USA, looking west from the top of Diamond Head.*

Fig. 2. Sand dune and beach facing the North Sea in The Netherlands.

*All photographs in this chapter by H. J. Walker.

it is flexible, does not result in intractable decisions, and is, in many cases, the only thorough solution that does not involve the possibility of periodic repetition.

In order to partly compensate for sand losses along the Dutch coast during the period 1952-1985, approximately $25x10^6$ m^3 of sand was supplied. Sand for shore protection was acquired from civil engineering projects, such as the coastal extension at Hoek van Holland, in the coastal zone, or from the seabed offshore (Table 1). Acquisition of sand was mainly accomplished using suction dredges that spouted immediately at the construction site (with or without the help of booster stations). Hopper dredges were used for relatively small projects where up to about $1x10^6$ m^3 of sand was placed. Small amounts of sand were obtained using trailing hopper dredges when the distance from the supply site was 10 to 30 km.

Most of the periodic replenishment conducted as part of the overall coastal protection efforts occur along the Dutch coast in the province of Zuid Holland. Sandfills in the rest of The Netherlands is mostly regarded as a one-time solution. The fills at Voorne and Goeree, for example, were combined with other projects that influenced cycle times. Fill frequencies for these and comparable coastal segments range from 5 to 8 years. Beach replenishments on the islands of Texel and Ameland have been repeated after six years of fill longevity.

2.3 BELGIUM

About half of the Belgian coast is protected by hard structures including seawalls and groins, most of which were built after the severe storm of 1953. The remainder of the coast consists of dune-belts and beaches. Because the hard structures did not provide the shore protection that was anticipated, the extension of the Zeebrugge harbor provided an opportunity to obtain beach fill from offshore dredging[17]. Although Belgian approaches are varied and often innovative, this brief review considers the effects of feeder berm construction (Table 1).

Enlargement of the harbor at Zeebrugge along the Belgian coast in 1976 included construction of new breakwaters extending about 3.5 km into the sea. Because of potential downdrift beach erosion resulting from the littoral drift blocker at the stabilized navigational entrance, it was decided to employ shore protection works at Knokke-Heist east of Zeebrugge. Over the period December 1977 through March 1979 about $8.4x10^6$ m^3 of beach fill was placed in this zone over a length of about 8 km. This was the largest artificial nourishment scheme attempted in Belgium[21][17]. Studies of potential impacts of the harbor construction, on coastal morphology anticipated the following sand losses: (1) 300,000 m^3 a^{-1} by tidal currents and waves in the offshore zones, (2) 430,000 m^3 a^{-1} by waves in the nearshore zone, and (3) 80,000 m^3 a^{-1} by wind erosion in the beach-dune area[22]. As it turned out, computational forecasts of the anticipated total losses were too pessimistic.

Europe's first feeder berm, approximately 2200 m long, was placed in 1991 near De Haan (also known as Le Coq) [17]. Two trailing-suction split-hopper dredgers placed 600,000 m^3 of sand in a longshore bar 600 m from the shoreline on a low water bar.

Mean high and low spring tides are respectively 4.74 m and 0.21 m. By mid-1992, profile nourishment restored the beach profile between the dike and feeder berm. Due to the long pumping distances involved, a booster station increased output in two lines that fed the beach. The experiment conducted at the eastern end of the Belgian coast appears to show that beach replenishment is an efficient and economical method of shore protection[17]. The effects of artificial nourishment are "durable" in this location providing the beaches are partially recharged with sediment on a periodic basis. Profile nourishment and imitation of natural sediment feeding processes were successful at the De Haan works.

2.4 GREAT BRITAIN

Beach replenishment in England, as part of a number of different coastal protection schemes (Fig. 3), takes place in the form of replacing shingle (an older practice) or placement of sand or mixed sand and gravel[9]. Rates of erosion along many beaches is estimated at about 1 m a^{-1}, but landslides along cliffs or individual erosional episodes have brought about shoreline retreats in excess of 2 m a^{-1} at individual points along the shore. Beaches range from sandy in protected locations to gravelly on some open coasts. Some of the earlier experimental replenishments took place along Romney Marsh, near Dungeness, Bournemouth, Portobello (east of Edinburgh), and Christchurch. Artificial replenishment of sand beaches was adopted later in Britain than in North America and even by the early 1970's was largely regarded as an experimental procedure. The first British use of sand injection occurred in 1970 when about 15,000 m^3 was added to an 0.5 km east-facing beach at Gorleston[9]. In the same year a larger injection of about 200,000 m^3 was added to 2 km beach at Portobello. Since 1975, beach replenishment has become an integral part of coastal protection programs of a number of Coast Protection Authorities.

Artificial beach renourishment at Bournemouth and Christchurch has taken place in two main phases, at Bournemouth in 1975 and 1976 and in several separate schemes in both towns during the late 1980s. In all cases, the additional beach material has been used to fill existing groin compartments or in association with the construction of new groins and walls. The sources of beach materials have varied greatly, partly because the natural beaches themselves vary so much and partly because of the costs involved. Most schemes used sand and gravel dredged from offshore banks (Table 1).

Maintenance dredging of both river and port channels provided sediments at low cost. Dredging and replenishment operations in 1989 for the largest beach renourishment scheme (800,000 m^3 near Bournemouth) was conducted during the winter half of the year, partly in association with deepening of a harbor entrance channel and a need to complete the beach reconstruction before the summer tourist season began. It was calculated that by using "material of opportunity" the cost of the Bournemouth project was reduced to one-eighth what it would otherwise have been[9].

Experience in Britain over the last two decades shows that cost remains a critical factor in the timing and degree of renourishment and that different sources of sediment need to be utilized to meet the size needs of different beaches. Beaches along high-energy coasts need to be replenished with gravel or mixed sand and gravel whereas along some of the more protected coasts, such as at Bournemouth, sand with some gravel has been more common. The amenity role of the beach also affects the choice ofreplenishment materials because tourists generally perceive sand beaches as more desirable than gravel beaches.

2.5 PORTUGAL

Beach nourishment along a cliffed coast in the Algarve, Portugal, provides an uncommon example of beach construction in an environment not normally associated with such activity. The beach at Praia da Rocha ("beach of the rock") was originally narrow and studded with stacks (Fig. 4). Small pocket beaches, interspersed and secluded niches among rocky outcrops (Table 1), were converted to a broad sweeping sandy beach that supported local infrastructure of restaurants, cabanas, amusement rides, among other activities[35]. Pocket beaches along this coast experience a general wave climate dominated by long, low swell from the southwest with wave heights of 1.0 to 1.5 m and periods greater than 7 seconds. Beach and cliff erosion is associated with the so-called *Levante*, a strong wind from the southeast that develops in conjunction with an atmospheric depression formed in the northern Sahara. Waves associated with the *Levante* average 2 to 3 m high but occasionally reach 5 m with very short wave periods (3 to 5 seconds). Psuty and Moreira[36] report that the pocket beaches and cliffs seem to have remained in a state of dynamic equilibrium for a century or more prior to 1970 when the main channel of the River Arade and the harbor at Portimão were deepened and widened. Construction of jetties at the harbor isolated Praia da Rocha from the shifting tidal deltaic shoal which was its sediment source. The volume of pre-fill beach sand was estimated to be about $5x10^4$ m^3, calculated from the subaerial beach, from 1968 engineering topographic survey and 1988 beach profiles.

In December of 1970, approximately $8.8x10^5$ m^3 of sediment was pumped through a pipeline building a beach 140-150 m wide along the front of the cliffs. The fill material over the next decade shifted about and presented a shoreline planform that was more orthogonal to the incoming southwest waves. During 1983, another dredging episode took place within the port and entrance of the harbor. Part of the dredging material was placed on the last 200 m of the western part of Praia da Rocha which was not infilled during the first nourishment. This second phase of nourishment contributed about $2.5x10^5$ – $3x10^5$ m^3 of coarse sand with gravel and a lot of broken shell[37]. The western portion of the beach was widened from 45 m to 100 m as approximately $0.75x10^5$ –$1x10^5$ m^3 of sediment was added. Field reconnaissance of the Praia da Rocha site conducted in 1983 showed that more than 80% of the subaerial filled remained. This observation suggested that the longshore transport capacity is limited and that the beach cell is a relatively closed system. Rock promontories provided protection for the renourished beach. Scarps in the beachfill westward of Praia da Rocha are geoindicators of active erosion, a situation resulting from high wave energies and more general southwesterly exposure.

Nourishment of this cliffed coastline in Portugal is regarded favorably. Although pocket beaches backed by cliffs represent specialized coastal conditions, the project

Fig. 3. Shingle beach backed by rock-filled gabions on the south coast of England.

Fig. 4. Coastline along the western shore of the Algarve, Portugal.

here illustrates the flexibility and usefulness of artificial beach renourishment in an unusual environment.

2.6 THE GEORGIAN BLACK SEA COAST

Along the northeast shoreline of the Black Sea (Fig. 5) are a number of towns and cities that are centers for recreation as well as other functions (Fig. 6). This section demonstrates well the relationships between landforms, river discharge, longshore currents and human activity such as beach mining, construction of dams on coastal rivers, groins and breakwaters and beach nourishment (Table 1). Along the Tuapse-Adler section, river discharge of beach-forming sediments was sufficient to form natural beaches 40-50 m in width. In the 1920s, railroad construction crews mined beaches for pebbles reducing beach width and dam construction on rivers further depleted the beaches[38]. In order to protect the railroad and other sections of the shoreline, seawalls and underwater breakwaters also aggravated erosion. As a result Peshkov wrote "The Caucasian Black Sea coast is a region which is suffering greatly from irrational economic activities"[39]. By 1980, 223 km of the 312 km of the Georgian coast had eroded. About $40x10^6m^3$ of material was removed from the coastline for construction purposes. Because these activities rendered the coast useless for recreation, a very elaborate plan was formulated in 1982 for renourishing the coastline. This plan included the use of $3.5x10^6m^3$ of nourishment materials in the regions of Gagres and Pitzunda.

The State Scientific-Industrial Association for Coast Protection (Krasnodarbe-regozashchita) which was established in 1989 planned for the removal of the "...ineffective coast-protecting structures..." and the emplacement of $10x10^6m^3$ of material along the beach restoring its width to 40-45 m. In Inal Bay (2.7 km long) north of Tuapse, beach fill was achieved by using pebbles shipped in by barge. By 1992, $1.83x10^5$ m^3 of the 20-25 mm size pebbles had widened two km of the bay's beach by an average of 35 m[41] .

Along much of the coast groins were employed to halt erosion, generally without success. It was found that periodic nourishment of pebble beaches without groins was as effective as building groins and cost only about half as much[42]. The accepted procedure is the placement of material updrift of the location to be renourished and allow the sea to transport it to the eroded sections. It was projected that the procedure was to be used along 300 km of coastline where needed. In 1982 and 1983, $7.25x10^5$ m^3 of rock debris was effectively placed along the Gagra-Pizunda shoreline in the surf zone.

One of the major nourishment projects on the Black Sea was at Poti which lies at the head of a submarine canyon. By 1988, some $6x10^6$ m^3 of sand were hydrotransported from a source seven km inland to the beach which was effectively widened[43].

Placement of sediment along the Black Sea Coast followed three basic procedures: (1) Dumping onshore to form a longshore berm by using bulldozers, (2) landside construction of sedimentary groins built out into the water by dump-trucks, and (3) nearshore dumping by split-hull barges to form a submarine bank.

The "experiments" on the Black Sea Coast led to the conclusion by Peshkov (1993) that there is a choice "...either to continue a senseless waste of money and to ruin finally the Black Sea coast by reinforced concrete, or to go by the way that is prompted by

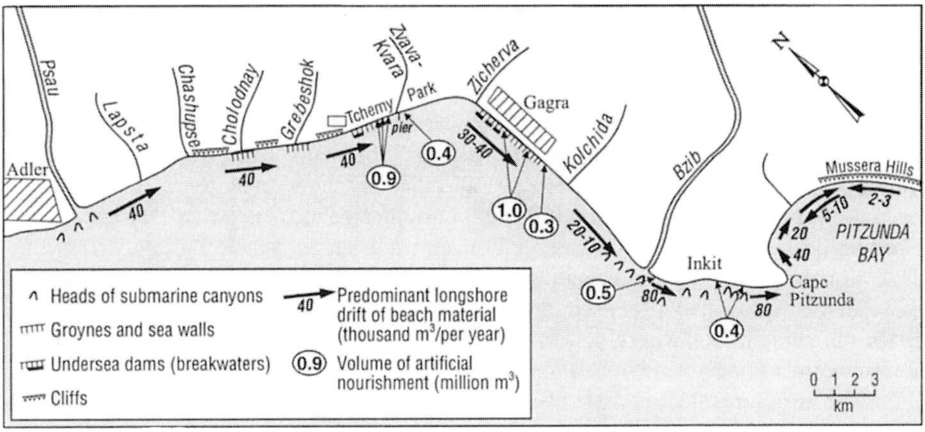

Fig. 5. The northern Black Sea coast illustrating coastal protection measures. Modified from Peshkov (1993)[24].

Fig. 6. The beach resort of Sochi, Russia with its rocky beach.

nature"[24]. He argues that the second way is (1) relatively simple and inexpensive, (2) that maintenance expenses on artificial beaches are "steady state", (3) that it creates an active sediment balance, and (4) that it is aesthetically and sanitarily meritorious.

3. Australia

Port Phillip Bay is one of Victoria's major boating and recreation areas, with a water area of about 2000 km^2 and 300 km of coastline. The major ports of Melbourne and Geelong are managed by their respective authorities but the remainder of he Bay is the Port of Port Phillip Bay and within this the Ports and Harbours Division has responsibility for protection of Crown Foreshores from erosion by wave action. A beach preservation program began in 1974 by the Ports and Harbours Division. Although engineers had proposed beach restoration since the early 1960s and had carried out some minor works, it was only after 1970 that there was political pressure for beach protection by sand nourishment.

After experimental beach nourishment projects with sand brought from inland quarries, pumped from offshore, or bulldozed from the shore, the Victorian Ports and Harbors Division decided to dredge coarse sand deposits from the floor of Port Phillip Bay for the nourishment of eroded beaches[16]. Between 1975 and 1987, dredged sand was pumped onshore to successfully nourish 18 beaches with a combined length of 19.3 km at a total cost of just over $A 4.5 million. The present coastline of Port Phillip Bay consists of cliffs or bluffs, fronted by shore platforms, and alternating with low-lying sandy plains and marshy areas. Narrow beaches of sand and gravel run intermittently along the east coast, widening in coves where the shore platforms are interrupted.

Some of the early attempts at beach nourishment in South Australia, using grain sizes smaller (pumped from the floor of Sandringham Harbour) than those on the native beach, failed when the sediment eroded away in a year or so[16]. In another early experiment, beach nourishment was attempted by bulldozing sand from nearshore shallows on the southern shore of Port Phillip Bay near Rosebud. A bulldozer was used at low tide to push sand from up to 100 m offshore onto the beach (Table 1). This procedure appeared successful and for the next 20 years several sectors of beach were widened in this way. Where transverse bars were present, they were used as an accessible source of sand for placement on the beach. Where they were not present, the nearshore area was deepened by up to 30 cm as the bulldozer pushed sand shoreward.

Between 1963 and 1984, from 3000 to 5500 m^3 of nearshore sand were delivered to the beach annually[16]. In 1984, bulldozing sand from the nearshore zone was discontinued because opportunistic seagrasses colonized the newly exposed sites. Growth of seagrasses in the nearshore zone was undesirable because they were unpleasant for swimmers, the beach was often obscured by rotting seagrass that was smelly and bred flies. The beach was nourished a year later in 1985 using sand pumped in from deeper water offshore. The sand was placed in transverse bars so that lateral dispersal of sand from these would suppress nearshore seagrass growth.

Beach nourishment began in 1977 at Mentone beach on the east coast of Port Phillip Bay. The project is of interest because it uses an interim dump site in 6 m of water to temporarily hold material from the primary borrow area 2 km offshore (Fig. 7). The 2000 ton dredging *vessel Mathew Flinders* (87.8 m long, with a draught fully loaded of 4.6 m, a hopper capacity of 1150 m^3) extracted sand from the offshore borrow area but its draught was too great to permit direct pumping ashore. A pump mounted on a raft was then used to pipe sand from the interim dump sit (400 m long by 150 wide by 1 m deep trench). The *Mathew Flinders* was able to dredge and dump one 800 m^3 load of sand per hour and eventually about 16,000 m^3 of sand was placed in the trench at the interim dump site. The transfer of coarse sand from the trench to the shore required the use of a cutter suction dredge. The pumps supplied about 100 m^3 of sand to the shore per hour and this was then bulldozed to form a beach terrace in the front of the seawall. The terrace was initially built 32 m wide, 2 m above chart datum (equivalent to low spring tide, or 1 m above calm sea level at high spring tide) with a seaward ramp graded to 1 in 3. The beach thus constructed extended 1800 m along the coast and contained 127,000 m^3 of coarse sand[16]. During the first year after completion of the project, the beach terrace width decreased from 32 m to 22 m due to cutting back and reshaping the seaward slope by wave action from 1 in 3 to 1 in 9 gradient. The beach terrace was subsequently cut back at an average rate of about 1 m per year, recession occurring mainly during brief periods of storm wave activity, especially at high tides. Depletion of the northern end of the artificial beach which had occurred by 1984 was offset in that year by pumping in a further 18,500 m^3 of coarse sand to restore the profile to its 1977 dimensions[14].

4. New Zealand

In New Zealand, a distinction is made between beach 'renourishment' and 'reconstruction.' The former term is used in reference to any sediment that is artificially emplaced within an existing beach system from the offshore bar to the dune, but without regard for morphological implications. Beach reconstruction (Table 1), on the other hand, implies that the beach fill is molded into site-specific beach-dune morphology[16]. Most cases of beach renourishment have been implemented in the last two decades under the realization that soft engineering practices are more flexible than stationary structures that interrupt coastal dynamic systems. The 'working with nature' approach has the additional advantage that coastal morphodynamic systems are not as perturbed by steep gradients that are typically set up with hard structures. The range of beach types that have been renourished, and sometimes reconstructed, in New Zealand includes: (1) open ocean sandy beaches (Fig. 8), (2) mixed sand-gravel beaches on littoral drift coasts, (3) deeply embayed, moderate wave energy barrier spit and pocket beaches, (4) sand veneer beaches over rocky shore platforms, (5) sheltered estuarine-lagoonal beaches and ebb tidal delta adjacent beaches, and (6) sheltered harbor mixed sand and gravel beaches[16].

One of the best known cases of beach erosion occurred on the northeast coast of New Zealand along a moderate wave energy, meso-tidal, embayed barrier spit beach known as Omaha Beach. During the 1970s the spit was developed for a housing subdivision without regard for geoindicators of active shore erosion, viz. steep faceted frontal dunes, numerous blowouts, a narrow low angle beach, and shoreline instability.

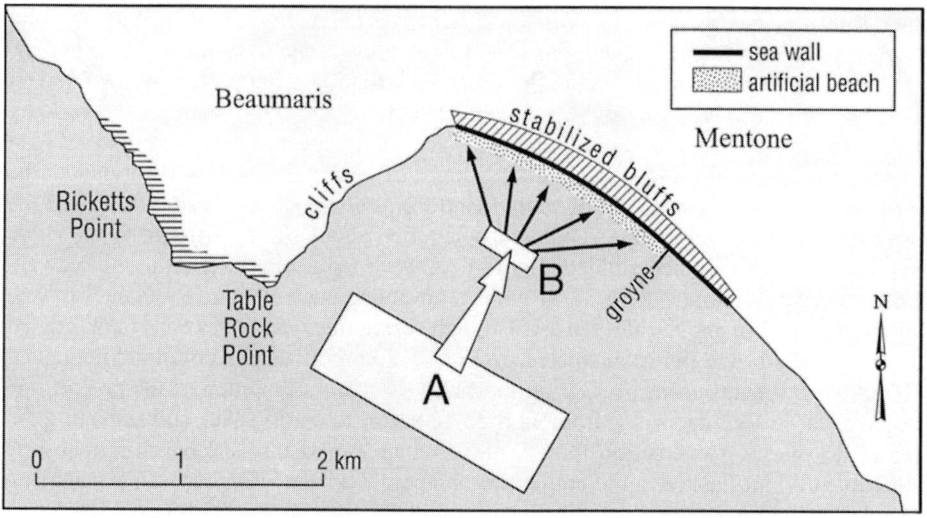

Fig. 7. Renourishment technique at Mentone, Port Phillip Bay, Australia. A = primary borrow area, B = trench emplacement from which it was pumped onshore. Modified from Bird (1990)[16].

Fig. 8. New Zealand shoreline.

Although the spit retained hummocky dune topography after erosion in the preceding two decades, the dunes were mechanically recontoured into a series of sequentially lower terraces so that property owners would have oceanfront views. Studies (e.g. Schofield[26]) indicated that about $1.3x10^6$ m^3 of sand was dredged from the site in addition to natural sand loss. Construction of a seawall in front of the development may have exacerbated the natural erosive processes. After a major storm catastrophically damaged the seawall in July 1978 and erosion progressed 40 m inland, the local authorities and developers agreed to: (1) beach reconstruction and nourishment, (2) construction of a rip-rap dike-type wall veneered with sand to replace the graded frontal dune, (3) construction of a rip-rap hooked groin to stabilize the tip of the spit, and (4) construction of two groins to contain artificially placed beach fill and to protect the hooked groin[16].

Remedial works were commenced in 1979. Beach nourishment and reconstruction was supported by dredging $4.5x10^5$ m^3 of medium-grained sand and shell material from the inner side of the tidal inlet gorge and pumping the slurry to the ocean beach where it was distributed by earth moving machinery. The hard structures have generally held up but the beach lacks a positive sediment budget.

Mixed sand and gravel beaches are on the east coast of New Zealand where their distribution is attributed to a provenance in graywacke rocks in nearby ranges and to Quaternary outwash deposits derived from them[27]. These beaches show distinctive dynamics in microtidal, east-coast swell type wave environments so that most knowledge and management techniques are derived locally. With respect to nourishment technology, these beaches typically show moderate to severe, natural and human-induced, regional long-term erosion rates between 0.5 to more than 8 m a^{-1}. Their distinctive two-part cross-shore profile corresponds with a binary sediment transport system[27]. There is an abrupt slope break at the submerged base of the beach with coarser materials above this line with flatter seaward slopes dominated by fine sand transport that is independent of beach sediment transfers. There is no onshore-offshore cycling of beach sediments as occurs on most sandy beaches. Instead of a surf zone, these beaches have a single breaker on the steep nearshore face. This topography results in a runup-dominated beachface that is frequently overtopped. Thus, washover of beach sediment to the back beach is an important component of long term erosion mechanism and of attempts to control it by nourishment.

In an effort to manage coastal erosion around the port and city of Timaru on the east coast of the South Island, a combined reconstruction and nourishment project was initiated on 300 m of beach in 1980. Bulldozing relocated about 6600 m^3 of storm washover materials from the back beach area to the berm crest and foreshore. A coarser capping of river gravels comprising some 8900 m^3 was then added to raise the crest by 2.0 - 2.5 m, and the foreshore was coarsened and steepened. The purpose of this technology was to reduce overtopping by storm waves and provide a beach surface that was more absorbent of wave energy and resistant to erosion[16]. Monitoring of the trial as conducted for the next five years using standard methods of grain-size assessment and surveyed profile change. During the trial there were eight further injections of sediment on the beach. Because the adjacent untreated beach segments retreated 11 - 17.5 m, the renourished trial formed a headland that was subjected to increasing erosional attack. Nevertheless, in 1985 the trial site profiles contained from 60 - 82% of the volumes originally placed. In terms of landward retreat of the shoreface contours, reductions in erosion ranged from 50 to 90%[16]. This project demonstrates effective management procedures for eroding beaches on high energy coasts.

In 1990, nourishment had been used at two locations for the primary purpose of building up beaches in anticipation of the increase in erosion that will accompany the predicted rise in sea level. At Westshore beach, one of the two locations, $9x10^4$ m^3 of dredge spoil from the harbor of Napier was dumped into 5 – 7 m of water just offshore (Fig. 9) with the objective of inducing down drift nourishment [16].

5. Japan

Eight artificial beaches, with a combined length of nearly 13 km, have been constructed along the northern shores of Tokyo Bay since the early 1970s (Fig. 10). The natural shoreline length of Tokyo Bay used to be about 180 km, but the total length of waterfront is now nearly 800 km, only 60 km of which is easily accessible for pedestrians. The Ministry of Transport plans, however, to triple the waterfront length, with safe access for pedestrians, to 180 km within the next decade. Rapid development polluted the Bay with industrial effluents and household wastes, making the water unsuitable for swimming. Under conservation policies guided by the Environmental Agency, public authorities are trying to purify the water, but water quality has not returned to its pre-1955 state which was ideal for swimming and commercial fishing[28].

Tokyo Bay, with a water area of 1200 km^2 and an average depth of 15 m, is practically surrounded by reclaimed land and artificial beaches[13]. The urban population around the Bay exceeds 15 million people. Tokyo (8.4 million), Yokohama (2.8 million), and Kawasaki (1.1 million) are the largest urban centers fronting the Bay. City dwellers wanted the beaches restored for swimming and gathering shellfish close to their residential areas. Artificial beach construction projects have been carried out along margins of the Bay adjacent to landfill areas since the early 1970s. Many seaside parks were developed in association with artificial beaches in an effort to promote waterfront accessibility. These waterfront parks feature board sailing centers, cycling paths, artificial beaches, wild bird sanctuaries, and other similar facilities that appeal to citizens of all generations[29].

Most of the landfills on the eastern coasts of Tokyo Bay were completed after 1966. The southern part of the landfill is occupied by heavy industry but the northern part was planned for artificial beaches and seaside parks. One of the earliest projects was Inage Sea Side Park which contained a 1200-m long beach (200 m wide) at a cost of $US 850 million in 1975[13]. Sand was dredged from 3 km offshore and pumped directly onshore to nourish the seafloor between two groins 200 m in length. The surface of the nourished beach was covered with a 30-cm thick layer of coarser material (3-8 mm mean diameter) in an effort to reduce beach erosion[30]. The Makuhari Beach in the Chiba Prefecture by the same method of sand nourishment but an underwater breakwater was built in front of the beach. In the decade immediately following construction, material gradually eroded away but the beaches were maintained by

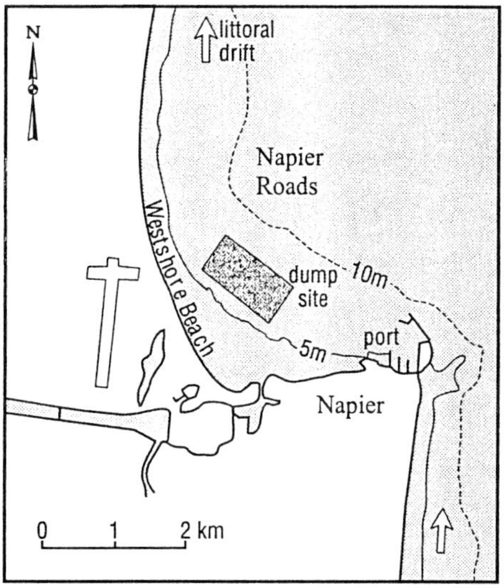

Fig. 9. Nearshore dredge spoil dump at Westshore Beach, New Zealand. Modified from Healy et al. (1990)[62].

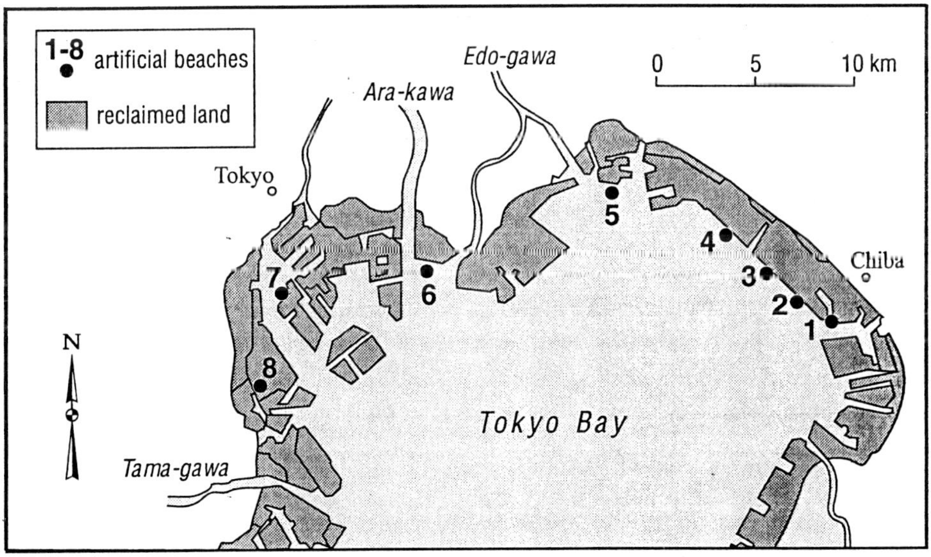

Fig. 10. Location of artificial beaches at the north end of Tokyo Bay. Modified from Koike (1990)[63].

adding 15,000 m^3 of sand every year to the Makuhari Beach and 45,000 m^3 to Inage Beach in 1983. In contrast to dredging offshore of the artificial beach project, some beaches were nourished with sand that was shipped from other coastal sectors (Table 1) and supplemented by offshore dredging [13].

6. Brazil

As Dieter Muehe[31] states, the trend in recent years in Brazil has been to utilize hard engineering rather than beach nourishment for combating coastal erosion. Hard structures have been mainly used for protecting coastal roads.

Nonetheless, there are a few locations in Brazil where beach nourishment has been used. Some of them are the Camburi Beach at Vitória (500 km north of Rio de Janiero), São Braz Beach at Mangaratiba Bay (west of Rio de Janiero), a beach in Araruama Lagoon (near Cabo Frio), and Copacabana Beach near Rio de Janiero.

The Copacabana project was the most important of Brazil's renourishment efforts. Copacabana Beach is a crescent shaped beach composed of fine to medium size sand. The Beach, protected at both ends by rocky promontories, is 4.2 km long and under natural conditions about 55 m wide (Fig. 11). It was decided in the late 1960s that the beach should be widened to 140 m. Before beginning construction, however, the authorities authorized a feasibility study which was conducted at the National Laboratory in Lisbon, Portugal[32]. The modelers experimented with two techniques, namely: stockpiling sand on the foreshore and offshore dumping at water depths of –5 m so that a hopper dredge could be used. Both methods were deemed workable, so it was decided to use both in combination. Two different sources of sand were used. That for stockpiling came from a protected bay and was piped 5 km to the beach (Fig. 11). That for dumping was dredged from depths of 10 to 15 m at a location some 4.5 km offshore and transported by hopper as close to shore as possible. The amounts of sand added to the beach totaled 3.5×10^6 m^3 about equally divided between the two methods. The results mimicked the model tests quite closely.

Because Copacabana Beach is one of the most utilized beaches in the world, its quality is critical to tourism. Storm water and uncontrolled sewage runoff occasionally discolors and contaminates sections of the beach. Because this condition, referred to as "black tongues", is both unsightly and unhealthy, a procedure for dealing with it has been developed. When a black tongue occurs, the contaminated sand is dug up and transported to the beach front at low tide and replaced by clean sand from the lower beach. In 1989-90, this operation was performed 20 times and involved the exchange of 5000 m^3 of sand[33]. Such a procedure will no longer be necessary once the sewer and storm drainage systems are upgraded.

7. The United States

The open ocean shoreline of the conterminous United States is about 8000 km long*.

*Several values for the length of the shoreline of the conterminous United States have been calculated. The U.S. Department of Commerce lists the General Shoreline as 8035 km long and the Tidal Shoreline as 86,382 km long[34] whereas the U.S. Army Corps of Engineers uses the value of 52,051 km based on the tidewater position[35].

Dominant coastal landform characteristics include sand beaches and barrier islands, pocket beaches between rocky headlands, rocky and cliffed coasts, and mudflats and marshes[36]. Beaches are common along shores backed by coastal plains along the Atlantic Ocean and Gulf of Mexico. Glaciated coasts in the northeast contain numerous pocket beaches, coves, and embayments interspersed along a predominantly rocky shore. Even though much of the Pacific coast is characterized by cliffy shores, beaches frequently lie at the base of cliffs. Great Lakes beaches are a mixture of several shore types including cliffed coasts and small coastal plains. In spite of the varied distribution of coastal landforms around the United States, many beaches are diminishing in width as retreating shorelines move landward. Of the 20 states in the conterminous United States that face the ocean about 30% is classified as beach. Along the shorelines of the Great Lakes it is about 57%[35].

States with the longest oceanfront shorelines are California and Florida (Table 2). The California shore, which stretches for 2900 km, is dominated by rocky and cliffed coasts (70% of the total shoreline length, including parts of San Francisco Bay and offshore islands) but there are significant segments of pocket and fringing sandy beaches below the cliffs[52]. About 85% of the California coast is actively eroding.

Florida, on the other hand, has some 1300 km of sandy beaches that represent about 25% of the total U.S. sandy shores. About 368 km (30%) of Florida's beaches are in a 'critical' state of erosion[42]. Renourished beach length in Florida (Table 2) amounts to 171 km and 117 km for Gulf and Atlantic coasts, respectively. The percentage of total sandy beach length that has been renourished is about 22% for both coasts of the Florida peninsula. Renourished shores in the glaciated coastal environment of New England average about 7% for the total shoreline length of New Hampshire, Massachusetts, and Rhode Island combined and about 25% for Connecticut (Table 2). Rock outcrops and cliffs developed in glacial outwash and morainic materials restrict the occurrence of sandy beaches in northern New England whereas unconsolidated sedimentary materials are more prevalent along the Connecticut coast.

According to Leonard et al.[51], as of 1988 approximately 90 Atlantic coast beaches, 35 Gulf coast beaches, and 30 Pacific coast beaches had been replenished. Beach nourishment efforts in the United States have resulted in more than 300 million m^3 of sand placed along more than 645 km of shoreline. The Atlantic coast has the greatest length of replenished shoreline (> 435 km), followed by the Gulf coast (> 160 km), and the Pacific coast (> 48 km).

The United States federal shore protection program started in 1922 in New Jersey[53]. In recent years there has been a shift in the philosophy of coastal protection from primarily hard engineering structures to beach nourishment as the preferable technique to mitigate beach erosion[54]. These changes occurred because of the realization that hard structures usually aggravate erosion problems, are expensive to deploy, cause the loss of aesthetic value on beaches, and require never-ending modification or new construction[55]. The U.S. Army Corps of Engineers (USACE) shoreline protection program covers about 362 km (8%) of the nation's 4330 km of critically eroding shorelines[54]. Presently the federal role in new beach nourishment projects is limited by budget constraints but that situation may change, as beach erosion problems become more widespread and difficult for local communities to cope with.

Table 2. Length of renourished shoreline segments for selected states.

	Total Shoreline Length (km)	Renourished Beach Length (km)[1]	Percent of Shore
New England[2]	**586**	**73**	**12.5**
New Hampshire	21*	2	8.8
Massachusetts	310	18	5.9
Rhode Island	64(Fisher[41])[3]	4	6.8
Connecticut	193*	49	25.4
Gulf Coast[2]	**2020**	**256**	**14**
Florida (West Coast)	645(Clark[42])	172	26.6
Alabama	75(Davis[43])	3	3.8
Mississippi	71*	97	136
Louisiana	639*	35	5.5
Texas	590(McCloy[44])[4]	10	1.6
Atlantic Coast[2]	**2167[1]**	**797**	**39.5**
New York (LI)	140(Psuty[45])	105	75
New Jersey	205(Psuty[45])[1]	105	51
Delaware	45*	47	104
Maryland	50(Monte[46])[5]	35	70
Virginia	180*	192	106
North Carolina	484*	103	21
South Carolina	320(Kana[47])	29	9
Georgia	160*	15	9
Florida (East Coast)	583(Clark[42])	117	20
Pacific Coast[2]	**3430**		
Washington	1265 (Schwartz and Terich[48])		
Oregon	500 (Stembridge[49])		
California	1665 (Orme&Orme[50])[6]	44[51]	

[1] Abstracted from Esteves and Finkl (1998)[37]. For the Gulf of Mexico renourishment episodes, project lengths were calculated for the time frame 1949 to 1998. The period of study for Atlantic coast beaches range from 1923 to 1996. Because some beaches were artificially renourished many times, the percentage of renourished coastline may exceed the total length of coast (i.e. Delaware and Virginia) when, in fact, many coastal segments remain in a natural condition.

[2] Compiled from Haddad and Pilkey (1998)[38], Trembanis and Pilkey (1998)[39], Esteves and Finkl (1998)[37], and Valverde, Trembanis and Pilkey (1999)[40].

[3] Includes the Atlantic coast and Narragansett Bay.

[4] Exclusive of the upper Gulf coast or low energy, marsh coast of the Big Bend area.

[5] Maryland has 6089 km of tidal shoreline of which 50 km is along the Atlantic oceanfront, 322 km on back bays, and the rest on the northern portion of Chesapeake Bay.

[6] Of the total, 70% is rocky or cliffed with pocket beaches commonly occurring below the cliffs. About 30% are either sandy fringing beaches backed by dunes or barrier beaches fronting lagoons and wetlands. There is an additional 1200 km of tidewater coast around San Francisco Bay, lesser bays, and several offshore islands but these shores are beyond the scope of this analysis.

*National Ocean and Atmospheric Administration, National Ocean Service (Washington, DC)[34]. Lengths of shoreline were calculated from general open coast and do not include total lengths of tidal shorelines.

7.1. THE EAST COAST

7.1.1. *Avalon, New Jersey.*

The southern coast of New Jersey is lined with a number of barrier islands that average about 18 km in length separated by tidal inlets. Avalon, located at the north end of Seven Mile Island (Seven Mile Beach) is adjacent to Townsends Inlet (Fig. 12). Widest at its northern end, Seven Mile Island narrows to the south ending in a spit at Hereford Inlet. The entire coastline is impacted by dominant northeast storm waves which in general cause a southward drift of sediment. The major exception to this trend is near Townsends Inlet where a long-shore current reversal occurs. The erosive/depositional history of the northern part of Seven Mile Island is therefore complicated and erratic[56].

Over the years, the Townsends Inlet channel migrated southward into the town of Avalon. Subsequent to a major storm in 1962, a bulkhead, rock revetment and three groins were constructed along the south side of the inlet. A large ebb-tidal shoal developed off Townsends Inlet. When Sea Island City, located on the barrier island to the north, needed beach renourishment in 1978 and again in 1984, the shoal served as the sand source. Subsequent to altering the morphology of the shoal, the shoreline south of Avalon eroded rapidly and by 1985 ocean-front property was being threatened.

Between 1987 and 1993, four renourishment projects were completed along the Avalon shoreline. A total of 1.8×10^6 m^3 of sand were used. The 1993 project, partially in response to a severe storm in December 1992, involved the establishment of a submerged concrete breakwater along a 300 m section of coast and the installation of sand-filled geotextile fabric tubes at Townsends inlet. Although the northern 500 m of shoreline continue to have problems because of northeast storms, Farrell reports that the area of beach to the south is stable[58].

7.1.2. *Tybee Island, Georgia.*

Tybee Island, an Atlantic coast barrier island, extends 4.4 km south of the Savannah River in Georgia, USA. The shoreline has an extensive history of erosion, some of it natural, some of it human induced. It is believed that dredging of the Savannah River since before the end of the last century has aggravated the shoreline's erosion. Traditionally, the spoil was dumped in impoundments or taken offshore to depths exceeding 10 m. Over the years, more than 130 groins and many bulkheads and seawalls were constructed along the island's shoreline[59].

The first nourishment project on Tybee Island was in 1975 when 1.7×10^6 m^3 of sand fill was placed by hydraulic pipeline dredge along 4150 m of the shoreline south of a terminal groin that had been built the year before. The fill came from shoals 600 m off the mouth of Tybee Creek (Fig. 13). Erosion at the south end of Tybee Island (south of the area of fill) accelerated because of the change in current patterns created by flow over the borrow sites. The Federal Government built a groin toward the southern end of the Island in 1986 and further renourished the area. The 1994 renourishment project utilized 0.75×10^6 m^3 of spoil from Savannah River dredging which proved to be of inferior quality and most of it was soon lost to the beach.

The southern end of the island, the part most important from a recreational standpoint, continued to suffer erosion. Therefore, the State of Georgia, authorized a

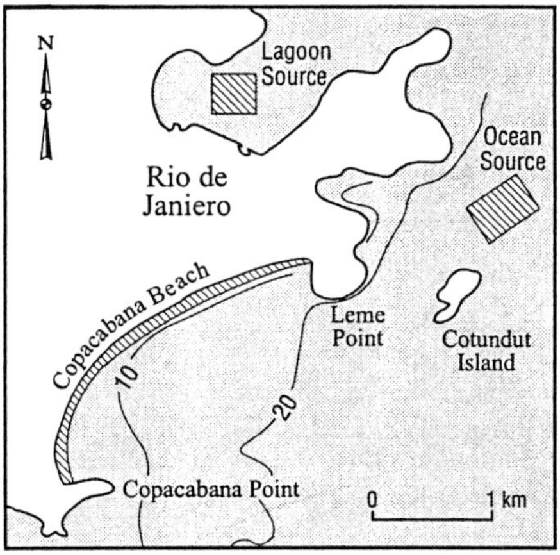

Fig. 11. Copacabana Beach, Brazil showing location of source areas for renourishment projects. Modified from Vera-Cruz[98].

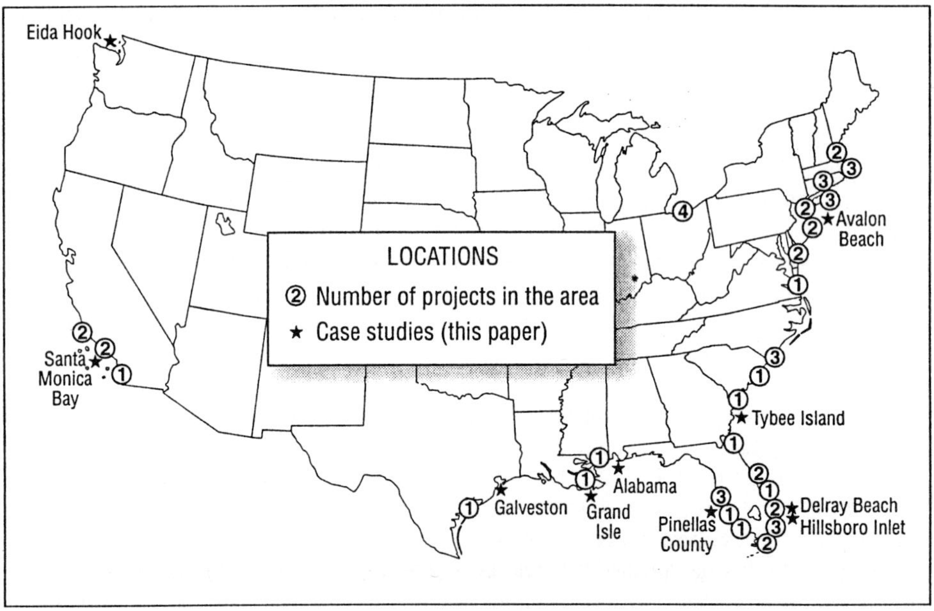

Fig. 12. U.S. map with nourishment locations, including the number and general location of those major projects sponsored by the USACE, modified from NRC[99].

stabilization project for the area between the South Federal groin and Tybee Creek (Fig. 13). The project included the construction of three 85-m-long groins and the placement of 38,000 m^3 of fill between them to form crenulate beaches. At the same time, 0.19 x $10^6 m^3$ of fill was placed north of the federal groin. Subsequent to construction, some of the additional sand has moved into the groin field and, by January, 1996, it was nearly full[59].

7.1.3 *Delray Beach, Florida.*
The town of Delray Beach, about 80 km north of Miami, Florida, is separated from the Atlantic Ocean by a beach and coastal road both of which were subject to heavy erosion prior to 1970[60]. The revetment that the city constructed earlier increased erosion so that, in 1971, it was decided to renourish the beach. In 1973, 1.22×10^6 m^3 of sand was placed along 4.34 km of the city's shoreline. By 1992 (i.e. within a 19-year period) the beach was renourished an additional three times with a total of 3.5×10^6 m3 of sand of which 0.75×10^6 m^3 was added in the 1992 project.

Because the Delray Beach shoreline was surveyed every year after 1973, good data about sand loss from the system is available. These data show that 1.15×10^6 m^3 of sand was lost between 1973 and 1990. This amount represents a 41% loss from the beach prior to the 1992 renourishment. These data also show that 0.51×10^6 m^3 (or 45%) of the sand was transferred to adjacent beaches widening them in the process. To the north (the Gulf stream side) of the renourished area the beach widened by an average of 18.6 m whereas to the south it was 16.8 m. The textural difference, finer to the south than to the north, led to a flatter profile to the south compared to the north.

The Delray Beach renourishment project, where fill was placed along a 4.34 km stretch of beach, actually benefited an 11.3 km stretch of the coastline along this section of Florida's east coast[60].

7.1.4 *Hillsboro Inlet And Its Downdrift Beaches, Florida.*
The east coast of Florida has numerous inlets (both natural and man-made) some of which have been jettied. The jetties at Hillsboro Inlet are short and parallel to the inlets entrance. In addition, a curved detached breakwater has been placed off the center of the entrance and, in essence, serves as a hard-structure downdrift spit (Fig. 14). The zone between the northern jetty and the detached breakwater serves as a conduit at high water and allows sediment to be passed into the north side of the inlet instead of trapping it to the north.

Between Hillsboro Inlet and Port Everglades Inlet (which is man-made) is a beach system that is 18.5 km long (Fig. 15). The northern part has been artificially renourished, with federal funds, twice, first in 1970 with $0.8 \times 10^6 m^3$ along 5.1 km of beach and second in 1983 with $1.46 \times 10^6 m^3$ along a 5.1 km section[61]. The source areas for these two renourishment projects has been offshore, seaward of a long-shore hard pan zone. In addition, to the $2.28 \times 10^6 m^3$ were used along another 11 km of Broward County's shoreline. Dredging has been so intensive that little quality sand is left in the county's offshore area[61].

Therefore, a management study was conducted in order to make recommendations on how to better utilize Hillsboro Inlet and eliminate the need for renourishment.

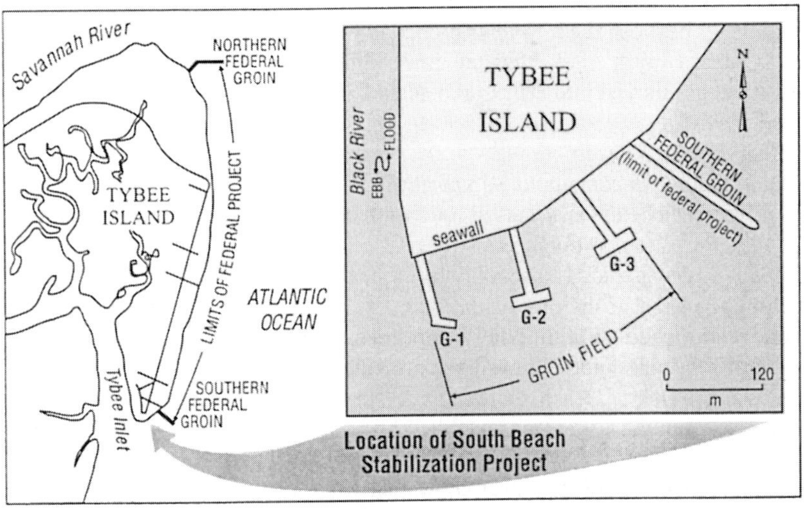

Fig. 13. Tybee Island, Georgia, stabilization project. Modified from Olsen[59].

Fig. 14. Hillsboro Inlet with its detached breakwater (see Fig. 16).

Although the official inlet dredging project was initiated in 1931, it was in 1947 that dredge materials for Hillsboro Inlet were bypassed to the beach north of the inlet. From 1948 to 1993, $2.33x10^6$ m^3, an average of 52,000 m^3 a^{-1}, was bypassed. The amount was increased after 1980 as a result of improved dredging capabilities. The pre-1993 sediment budget (Fig. 16) indicates that bypassing made 52,800 m^3 a^{-1} (i.e., 60,400 m^3 minus the 7600m^3 washed into the inlet from the south) available to the beaches to the south.

The pre-1993 sediment budget also shows that 27,000 m^3 was lost to the sea. A 1995 plan calls for the establishment of a 40,000 m^2 fan-shaped sand trap out from the navigation channel in order to trap the sand that is usually lost offshore. Such sand trapping would increase the volume of sand available for mechanical bypassing. The plan calls for increasing the amount by-passed to 89,000 m^3 a^{-1}.

It is contended that the "...already successful bypassing program at Hillsboro Inlet, once further improved, is expected to completely eliminate the need for continuous renourishment of the Pompano/Lauderdale-by-the-Sea area"[61].

7.2 THE GULF COAST

7.2.1 *Pinellas County, Florida.*

Between 1988 and 1992, three projects of beach nourishment were completed along a 16 km-long section in Pinellas County on Florida's Gulf Coast[62]. The three beaches, about equal in length, are Redington Beach, Indian Shores Beach, and Indian Rocks Beach (Fig. 17). Although the projects were similar in size, there were differences in the source and type of sediment used, construction methods employed, and subsequent performance of the renourished areas.

Because of serious erosion along this section of Pinellas County's beach, breakwaters and groins were built starting in the 1950s. Erosion with beach loss continued and decisions were made to "restore" a protective beach along 16 km of shoreline (divided into three sections) by renourishment (Fig. 17).

Two source locations were used, both being ebb-tide deltas--one at the mouth of St. John's Pass; the other, known as Egmont Delta, off the mouth of Tampa Bay. Because St. John's Pass is close to Redington Beach, the shelly sand was pumped directly onto the beach using a suction dredge. The other two locations were renourished from the Egmont source. For Indian Shores Beach the material was obtained by clamshell dredge, barged to the site, and transferred to the beach by conveyor belt. For the Indian Rocks Beach, the material was obtained by suction dredge, barged to the site and then pumped onto the beach.

Monitoring of the three beaches during the first six months after completion showed that there was a differential landward movement of the beaches with the Indian Shores Beach (the one for which the clamshell dredge and conveyor belt were used) experienced the greatest retreat. There was even greater difference between the volume changes along beach profiles. Indian Rocks Beach (the one where the suction dredge, barge, and pump were used) had the greatest loss.

Davis et al[62] concluded that much of the difference in response was because of difference in wave action. The six months after Redington Beach and Indian Shores Beach

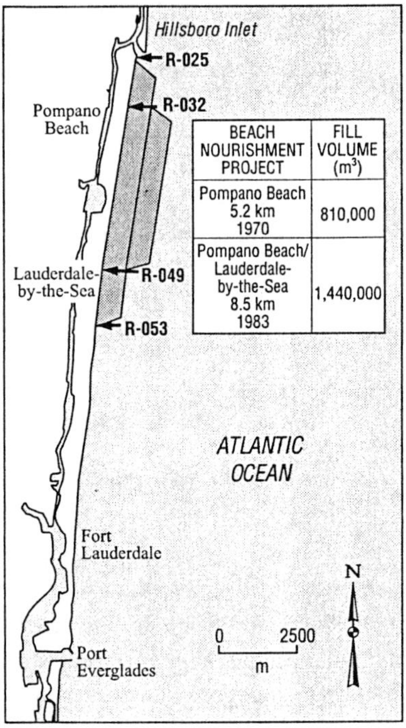

Figure 15. Map of Hillsboro nourishment budget. Modified from Lin et al[61].

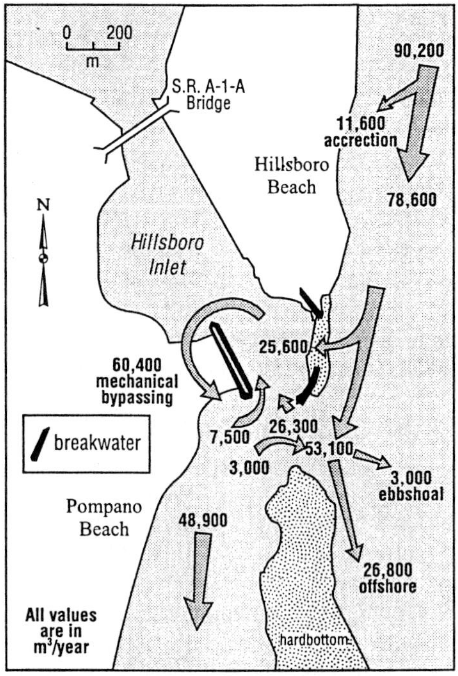

Figure 16. Hillsboro Inlet sediment project. Modified from Lin et al[61].

were renourished, low energy conditions prevailed whereas in the case of Indian Rocks Beach, high energy conditions helped cause it to have the largest loss of the three beaches. The authors concluded that variations in beach performances were related to initial profile, emplacement method, and angle and intensity of wave attack[62].

7.2.2. *Alabama.*

Davies et al report that most of the Alabama Gulf and estuarine shoreline is eroding and that replenishment and maintenance will be needed[63]. If such replenishment is undertaken on a large scale, more sand than is available onshore locally will be needed. Therefore, a large area offshore was surveyed for probable future utilization. Although there is an extensive area of mud out from the outfall of Mobile Bay, most of the Alabama shelf area is sandy. The survey included vibracoring at 59 locations and collecting bottom samples. Five "target" areas were examined in detail. For each, the area size, water depths, bottom surface relief, ridge and trough numbers, and potential volume of usable sand were determined. For example, Target Area 3 (Fig. 18) south of the west end of Morgan Peninsula, has an area of 92 km^2 and a maximum water depth of 18m. There is a large shoal area about 15 km offshore with fine to medium sand

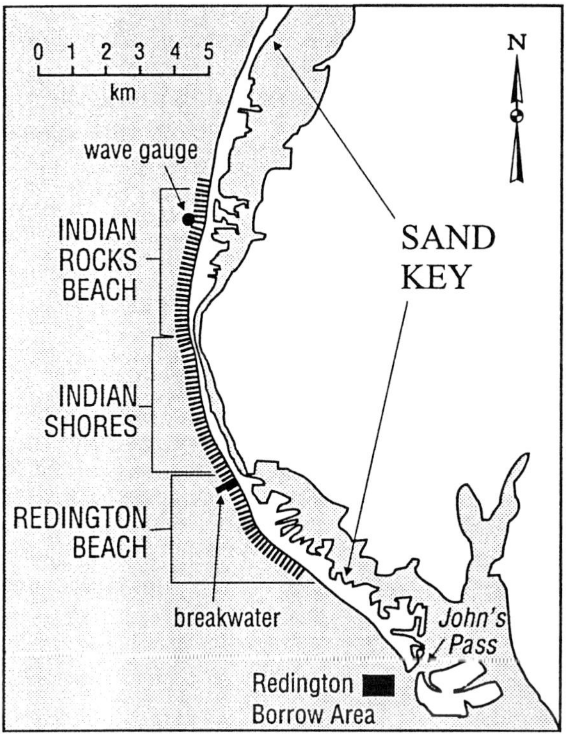

Fig. 17. Pinellas County, Florida showing the location of its three renourishment projects. Modified from Davis et al[62].

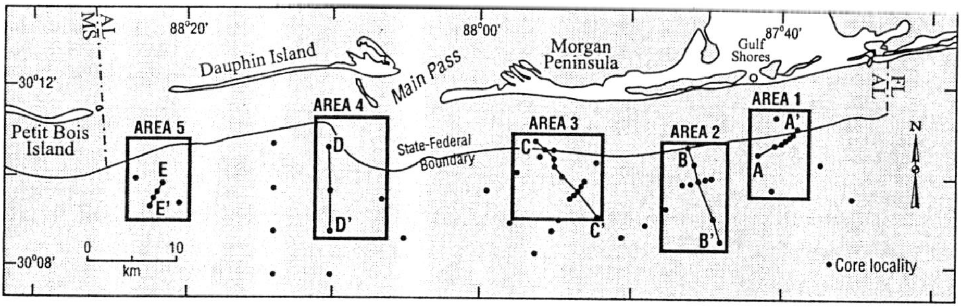

Fig. 18. Alabama coast showing the location of sand resource target areas. Modified from Davis et al[63].

making up 96% of the total deposit. It was calculated that Target Area 3 contains 151×10^6 m^3 of sand suitable for beach replenishment. The other four areas combined have another 398×10^6 m^3 of potentially usable sand. The specific area used may well depend upon the distance from the sand source to the specific beaches to be renourished[63].

7.2.3 *Grand Isle, Louisiana.* Grand Isle, Louisiana, is a barrier island located about 100 km south of New Orleans. Facing the Gulf of Mexico, Grand Isle is subject to erosive waves from occasional hurricanes and winter frontal passages. The sand of Grand Isle consists of the coarse grained fractions that accumulated as former Mississippi River delta deposits eroded. Several hurricanes occurred in the 1960s and 1970s causing much damage. Two years after Hurricane Carmen (1974) destroyed the dune ridge that backed the beach at the time, an artificial dune was constructed to a height of 2.2 m NGVD, or at an elevation about twice that of the height of the town itself (Fig. 19).

In 1983, a major project of renourishment was begun[64]. The design was to raise the dunes to 3.5 m and expand the beach by utilizing 2.3×10^6 m^3 of fill. By the time the construction was finished in the summer of 1984, more than 4.2×10^6 m^3 of sand had been used. Most (85%) of a the fill was obtained from two source areas 900 m in the Gulf of Mexico south of the center of Grand Isle. A hydraulic dredge transferred the sand from the borrow areas to compartments on the beach where it was positioned by bulldozers. The dunes were planted with sea oats and bitter panicum to help control erosion, enhance wildlife, and beautify the beach[64].

During the 1984-85 winter, several frontal passages caused erosion along some 3200 m of shoreline with a loss of 0.18×10^6 m^3 of sand. Further, two cuspate bars developed at locations on the beach that suggested they were the result of a wave climate that had been modified by the location of the two large holes created during hydraulic pumping.

Further, during 1985 three hurricanes struck Grand Isle. The first two caused erosion to reach the artificial dune at several locations. The third hurricane, Juan, in late October, caused severe damage to the beach and leveled 1800 m of the dune. The erratic nature of Juan caused storm waves to beat on the shore for several days.

The sequence of events at Grand Isle during a period of two years illustrates the "sacrificial" nature of beach and dune nourishment. It was estimated by the USACE that the project saved $12 million in damages that otherwise would have occurred in the town of Grand Isle.

7.2.4. *Galveston, Texas.*

Galveston, a city of 60,000 population, faces the Gulf of Mexico on its southeast side. Galveston Island is 45 km long and lies in the path of an occasional hurricane. Soon after a powerful hurricane struck Galveston (killing some 6000 people) in 1900, the USACE constructed a 16-km long seawall in front of the city. Like many seawalls, the one at Galveston aggravated shoreline erosion to such an extent that between 1936 and 1939 a number of groins were constructed perpendicular to it (Fig. 20). The groins were successful in trapping sand so that the shoreline accreted slightly in some locations[65]. In 1968 the groins were improved.

Fig. 19. Artificial sand dune at Grand Isle, Louisiana.

Fig. 20. Sea wall and sand fill at Galveston, Texas.

Periodic surveys since 1968 indicated that beach width fluctuated within the groin field. The beach area between the northern-most groin and the south jetty grew nearly 1500 m since the jetty was built in 1893. About two-thirds of that growth occurred during the first ten years after jetty construction. In contrast, shoreline erosion southwest of the seawall averaged 4 m a^{-1} between 1959 and 1985[66].

Because the beach at Galveston was not meeting its recreational potential, the city decided to renourish it and authorized a project to begin in 1994. The nourishment was within the groin field and extended for a distance of 5.8 km. The original sand source selected by the city was the shoal known as "Big Reef" in the Galveston Harbor ship channel. However, the presence of a hardpan layer in the shoal and its distance from the groin field prompted a search for another source area. Three promising locations were found although the surveys for two of them turned up anomaly patterns that were probably telegraph cables placed there in the late 19th century.

Renourishment was begun on December 24, 1994 and completed by June 1, 1995. The original contract had stipulated completion by April 30, 1995 so as not to interfere with the spawning season of white shrimp and red drum. Delays, because of adverse weather, necessitated an additional month of work. A total of 0.54×10^6 m^3 of sand was emplaced within the groin field[65]. As of June 1998, the beaches are loosing sand. It is being replaced by truck from the area near the jetty at the east end of Galveston[67].

7.3. THE WEST COAST

7.3.1. *Dockweiler and El Segundo Beaches, California.*

The 450 km long coastline of southern California that extends from Point Conception to the United States-Mexico border (Fig. 21) experiences a variety of wave directions and nearshore circulation patterns as well as distinctive sand sources and beach dimensions[68]. Among the most critical features of southern California's coastline are the littoral cells that divide it into six distinct subregions (Fig. 21). Each cell has its own sand sources, circulation pattern, and terminal sink[69]. Because many of the sand sources, such as updrift cliffs and inland streams, have been modified by engineering projects, natural sources of sand have been decreased to the point where beach erosion is severe. It is not surprising, given the importance of beaches to southern California residents and visitors, that beach nourishment has received much attention. Flick reports that more than 100×10^6 m^3 of sand has been emplaced on the beaches of southern California[69].

The Dockweiler and El Segundo beach replenishment projects illustrate one of California's many nourishment efforts and also serves as an example of opportunistic renourishment. These two beaches located on Santa Monica Bay near Los Angeles (Fig. 21) have suffered sand loss over the years like most other southern California beaches because of the elimination of most natural sources especially the streams that used to bring sand from inland.

Inland from the coast in the El Seguado sand dunes is the Hyperion Sewage Treatment Plant. It has been the source of sand for renourishment projects since the late1940s. In 1947 and 1948 when the plant was being constructed, more than 10×10^6 m^3 of dune sand was transferred from the excavation site to beaches at Santa Monica, Venice, and Dockweiler among others. At that time it was pumped to the beaches as a slurry by pipeline. The beaches were widened on average by 180 m.

In 1988, during an expansion of the treatment plant, excavated sand was again used for beach nourishment. This time the method of transfer was by conveyor belt from the excavation site. The sand was moved by bulldozer to a hopper that fed it onto a conveyor belt which could move 750 m^3/hr. At the beach it was placed in mounds and eventually bulldozed into the water widening the beach[70].

The actual amount of sand placed on these beaches seems to be in doubt although it apparently was on the order of $0.75x10^6 m^3$[68]. This was sufficient to widen the beaches by 33 m at Hyperion and 65 m at El Segundo.

Further, Goldman quotes an official as saying that in addition to recreational benefits, the fill "...is really the only defensive structure we have against storms. It protects public and private property from costly damage"[70].

7.3.2 *Ediz Hook, Washington.*

Virtually all of the beach nourishment projects on the west coast of the United States are to be found in southern California, the only major project along the ocean coasts of Oregon and Washington is at Ediz Hook[68]. Ediz Hook, at the northern end of the Olympic Peninsula, is comprised of sand, gravel, and cobbles. It is a spit about 5.6 km long oriented east and west at the mouth of the Elwha River. As Ediz Hook grew eastward, subsequent to sea level reaching its virtual still-stand position, the eastern end became enlarged, especially in relation to its long, narrow neck (Fig. 22). Frequently exposed to storm-wave attack this neck is subject to overrunning and breaching. In the past, breaks healed rapidly because of a strong littoral drift. Updrift cliff erosion originally supplied about 85% of the total annual volume of $0.26x10^6 m^3$ of material to the spit[71]. The other 15% came from the Elwha River.

In 1911 a dam constructed on the Elwha River eliminated 38,000 m^3 of material from reaching the river mouth and between 1930 and 1961 several structures built along the sea cliffs which reduced the annual contribution of detritus from 206,000 m^3/a^{-1} to only 31,000 m^3/a^{-1}. This reduction in potential beach material accompanied by shoreline erosion resulted in facility damage on the spit. Because the numerous hard structures used to combat shoreline erosion were not completely satisfactory, other techniques such as detached breakwaters, groins, and beach nourishment were suggested.

In deciding among the various options of shore protection a Beach Feed Evaluation Test (BFET) was conducted. One of its determinations was to decide the most appropriate locations for nourishment placement as well as the type materials to be used. Experiments included the use of gravel and cobbles, quarry rock, and pit-run sand and gravel[71]. The gravel/cobble combination proved most cost effective and was used at five locations along the spit (Fig. 22). In addition a rock blanket was placed near the distal end of the spit. The beach-nourishment materials came from glacial drift which is composed of different rock types. The project, completed in 1977-78, was subsequently monitored using direct beach observations, aerial photographs,

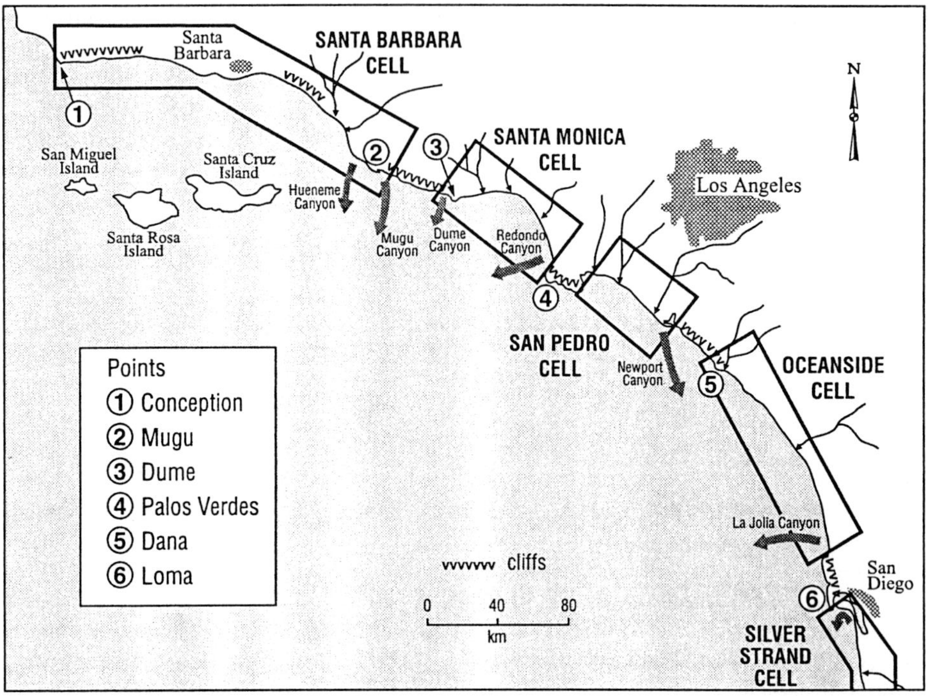

Fig. 21. Littoral cells along the southern California coastline. Adapted from Flick[69] and Wiegel[68].

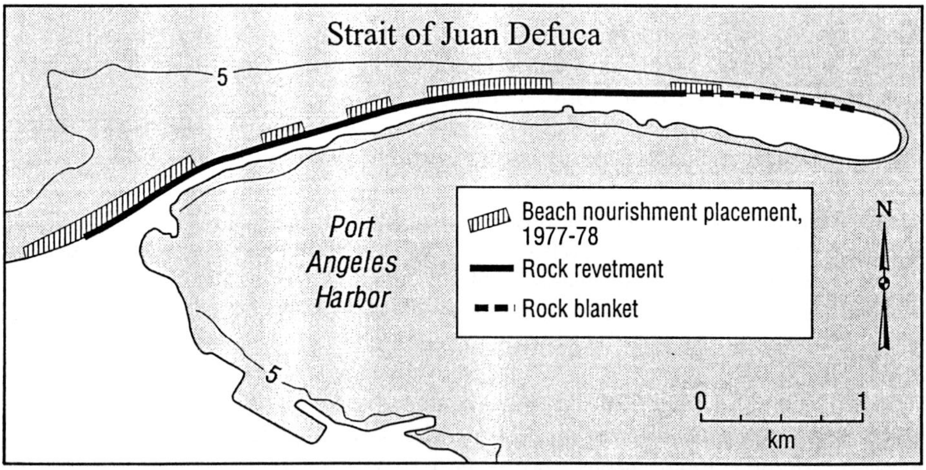

Fig. 22. Ediz Hook, Washington. Modified from Galster and Schwartz[71].

hydrography, side-scan sonar, bottom samples, and tracer cobbles. Soon after nourishment was completed there was a general merging of the materials downdrift from depositional sites. By 1984, "...the mid-tide shoreline was impinging directly on the revetment toe over much of the length of Ediz Hook;..."[71]. It had been calculated that maintenance rehabilitation would be needed about every five years. Thus from that standpoint, the project was considered successful and, in 1985, the first of the projected rehabilitation nourishments was begun. It is believed that such maintenance will continue to be needed in the future. However, studies show that if the two dams on the Elwha River should be removed the quantity of material needed could be reduced.

7.4. HAWAII

7.4.1 *Waikiki Beach.*
One of the most famous beaches in the United States is Waikiki Beach. Backed by the city of Honolulu, Hawaii, it is about 3.5 km long and extends northwest from Diamond Head to Ala Wai Yacht Club (Fig. 1). Prior to the 1920s it was a barrier beach which was used for recreational purposes by Hawaiian ruling families and affluent non-natives. In the 1920s the lagoon and wetlands (duck ponds, taro fields, and swamps) behind the bar were drained and filled with coral from offshore. The Waikiki beach area today is almost completely artificial in that it is composed of imported sand, compartmentalized with groins, and backed by seawalls[72].

The 3.5 km stretch of shoreline is composed of a number of distinct, mostly narrow and short, sections of beach. Each has had its own history of modification. In some locations, former beaches are entirely gone because of erosion associated with sea walls. Such is the situation at Kaluahole Beach near Diamond Head. Clark writes that "Ironically enough, most of the sand that remains from this once beautiful beach is contained *behind* the seawalls..."[72].

The first beach nourishment project was before World War II (1939) along 210 m of Kuhio Beach, and then between 1951 and 1957, 120,000 m^3 of sand was added to the beach between two groins[73]. Much of these early deposits were lost offshore where they filled in many of the depressions in the reefs offshore affecting surfing waves[72]. There is a dispute about some of the sources of sand although most of it seems to have come from Yokahama (Keawa'ula Beach) northwest of Honolulu. An example of another source of sand is found at the Outrigger Canoe Club. The sand was obtained from the Club's property including that excavated during building construction. Because sand in Hawaii is relatively scarce, due partly to the relatively recent geologic history of the Islands and their small size, some consideration has been given to the idea of mining offshore sand from shorelines that existed during periods of lower sea level.

8. Conclusion

Beaches, which are present along about 40% of the world's coastline[74] are among the most dynamic of environments on earth. During human history many of them have become utilized for residential, commercial, industrial, and recreational purposes.

Although they serve as barriers to storm waves, today some 70% of them are subject to erosion. Over the years many techniques have been devised and attempts made to mitigate erosional events. Traditionally, hard structures (e.g. seawalls, break waters, jetties, groynes) have been relied upon but with only partial success. During the past half-century or so, beach nourishment has come into prominence especially along the coasts of the western world.

The 21 case studies outlined here show the many factors involved in the implementation of beach nourishment projects. Various natural causes of erosion are documented although emphasis has been placed on those that have a human dimension such as beach mining, inlet dredging, seawall construction, levee building, and river damming. Various sediment sources, including onshore (often at some distance from the coast), inlets and lagoons, and offshore as well as methods of transporting the sediment to the beach to be nourished have been exampled. Often the fill is dumped directly onto the beach (by truck or as a piped slurry) and then 'bulldozed' into place. Frequently, the material to be used is left offshore (updrift) as a 'feeder bar' so that waves and currents can transport it to the desired location. The case studies also provide both general and specific examples of the durability of nourished beaches. Although durability varies greatly from one project to the next it is generally on the order of 5 to 10 years.

Despite the fact that beach nourishment has been used in many hundreds of locations, under a variety of environmental conditions, and has been frequently integrated with hard structures, there is much debate about whether it is the best solution to specific shoreline problems. Although many arguments against beach nourishment are extant, it appears to remain the most acceptable method of protecting against storm surges, advancing the shoreline seaward, and widening recreational beaches.

9. References

1. Leatherman, S.P. 1991. Coast and beaches. In: Kiersch, G.D., Ed. *The Heritage of Engineering Geology*; The first hundred years. Geological Society of America. Centennial Vol. 3, 183-200.
2. Dent, E.J. 1916. *The preservation of sandy beaches in the vicinity of New York City*. American Society of Civil Engineers.
3. Hall, J.V. Jr. 1953. Artificially nourished and constructed beaches. *Proceedings, 3rd Conference on Coastal Engineering, Cambridge, MA*, 119-136.
4. Swart, D.H. 1996. The history of coastal engineering in South Africa. In: Kraus, N.C. Ed. *History and Heritage of Coastal Engineering*. American Society of Engineers, New York, New York. 429-464.
5. Kunz, H. 1993. Sand losses from an artificially nourished beach stabilized by groynes. In: Stauble, D.K. and Kraus, N.C. Eds. *Beach Nourishment Engineering and Management Considerations*. American Society of Civil Engineers, New York, New York. 191-205.
6. Møller, J.T. 1990. Artificial beach nourishment on the Danish North Sea coast. *Journal of Coastal Research*, SI 6, 1-9.
7. Roelse, P., 1986. Artificial nourishment as coastal defence in The Netherlands: previous fills, future development. In: Pilarczyk, K.W. and Overeem, J. van (eds.*), Background Information on Artificial Beach Nourishment*. Emmeloord, The Netherlands: Delft Hydraulics Laboratory, Annex IV.
8. De Ruig, J.H.M., 1998. Seaward coastal defense: limitations and possibilities. *Journal of Coastal Conservation*, 4(1), 71-78.
9. May, V. 1990. Replenishment of resort beaches at Bournemouth and Christchurch, England. *Journal of Coastal Research*, SI 6, 11-15.

10. Psuty, N.P. and Moreira, M.E.S.A., 1990. Nourishment of a cliffed coastline, Praia da Rocha, the Algarve, Portugal. *Journal of Coastal Research*, SI 6, 21-32.
11. Zenkovich, V.P. and Schwartz, M.L., 1988. Protecting the Black Sea - Georgian S.S.R. gravel coast. *Journal of Coastal Research*, 3(3), 201-209.
12. Kiknadze, A. G., Sakcarelidze, V. V., Peshkov, V.M., and Russo, G.E. 1990. Beach forming process management of the Georgian Black Sea coast. *Journal of Coastal Research*, SI 6, 33-44.
13. Koike, K. 1990. Artificial construction on the shores of Tokyo Bay, Japan. *Journal of Coastal Research*, SI 6, 45-54.
14. Jones, J.C.E. and Schaffer, I.F., 1986. Case studies of beach nourishment in Port Phillip Bay, Victoria, Australia. *Delft Hydraulics Laboratory, Manual on Artificial Beach Nourishment*. Gouda, The Netherlands: CUR/Rijkswaterstaat/Delft Hydraulics, Report 130, 26-36.
15. Bird, E. C. F. 1990. Artificial beach nourishment on the shores of Port Philip Bay, Australia. *Journal of Coastal Research*, SI 6, 55-68.
16. Healy, T.R., Kirk, R.M., and deLange, W.P. 1990. Beach renourishment in New Zealand. *Journal of Coastal Research*. SI 6, 77-90.
17. Charlier, R. H. and De Meyer, C. P., 1995. New developments on coastal protection along the Belgium coast. *Journal of Coastal Research*, 11(4), 1287-1293.
18. Charlier, R. H. and De Meyer, C. P., 1999. Ask Nature to protect and build-up beaches. *Journal of Coastal Research*, 15(4), in press.
19. De Ruig, J.H.M. & Hillen. R., 1997. Developments in Dutch coastline management. *Journal of Coastal Conservation*, 3, 203-210.
20. Hillen and Roelse, P. 1995. Dynamic preservation of the coastline in The Netherlands. *Journal of Coastal Conservation*, 1, 17-28.
21. Malherbe, B. and LaHousse, B. 1998. Building coastal protection with sand in Belgium. *Journal of Coastal Research*, SI 26, 101-107.
22. Roovers, P.P.J.; Kerckaert, P.; Burgers, A.; Noordam, P., and de Candt, P., 1981. Beach protection as a part of the harbour extension at Zeebrugge, Belgium. *Proceedings of the 25th Permanent International Association of Navigation Congresses* (PIANIC) (Edinburg, Scotland), 2(5), 755-769.
23. Zenkovich, V. and Shuysky, Y. 1988. USSR—Black, Azov, Caspian, and Aral Seas. In: Walker, H.J. Ed. *Artificial Structures and Shorelines*. Kluwer Academic Publishers, Dordrecht. 223-240.
24. Peshkov, V.M. 1993. Artificial gravel beaches in the coastal protection. In: Kosiyan, R. and Magoon, O.T. Eds. *Coastlines of the Black Sea*. Coastal Zone '93. American Society of Engineers, New York. 82-102.
25. Kirillov, V.G., Ivashkov, G. I. and Peshkov, V. M. 1993. The Azov-Black Sea coast of Russia. An experience of artificial beach formation. In: Kos'yan, P. and Magoon, O. T. Ed. Coastlines of the Black Sea, *Coastal Zone '93*, American Society of Civil Engineers, New York. 355-365.
26. Schofield, J.C., 1985(check date). Coastal change at Omaha and Great Barrier Island. *New Zealand Journal of Geology and Geophysics*, 28, 313-322.
27. Kirk, R.M., 1980. Mixed sand and gravel beaches: morphology, processes and sediments. *Progress in Physical Geography*, 4(2), 189-210.
28. Yasuda, Y., 1988. Tokyo on and under the bay. *Japan Quarterly*, 35, 118-132.
29. Tokyo Metropolitan Government Staff, 1988. *Planning for Tokyo 1988*. Tokyo Metropolitan Government, 86p.
30. Sugawara, K., 1980. On the construction of Inage artificial beach (named "Inage no Hana"). *Fisheries Engineering, Japan*, 13(2), 29-35. (in Japanese)
31. Muehe, D. 1999. Personal Communication.
32. Vera-Cruz, D. 1972. Artificial nourishment of Copacabana Beach. *13th Coastal Engineering Conference*, Vol. II, 1451-1463.
33. Filippo, A. M. and Zee, D. M. W. 1993. Sandy beach restoration in heavily populated areas. In: Magoon, O.T. Ed. *Coastal Zone '93*. American Society of Civil Engineers. New York, 753-761.
34. NOAA 1975. *The Coastline of the United States*. U.S. Department of Commerce, Washington, D.C.
35. USACE 1971. *The National Shoreline Study*. U.S. Government Printing Office, Washington, D.C.
36. Dolan, R. 1970. Coastal landforms. In: Gerlach, A.C. (ed.), *National Atlas of the United States of America*. Washington, DC: Government Printer, pp. 78-79.
37. Estves, L.S. and Finkl, C.W., Jnr., 1998. The problem of critically eroded areas (CEA): An evaluation of Florida beaches. *Journal of Coastal Research*, SI 26, 11-18.

38. Haddad, T.C. and Pilkey, O.H., 1998. Summary of the New England beach nourishment experience (1935-1996). *Journal of Coastal Research*, 14(4), 1395-1421.
39. Trembanis, A.C. and Pilkey, O.H., 1998. Summary of beach nourishment along the U.S. Gulf of Mexico shoreline. *Journal of Coastal Research*, 14(2), 407-417.
40. Valverde, H.R.; Trembanis, A.C., and Pilkey, O.H., 1999, Summary of beach nourishment episodes on the U.S. East Coast Barrier Islands. *Journal of Coastal Research*, 15(4), 1100-1118.
41. Fisher, J.J. 1988. USA—Rhode Island. In Walker, H. J., Ed. *Artificial Structures and Shorelines*. Kluwer Academic Publishers, Dordrecht. 561-572.
42. Clark, R.R. 1993. Beach conditions in Florida: A statewide inventory and identification of the beach erosion problem in Florida. Tallahassee: Florida Department of Environmental Protection, *Beaches and Shores Technical and Design Memorandum 89-1* (5th edition), 202p.
43. Davis, D. W. 1988. USA—Mississippi and Alabama. In Walker, H.J. Ed. *Artificial Structures and Shorelines*. Kluwer Academic Publishers, Dordrecht. 615-628.
44. McCloy, J. 1988. USA—Texas. In Walker, H.J. Ed. *Artificial Structures and Shorelines*. Kluwer Academic Publishers, Dordrecht. 641-648.
45. Psuty, N.P. 1988. USA—New Jersey and New York. In: Walker, H.J. Ed. *Artificial Structures and Shorelines*. Kluwer Academic Publishers, Dordrecht. 573-580.
46. Monte, J.A. 1988. USA—Maryland. In Walker, H.J. Ed. *Artificial Structures and Shorelines*. Kluwer Academic Publishers, Dordrecht. 581-592.
47. Kana, T. W. 1988. USA—South Carolina. In Walker, H.J. Ed. *Artificial Structures and Shorelines*. Kluwer Academic Publishers, Dordrecht. 593-606.
48. Schwartz, M.L. and Terich, T.A. 1988. USA—Washington. In: Walker, H.J. Ed. *Artificial Structures and Shorelines*. Kluwer Academic Publishers, Dordrecht. 499-506.
49. Stembridge, Jr., J.E. 1988. USA—Oregon. In: Walker, H.J. Ed. *Artificial Structures and Shorelines*. Kluwer Academic Publishers, Dordrecht. 507-512.
50. Orme, A.R. and Orme, A.J. 1988. USA—California. In: Walker, H.J. Ed. *Artificial Structures and Shorelines*. Kluwer Academic Publishers, Dordrecht. 513-528.
51. Leonard, L.A.; Dixon, K.L., and Pilkey, O.H., 1990. A comparison of beach replenishment on the U.S. Atlantic, Pacific, and Gulf coasts. In: Schwartz, M.L. and Bird, E.C.F. (eds.), Artificial beaches. *Journal of Coastal Research*, SI 6, 127-140.
52. Orme, A.R., 1985. California. In: Bird, E.C.F. and Schwartz, M.L., (eds.), *The World's Coastline*. New York: Van Nostrand Reinhold, pp. 27-33.
53. Balsillie, J.H., 1996. Florida's history of beach nourishment and coastal preservation: the early years, 1910-1974. *Proceedings of the 1996 National Conference on Beach Preservation Technology. Tallahassee, Florida*, ASBPA, 350-368.
54. Hillyer, T.M.; Stakhiv, E.Z., and Sudar, R.A., 1997. An evaluation of the economic performance of the U.S. Army Corps of Engineers Shore Protection Program. *Journal of Coastal Research*, 13(l), 8-22.
55. Silvester, R. and Hsu, J.R.C., 1993. *Coastal Stabilization: Innovative Concepts*. Englewood Cliffs, New Jersey: Prentice-Hall, 578p.
56. Farrell, S. C. and Sinton, J. W. 1983. Post-storm management and planning in Avalon, New Jersey. *Coastal Zone '83*, 1:662-681.
57. National Research Council (NRC), 1995. *Beach Nourishment and Protection*. Washington, DC: National Academy Press, 334p.
58. Farrell, S. 1995. Beach nourishment at Avalon, New Jersey: A comparison of fill performance with and without submerged breakwaters. In: Tait, L. S. Ed. *Sand wars, sand shortages & sand-holding structures*. Tallahassee, Florida. 149-164.
59. Olsen, E. J. 1996. South Beach stabilization project. In: Tait, L. S. Ed. *The future of beach nourishment*, Tallahassee, Florida, 5-16.
60. Beachler, K. E. 1993. The positive impacts to neighboring beaches from the Delay beach nourishment program. In: Tait, L. S. Ed. *The State of the Art of Beach Nourishment*, Tallahassee, Florida, 223-238.
61. Lin, P. C.-P., Hansen, I. and Sasso, R. H. 1996. Combined sand bypassing and navigation improvements at Hillsboro Inlet, Broward County, Florida: the importance of a regional approach. In: Tait, L. S. Ed. *The future of beach nourishment*, Tallahassee, FL, 43-59.
62. Davis, Jr. R. A., Inglin, D. C., Gibeaut, J. C., Creaser, G. J., Haney, R. L. and Terry, J. B. 1993. Performance of three adjacent but different nourishment projects, Pinellas County, Florida. In: Magoon, O. T.

Ed. *Coastal Zone '93*, American Society of Civil Engineers, 379-389.
63. Davies, D. J., Parker, S. J. and Smith, W. E. 1993. Geological characterization of selected offshore -sand resources on the OCS, offshore Alabama, for beach nourishment. In: Magoon, O. T. Ed. *Coastal -Zone '93*, American Society of Civil Engineers, 1173-1186.
64. Combe, A. J. and Soileau, C. W. 1987. Behavior of man-made beach and dune, Grand Isle, Louisiana. - *Coastal Sediments '87*, 2, 1232-1242.
65. Spadoni, R. H. 1996. The 1994/1995 Galveston Island beach nourishment project. In: Tait, L. S. Ed. *The future of beach nourishment,* Tallahassee, Florida, 17-26.
66. Beumel, N. H. and Beachler, K. E. 1994. Beach nourishment design within an existing groin field at Galveston, Texas. In: Tait, L. S. Ed. *Alternative Technologies in Beach Preservation.* Tallahassee, Florida. 183-197.
67. McCloy, J. 1999. Personal Communication.
68. Wiegel, R.L. 1994. Ocean nourishment on the USA Pacific coast. *Shore and Beach*, 62(1), 11-36.
69. Flick, R.E. 1993. The myth and reality of southern California Beaches. *Shore and Beach*, 61(3), 3-13.
70. Goldman, J. 1988. *Sand on the run*. Los Angeles Times, Mar. 2, 1988.
71. Galster, R. W. and Schwartz, M. L. 1990. Ediz Hook – a case history of coastal erosion and rehabilitation. *Journal of Coastal Research*, SI-6, 103-113.
72. Clark, J. R. K. 1977. *The beaches of O'ahu*. University Press of Hawaii, Honolulu.
73. Wiegel, R.L. 1995. Waikiki Beach, Oahu, Hawaii. *Shore and Beach*, 63(4), 34-36.
74. Bird, E. C. F. 1996. *Beach Management*. John Wiley & Sons, Chichester.

THE MISSISSIPPI RIVER: ENGINEERED ROUTES TO THE SEA

H. JESSE WALKER
Department of Geography and Anthropology
Louisiana State University
Baton Rouge, LA 70803-4105

DONALD W. DAVIS
Energy Center/Oil Spill Coordinator's Office
Louisiana State University
Baton Rouge, LA 70803-4105

1. Introduction

Among the major rivers in the temperate and tropical areas of the world the Mississippi River was one of the last to be densely populated and intensively utilized. Other rivers, such as the Nile, Tigris-Euphrates, Ganges-Brahmaputra and Yangtze, have been centers of dense population for millennia. The Mississippi River and delta only began to be utilized intensively after becoming a center for colonization in the 18th and 19th centuries and as a major transportation route only after the passes at the river's mouth were opened to commercial traffic in the mid-19th century.

The natural setting of the Mississippi River combined with the deltaic processes operating at its mouth have been the cause of serious navigational problems from the time of the earliest adventurers to the present.

The recency of utilization of the river and the extreme difficulties associated with modification of the river's natural system combine to provide an example of how engineering, scientific, political and commercial interests combined to convert the Lower Mississippi River into one of the most utilized and modified of the world's major river systems.

2. The Lower Mississippi River: Its Natural History

The Mississippi River has altered its routes to the Gulf of Mexico repeatedly by a process known as delta-switching[1]. Since sea level reached its virtual-still-stand about 6000 years ago, this process has occurred numerous times. There have been at least seven major delta-switching events, interspersed with a number of sub-deltas associated with each course change (Fig. 1). Each time the Mississippi began a shorter and more direct route to the

J. Chen et al. (eds.), Engineered Coasts, 61–83.
© 2002 *Kluwer Academic Publishers. Printed in the Netherlands.*

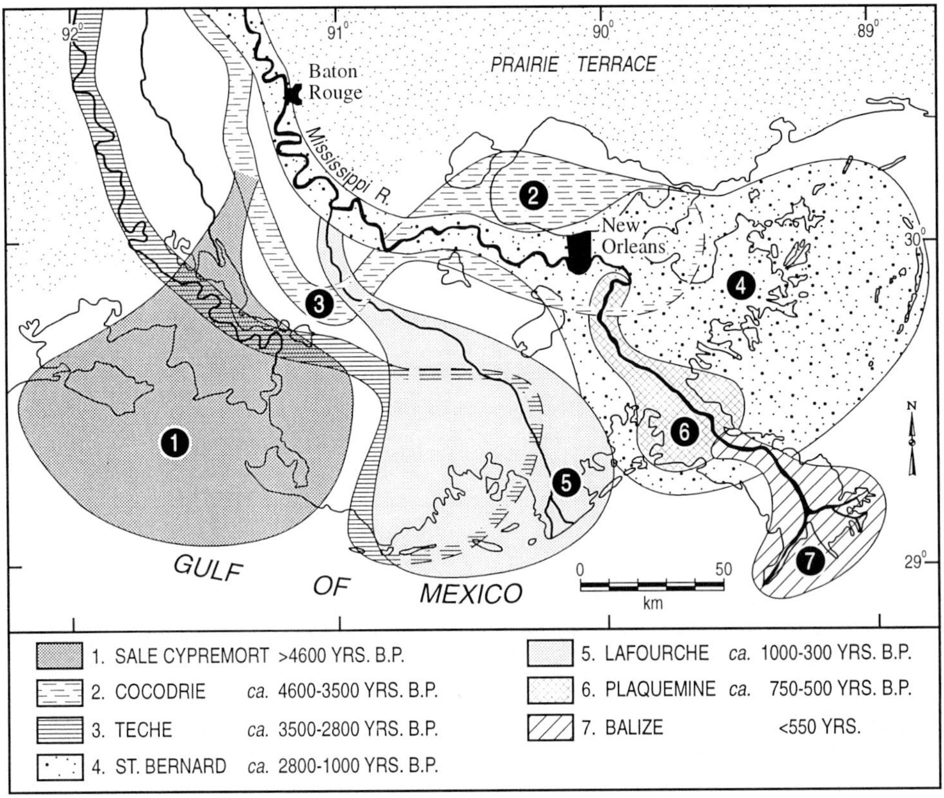

Fig. 1. Deltaic lobes of the modern Mississippi River[2].

Gulf, the old delta was gradually abandoned. Combined, these lobes have extended the coastal plain seaward along an arc more than 300 km long (Fig. 1). However, as each lobe was abandoned, subsidence and shoreline erosion caused coastal retreat. The most recent of these events, which began about 700 years ago, formed the Balize (from the French word balise, meaning "beacon") or present, bird's foot delta. The bird's foot, perched at the end of a 70-km long protrusion, is composed of several major distributaries (locally known as passes), subaerial and subaqueous levees and interdistributary marshes and bays (Fig. 2).

As the Balize delta encroaches on the continental slope, it is in danger of being abandoned for a route down the Atchafalaya River which already is carrying enough sediment to be building one of the world's fastest growing deltas. The Atchafalaya River route to the Gulf is about 300 km shorter than the route via New Orleans and, if it was not for the Old River Control Structures, it is likely that the Atchafalaya would be the Mississippi's primary channel today[3]. These structures force the Mississippi to maintain its current course and obstruct nature's most "preferred" channel, i.e., the Atchafalaya route.

3. The River's Major Distributaries And Their Modification

The Mississippi's system of distributary channels serve as natural corridors enhancing the movement of river water to the Gulf of Mexico. As early as the 16th century, French explorers recognized the Atchafalaya River as a distributary of the Mississippi River. At that time, however, the Atchafalaya was partially blocked by a log-jam or log raft. In 1860, the log raft was cleared increasing water flow through the Atchafalaya. By 1950, about 25% of the Mississippi River's water—an amount that increased annually until controlled in the mid-1950s—was diverted to the Atchafalaya.

Other distributaries within the modern system include Bayou Manchac, Bayou Plaquemine and Bayou Lafourche (Fig. 3). Pirogues, flatboats and steamboats used all of these water courses for commercial purposes. Many, like the Atchafalaya, were blocked naturally by rafts. Because they served as connections into the Mississippi, they were cleared frequently of these obstructions. Subsequently, all were modified by the construction of dams, levees or locks (Fig. 4). In addition, a network of canals connected numerous other bayous to the Mississippi. Although not part of the original distributary network, they none-the-less became important transportation routes.

Fig. 2. Image of Lower Mississippi River delta. The Balize delta is in the lower right corner. Image provided by Harry Roberts.

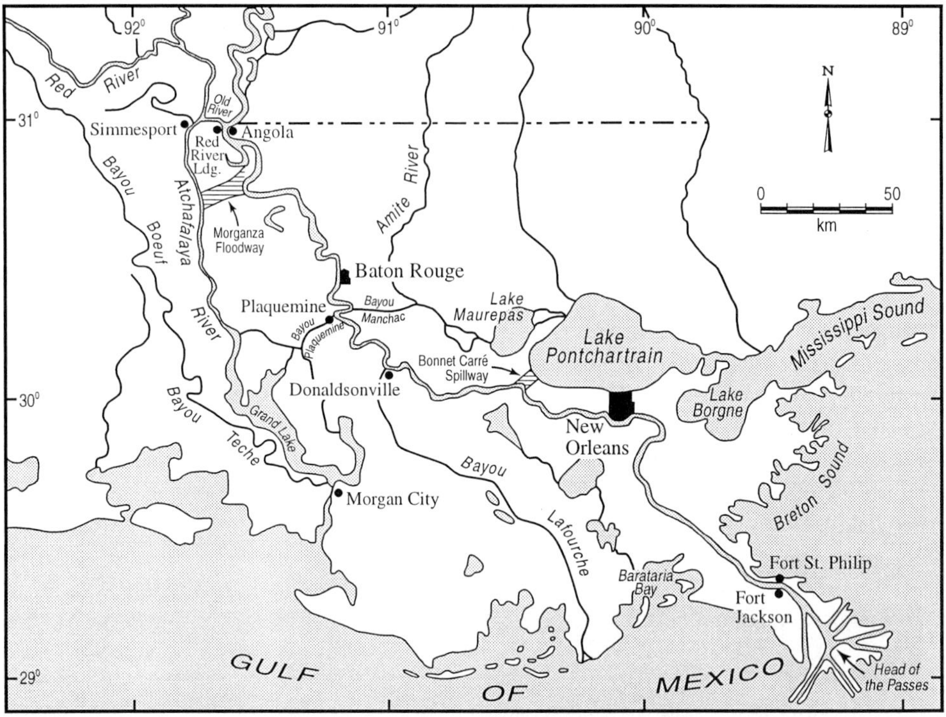

Fig. 3. The Lower Mississippi River and its distributaries and spillways.

Bayou Manchac linked the Mississippi River with what is now known as the Amite River and Lake Pontchartrain. Commercial traffic used this route to avoid the dangers associated with the "treacherous" passes through the Balize delta. This route was actively used by small boats until 1814, when General Andrew Jackson closed the bayou with an earthen dam, a dam that was made permanent in 1824[5]. Although various plans have been proposed to reopen Bayou Manchac, nothing has been done to date.

During the War of 1812, other distributaries, like Pass Manchac, were also potential routes for the British during the battle of New Orleans in 1814-15. Jackson, therefore, ordered them closed. For example, to his commander in the zone south of New Orleans, he ordered

> *... all the canals and bayous leading out of the Mississippi from a point thirty miles below the English turn, and on both sides of the river down to Fort St. Philips, filled with trees, so that no craft of any kind can pass through or navigate them*[6].

For people living along the distributary channels, such closures often caused hardships because of increased sedimentation and enhanced floods. Along Bayou Lafourche, for example, barricades caused "... great injury to the plantations threatening the lower portion of the settlement with destruction..."[7]. Such statements led Congress to appropriate funds for the removal of obstructions from some of the distributaries. Even so, by 1889 the

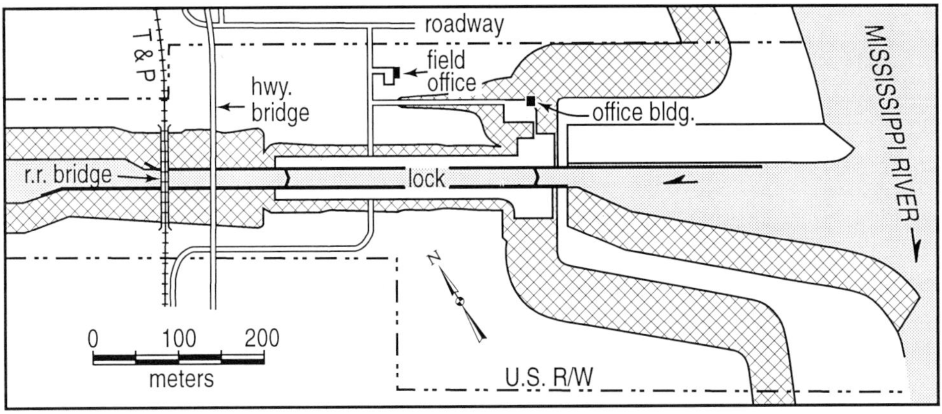

Fig. 4. The Port Allen lock[4].

Mississippi River Commission (formed in 1879) had come to the conclusion that General Jackson's actions had actually been beneficial. His blockages, they considered, were "...the first step towards closing outlets and thus confining the Mississippi to a single channel and forcing it to cut out and deepen that channel"[8].

4. The Balize Delta: Its Early Navigational History

Although Native Americans occupied the Lower Mississippi River Valley for most (if not all) of the delta-switching events, they had little direct impact on the river. Even so, the area's archaeology indicates that aboriginal peoples developed a close relationship with the land. Most landform changes in the delta occur at a rate that is measured on a human time scale[9]. Kniffen, for example, noted the consistent association of pre-historic sites with natural levees and that many lost because of subsidence[10, 11]. Shell middens, in association with natural levees, give evidence of major occupation whereas other mounds in the floodplain presumably served as refuge sites during high water. Whether the aboriginal peoples ventured into the Gulf or used the Balize delta passes intensively is unknown, although, based on explorer's accounts, Indian settlements were quite numerous in the delta plain at the time of discovery.

Also unknown is the name of the first European to see the mouth of the Mississippi River. It may have been in 1527 when the Spaniard Alvar Nunez Cabeza de la Vaca sailed along the coast and observed mud lumps[12]. These features are not only distinctive to the Mississippi River delta but also were important in the early utilization and modification of the river's outlets. It was not until 1684 (156 years after Cabeza de Vaca) that Sieur de La Salle entered the Gulf of Mexico by descending the Mississippi. La Salle was followed in 1699 by Pierre d'Iberville who successfully ascended the Mississippi from the Gulf of Mexico. This discovery and further trips, eventually led to the establishment of New Orleans, in 1717, more than 150 km upstream. Although New Orleans soon became the port city, the first overseas shipment from the Mississippi Valley was in 1705 via Bayou Manchac, not through the delta's passes.

Between 1684 and 1699 the Spanish sent at least 11 expeditions into the Gulf of Mexico to find the entrance reported by La Salle. The English were also in the area, but how many were involved is unknown. Some of those who tried concluded the entrance channels were not navigable and even gave the river such names as "Rio de la Palizada"[13]. The passes, once discovered, proved to be difficult to navigate and plagued commercial traffic from the beginning. One of the major impediments was the delta's unique mudlumps that blocked easy access to the main channel (Fig. 5). Attempts at channel improvements, were nearly continuous, politically controversial, technologically difficult, temporary and often ineffective.

Perhaps the first thoughts concerning channel maintenance were by Pauger who surveyed the Balize passes in 1722 for d'Bienville. He reported

> *... with little trouble, by giving a proper direction to the floating timber, dykes might be formed along one of the channels, and by sinking old vessels, so as to stop the others, the velocity of the water might be increased in the former, and a very great depth obtained in time... (Martin, 1882:141).*

Four years later (1726), French engineers attempted to deepen the channel by dragging iron harrows across the river-mouth bars to stir up the sediment so currents could carry it offshore. This technique was successful, even if temporary, and was used many times during the next 150 years. Improvement of the channel was a problem for engineers, as well as navigators, as suggested by M. Le Page Du Pratz. He wrote: "I know not why this entrance [Southeast] is left so neglected, as we are not in want of able engineers in France" (Du Pratz, 1774:51). The river-mouth bars were so changeable that river pilots

Fig. 5. Sketch of a 4.3 m high mudlump at the mouth of Northeast pass[14].

began to be used early in the 18th century. The fort at Balize, built in 1722, had its bar-pilots as well as its soldiers (Du Pratz, 1774). The site was utilized for 150 years. Eventually, a crevasse in 1865 diverted the Mississippi away from the channel and sealed the fate of the Balize community.

5. The Balize Delta And Its Passes

The modern delta is the product of fluvial and marine oceanic processes, with fluvial dominating. As the Balize delta extended seaward, several channels, or passes, evolved. Each of these channels has its own extension into the Gulf resulting in a planform that gives the appearance of a birdfoot (Fig. 6 a, b & c). Three passes predominate: Pass a'Loutre, South Pass and Southwest Pass. Bifurcation, a natural process (unless prevented by construction of jetties or engineered levees), is demonstrated by Pass a'Loutre which has divided into North, Northeast and Southeast Passes (Fig. 6). When a levee is breached, the associated crevasse furnishes a pathway for water and sediment to be discharged into the backswamps and interdistributary basins[17]. Before construction of levees, crevasses were a common occurrence. They directed sediment out of the main channel and reduced flood stage down river. Some of them remained open and functional for several hundred years[18].

Throughout the delta, numerous crevasses have been recorded. Cubit's Gap, Pass a'Loutre, The Jump, Batiste Collette, Grand Bayou, Joseph Bayou, and probably many other diversions within the delta were initiated by crevassing. A number of them, including Cubit's Gap, Grand Bayou and The Jump, were dammed. Such protective dams were often destroyed during flood[19]. The Jump south of Venice originated as a crevasse in 1839. By 1840, it was 0.8 km wide and 18 m deep and serving as a major sediment conduit[19].

Sometime in the 1860s, Cubit's Gap crevasse formed north of Head of Passes. Some authors speculate this course "...originated ... from a cut made by the Navy through the bulkhead of a fisherman's canal to provide a boat passage to oyster beds"[20].. Sediments moving through this outlet and deposited in Bay Rondo were responsible for creating a subdelta complex of more than 250 km^2[21]. The gap was originally quite small but enlarged to 914 m and at the time carried about 12% of the river's flow. Cubits Gap is in many ways a deltaic anomaly because it continues to distribute sediments into the interdistributary basin.

Because of the variability in river deposition and diversion, the passes vary with time in the amount of water they carry to the Gulf and in their navigability. An 1887-88 report noted: "Since LaSalle discovered the mouth of the river, two centuries ago, the deepest pass through which vessels plying to and from New Orleans have sailed, has changed no less than four times"[8]. In 1750, Northeast Pass was the main shipping route, followed in sequence by Pass a'Loutre, Southwest Pass and South Pass (Fig 6). Of these, Southwest Pass is the current (1998) principal navigation channel. South Pass and Pass a'Loutre are second and third respectively. These three passes diverge from the main channel at Head of Passes, a point in the river used for calculating distance up river and down the passes. Three additional passes out of the river, all north of Head of Passes are Baptiste Collette, Grand-Tiger Pass and Main Pass.

Because the Mississippi is a major transporter of sediment, shoaling at and near its many mouths is extensive. The quantity and location of shoaling is affected by coastal currents, tides, wind and waves, discharge, flow diversion, and the location of the salt water wedge, among other factors[22]. Shoaling (e.g. deposition within the channel itself and creation of river-mouth bars) has been the greatest problem to navigators attempting to use the passes.

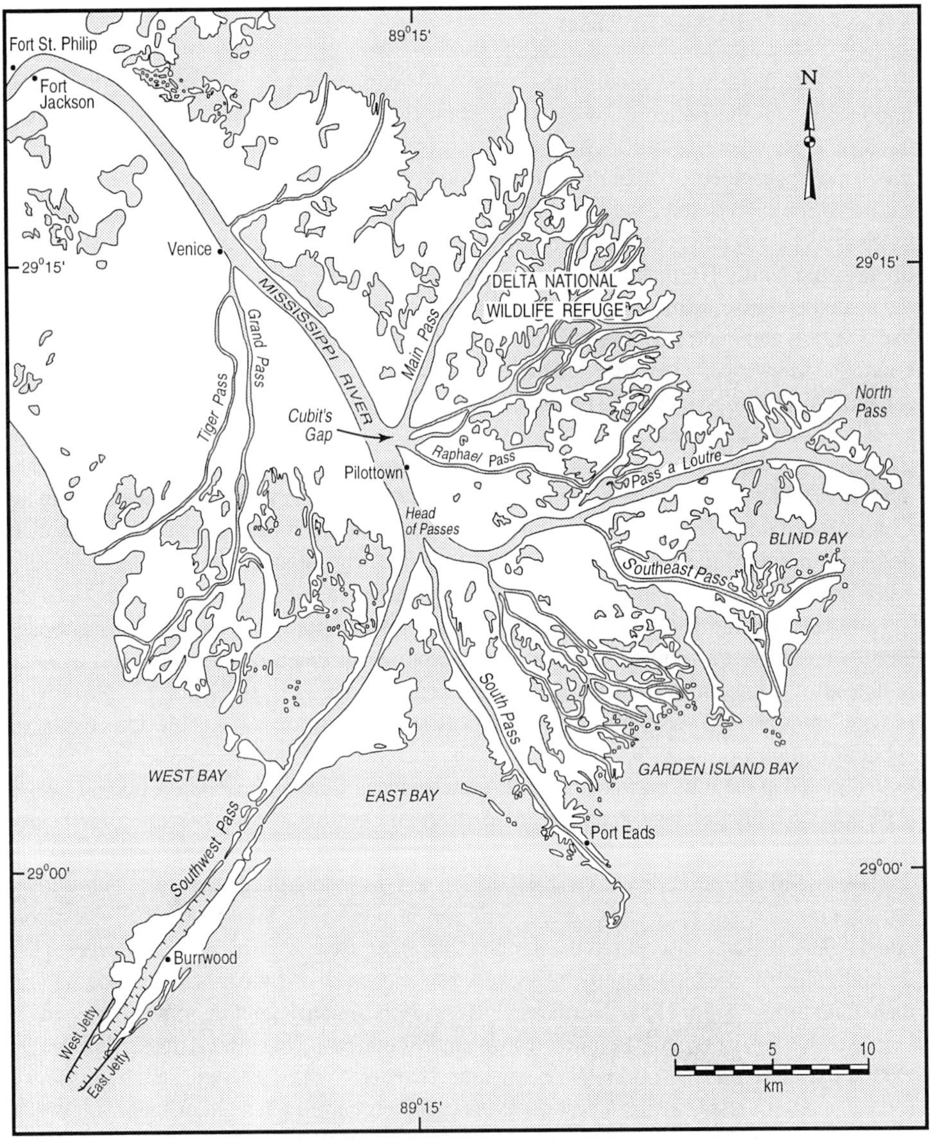

Fig. 6(a). The Birds Foot delta: map showing the passes.

Fig. 6(b). Photo looking north from the Jump.

Fig. 6(c). Photo looking south from Main Pass (Photos courtesy of the USACE, New Orleans District).

Even with early 18th century shallow-draft ships, the bars were difficult to navigate, as demonstrated by the use in 1726 of harrows to deepen the entrances to Southeast Pass. As vessel size increased, the problem was exacerbated. Ships were often delayed both outside the passes and in the river, delays that affected commerce in the entire Mississippi River basin. After the Louisiana Purchase, commerce grew rapidly as did pressure on shipping. Delays were common. For example, in 1852, 40 ships were stranded from two days to eight weeks trying to gain access to the Mississippi.

So severe were navigational issues, that river-dependent entrepreneurs lobbied Congress to appropriate the necessary funds to deepen the river's passes, since navigation was a matter of national concern. After the Civil War, the US Engineer Corps (now the US Army Corps of Engineers) tried to improve navigation in the passes. Enlarged harrows and later steam-driven propellers were used to stir up the muddy bottom. Explosives were also tried. These attempts, at best, were of limited value and, as Corthell notes, all of "...the methods employed from 1837 to 1875 were simply a contest with the depositing forces of the river and the wave forces of the gulf"[23].

6. James B. Eads And The South Pass Jetties

In 1852, after a number of "river improvement conventions" and several official surveys and reports, it was recommended that parallel jetties be fabricated at Southwest Pass and all lateral outlets closed. If these projects failed, a ship canal should be constructed between the Mississippi and the Gulf. On the east side of Southwest Pass, a 1.6 km long jetty was built that, along with harrowing and dredging, maintained a 5.5 m deep channel until maintenance was stopped because of the Civil War. After the War there were new demands for opening the passes.

In 1873, James B. Eads, a civilian engineer, proposed jetties be constructed to maintain an open channel through Southwest Pass. Eads' plan was adopted (over strenuous opposition by Andrew A. Humphreys, Chief of the US Engineer Corps) with two notable exceptions. Congress insisted Eads fund the project personally and the jetties be built at South Pass instead of Southwest Pass.

Objections to Ead's proposed jetty system included: 1. the jetties would be undermined by waves, 2. they would sink into the unstable sediments, 3. the mouth-bar would reform at the end of the jetties and 4. mudlump formation would destroy them[19].

Eads' engineering principles varied only slightly from those Pauger advocated in 1722. However, his method of forming the channel's "dykes" was different. Eads used broad, flat mattresses of willows, anchored to a row of piles as the jetty's foundation (Figs. 7 a & b). The mattresses were constructed upstream, towed to the location, and sunk by loading them with rocks. Several layers were laid on top of each other until the appropriate height was achieved. The Gulf ends of the jetties were subject to strong wave action and, therefore, were constructed of concrete[23]. Eads used a large mixer from which the concrete was transported to boxes (molds) on small rail cars.

In order to promote channel scour, the rate of flow and volume through South Pass was increased by placing wing dams in the channel and by constructing T-dams near Head of

Passes. In addition, low sills were built in Southwest Pass and Pass a'Loutre to divert more water toward South Pass[26]. To further increase channel deepening, Eads commissioned construction of a dredge boat outfitted with the largest centrifugal pump built to that time. Its suction pipe was 68.5 cm in diameter[27].

The jetties were a success (Fig. 8) and largely responsible for commercial expansion within the Mississippi River Valley. They also played an important role in the creation, in

Fig. 7(a). Ead's jetties: willow mattresses in process of construction [24].

Fig. 7(b). Inside willow sand-barriers[25].

Fig. 8. Diagram showing South Pass cross-section prior to and subsequent to dredging[25].

1879, of the Mississippi River Commission[26]. The importance of Southwest Pass to commerce is illustrated by Trowbridge who wrote that :

> *In total value of exports and imports coming in and going out each year, South Pass is second only to the entrance to New York Harbor among the commercial channels of the United States*[28].

Even after Southwest Pass was jettied and became the principal shipping channel, maintenance dredging was continued. The outlet was further improved in 1938 by adding a concrete cap over the jetties[29].

7. Southwest Pass

Although Eads believed Southwest Pass was better suited for jettying than the river's other passes, politics intervened and, reluctantly, he used South Pass. His work on South Pass was successful and served commerce well during the rest of the 19th century. However, as the century progressed, ship size and draft increased and South Pass became insufficient in width and depth. To counteract this problem, Congress recommended deepening Southwest Pass. Begun in 1903, work on Southwest Pass was essentially completed by 1908 (Fig. 9 a & b). However, Cowdrey[26] reported that work continued after that date; by 1923 the jetties had reached a length of 7597 m. Unfortunately the Southwest Pass jetties had been constructed too far apart to insure maximum channel scour. To correct this problem engineers, over the years, have added spur- or cross-dykes between the jetties (Fig.10).

Channel specifications were frequently changed. In the late 1980s, the US Army Corps of Engineers was charged with keeping the navigation channel of Southwest Pass at 12.2 m deep and 244 m wide from Head of Passes to mile -18 (-29 km). From mile -18 to the Gulf, the channel's width was reduced to 183 m[22]. Today the depths are maintained at 13.7 m.

Shoaling associated with Southwest Pass is a continuous and expensive maintenance issue, especially near the ends of the jetties but also within the channel itself. The shoals are directly related to discharge rates. During high water, shoaling occurs at Head of Passes and the tip of the jetties. At low water, shoals form further upstream. Maintenance is very expensive. In 1997, $37 million US was budgeted for Southwest Pass alone. Unusually large sediment years, such as the October tropical storm and severe flooding in the Ohio River Valley in 1997, tax dredging resources heavily.

Studies conducted in the late-1950s and again in the mid-1980s examined the best ways to reduce channel shoaling. Recommendations included: construction of lateral spur dikes, a sediment diversion channel, sediment basins, and a new entrance channel[22].

The river's jetties frequently need to be repaired because of subsidence and storm damage (Fig. 11 a & b). After a decade of neglect during the 1940s, portions of Southwest Pass' east jetty were below sea level. This jetty is subjected to more severe waves than its western counterpart. Therefore, in 1949 it was improved with the addition of more than 72,000 metric tons of granite shipped by rail to New Orleans from Stone Mountain, Georgia and then barged down river[29].

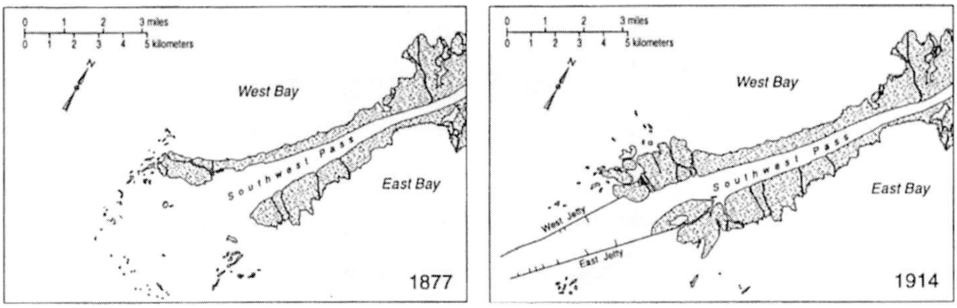

Fig. 9. Maps of Southwest Pass: (a) prior to jetty construction, (b) in 1914 after jetty construction. Coast and Geodetic Survey, Mississippi River. Charts 94(1877) and 194(1914).

Fig. 10. Photo of west channel showing dikes along left bank and midchannel dredge (Photo courtesy the USACE, New Orleans District).

In 1973, high water damaged structures and banks in the river south of Venice and in Southwest Pass. Bank restoration included foreshore dikes, bank nourishment, bulkheads, jetty heads and lateral pile dikes[30]. After this work in 1974, additional damage was done and additional restorations of both the east and west jetties has been needed. The extent of the work is suggested by the fact that by 1990 rock foreshore dikes and bank nourishment projects were completed along 80 km of channel. Some of the material used is dredged from the channel, the remainder is used in the creation of wetlands (Figs. 10 & 12). The average annual volume of dredged material needed to maintain the channel in Southwest Pass is 15 million m^3/a.

Fig. 11(a) and (b). Jetty repair, Southwest Pass (Photos courtesy USACE, New Orleans Distict).

Fig. 12. Dredge spoil, Southwest Pass (Photo courtesy USACE, New Orleans District).

There recently has been a change in the type of construction used for dikes. In 1994, "soft dikes" were used for the first time in Southwest Pass. They consist of geotextile bags filled with 230 m^3 of sand. They were placed perpendicular to the channel, thus narrowing it and increasing the rate of flow and reducing shoaling[30].

8. The Mississippi River Gulf Outlet (MRGO)

Although the passes provide natural outlets to the Gulf, they were difficult to use and, for most destinations, increased the distance traveled. Thus, it is not surprising that plans for excavating canals were formulated early. The first suggestion appears to have been in 1774 by Du Pratz who wrote:

> *...this navigation might be easily mended,...by means of a bay...lying to the west of the South Pass of the river...not above a mile from the Mississippi, above all shoals...where the river has one hundred feet of water. By cutting through that one mile then, it would appear that a port might be made there for ships of any burden;...* [16].

A somewhat more serious effort was made in 1832 when Chief Engineer Benjamin Bouisson of Louisiana proposed to build a canal from Fort St. Philip to Breton Sound (Fig. 13). The US War Department, however, declared it was too expensive. The notion of a canal was revived in 1858 and again in 1871—the same year Eads' jetty plan was being considered. If the jetties had failed, the canal would probably have been built at that time[31].

The Mississippi River Gulf Outlet (MRGO) was authorized by the River and Harbor Act of 1956, the Water Resources Development Act of 1976, and the Water Resources Development Act of 1986[32]. Construction was begun in 1958 and completed in 1968 (Fig. 13). Subsequent work included lengthening MRGO's jetties. The north jetty extends to mile 20.2 (32.5 km), the south jetty to mile 14.8 (23.8 km). For erosion abatement, plans call for adding 48 km of rock dikes along the channel's north bank. It is argued that this addition will reduce channel maintenance and also help preserve the adjacent marsh. In 1995, extension of the jetties was deferred, because maintenance dredging has proven more economical[30].

The MRGO route is 60 km shorter than that of the Mississippi River. The route extends from the Gulf Intracoastal Waterway (GIWW) past Lake Borgne and associated wetlands to the ll.6 m contour in the Gulf of Mexico.

9. Emergency Outlets

In flood, the Mississippi frequently created crevasses. Although crevassing is a natural process, once engineered levees were constructed on top of natural levees and population densities increased, flooding became increasingly more damaging with each crevasse occurrence. In the mid-19th century engineers suggested constructing floodways or spillways to help control flooding. In 1816, William Darby recommended using controlled crevassing as a possible means to reduce flooding. However, such plans were opposed during the latter half of the century by the powerful 19th century engineering duo of Humphreys and Abbott[33].

After the "great-flood" of 1927, a system of floodways and spillways was planned to "... duplicate the effects of the swamp reservoirs and natural outlets ..."[26]. Edgar Jadwin, Chief of Engineers, proposed that low levees (called "fuse-plugs") be built to emulate crevasses during extreme floods. It was suggested these structures be placed at locations that historically had been sites of prolonged flooding.

Three major flood outlets were constructed. All are designed to divert floodwater from the Mississippi River to the Gulf of Mexico. The three are: the Bonnet Carré Spillway completed in 1936, the Morganza Floodway completed in 1954, and the Old River system (Fig. 3). Combined, these three emergency outlets can carry half of the river's total flood discharge[30].

9.1. BONNET CARRÉ SPILLWAY

In 1850, a crevasse, 2 km wide, broke through the Bonnet Carré levee and flowed for more than six months at a rate of more than 4000 m^3/s[34,35]. As long as water passed through the gap with sufficient velocity, the opening continued to widen. Ellet wrote that the floodwaters did not cut a channel and that "...the furrows left by the plough and the roots of

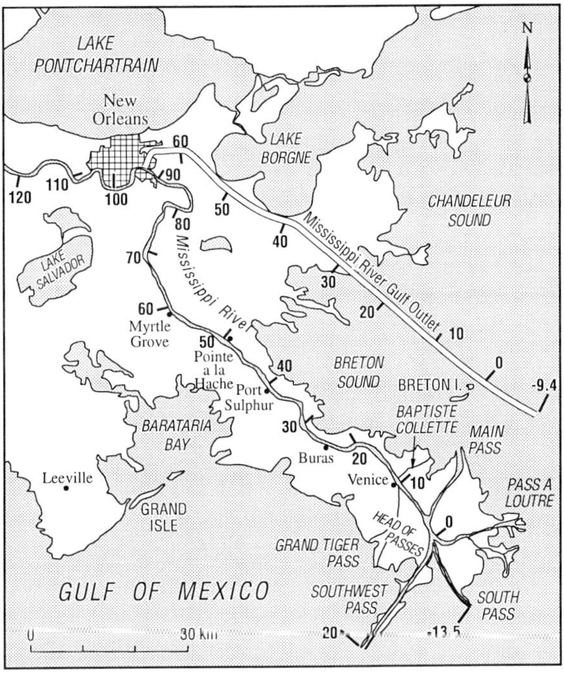

Fig. 13. Map of the lower delta showing the Mississippi River Gulf Outlet. Note mileage markers on both the MRGO and the main channel.

the crops remained on the field where it had been swept by the water, after the flood had subsided"[34]. An 1874 break remained open for ten years and raised the level of Lake Pontchartrain by 0.6 m[36]. Water flowed through a gap 417.5 m wide with an average channel depth of 6.8 m. The flow was so strong that the steamer Katie was drawn into the vortex and saved by the use of tugs[37]. Further, damage was caused to railroad tracks and to Lake Pontchartrain's oyster beds and saltwater fishery.

The Bonnett Carré Spillway, was begun in 1929 and completed in 1936. It can carry 7080 m^3/s (250,000 ft^3/s) from the river to the Gulf via Lake Ponchartrain. The spillway consists of a concrete control structure in the riverbank with guide levees between the river and the lake that range from 2350 m to 2780 m apart. It has 350 bays which are equipped with removable timbers called "needles." Because of the volume of concrete needed for the structure, a laboratory, to determine the most effective mixture, was established. The laboratory was successful in developing a concrete that nearly doubled the compressive capacity of concrete being produced in the 1930s. In operation, cranes move on tracks across the top of the structure to lift the needles from the bays allowing flood water to escape the river. It has been used eight times since being completed in 1936.

9.2. THE MORGANZA FLOODWAY

The Morganza Floodway, located about 425 km above Head of Passes has been used only once and then only partially. This was during the major flood of 1973. The floodway has 5395 m of levee and 128 gated openings each 8.6 m wide and is capable of carrying 18,400 m^3/s (650,000 ft^3/s). Flood waters entering the floodway flow into the Atchafalaya River basin and on into the Gulf of Mexico.

9.3. THE OLD RIVER COMPLEX

Before the Old River control project was built, the Mississippi River and the Atchafalaya River were connected by a short channel called Old River (Fig. 14). It had been determined that sometime between 1965 and 1975 about 40% of the Mississippi River's flow would be going down the Atchafalaya unless something was done to control further capture. In 1954, the 83rd Congress approved a plan to "...maintain the balance of flow from the Mississippi River into the Atchafalaya River and Basin by control structures on the right bank of the Mississippi River"[30]. The main features included a low-sill control structure, an overbank control structure, navigation locks, levees and an Old River closure structure (Fig. 14). It is designed to transfer as much as 17,560 m^3/s (620,000 ft^3/s) from the Mississippi River into the Atchafalaya.

The major flood of 1973 caused excessive turbulence adjacent to the low sill structure eroding a large scour hole beneath it and destroying an adjacent wing wall. Emergency work during the flood saved the structure. Subsequently the structure was strengthened and the wing wall was replaced by a rock dike. The structure is designed to withstand a difference in water levels between a flooding Mississippi and the lower Atchafalaya River of 6.7 m (22 ft). In recent years an auxiliary control structure and a 192-megawatt power plant have been added to the complex (Fig. 15).

10. Other Mississippi River Delta Channels

The wetlands of Louisiana are a maze of artificial canals. The earliest may well stem from Indian days when pirogues were guided through the marshes in search of game. These increased in number with the early European settlers and were later added to by oil companies in their exploitation of petroleum products. Many of these canals have enlarged, especially because of erosion caused by passing motorized vessels and by dredging and have contributed to land loss in coastal Louisiana.

Venice, located 18 km above Head of Passes, is a main center for much of Louisiana's offshore oil industry. Because of the importance of this activity, a number of Mississippi River outlets were dredged and jettied in order to reduce transportation costs. For example, Baptiste Collette Bayou and Grand-Tigre Pass are maintained with depths of 4.25 m and have jetties built out to the -1.8 m contour[30]. Dredging materials are used to create islands and wetlands.

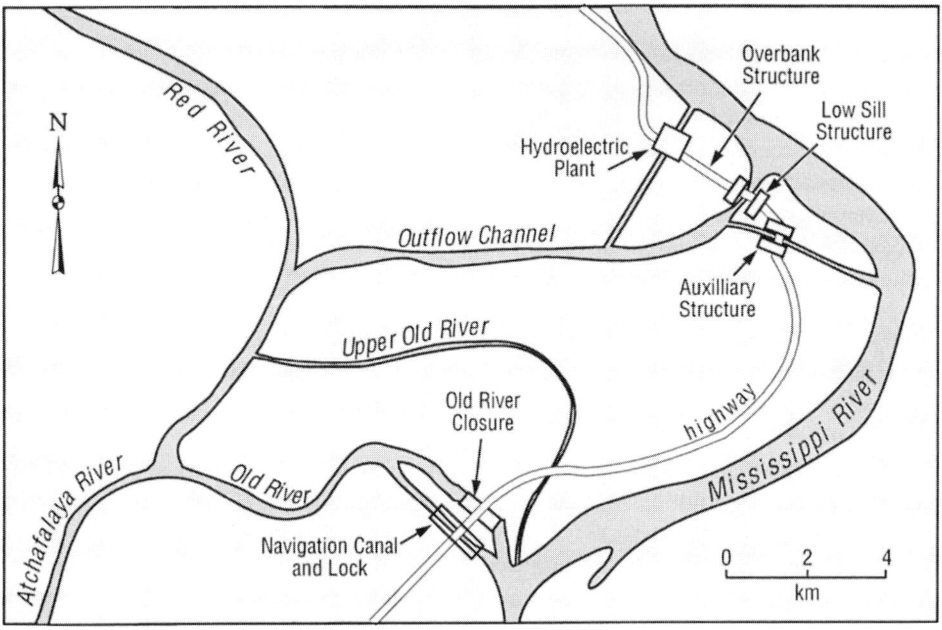

Fig. 14. Old River control structures

Fig. 15. Photo of the 192-megawatt power plant at the Old River Control Structure (Photo courtesy J. B. Lancaster, Jr.).

11. The Mississippi River And Other Engineered Structures

The rapid rate of land loss in coastal Louisiana began receiving serious attention in the 1970's and various government and private organizations began to take action to curb such loss. The loss, to a large extent, is the result of the reduction in the amount of sediment reaching the coastal marshes due to trapping in upriver dams and funneling into deep water by artificial levees. Actions to help mitigate this loss include construction of freshwater diversions, regulating dredge and fill activities, and rebuilding barrier islands[38]. Constructed diversions are of two types, those that siphon freshwater into the marshes and sounds in order to promote biologic activity and those designed to be like crevasses in order to promote sediment accretion in marshes that are under erosional stress (Fig.16). Between 1983 and 1995, 24 crevasses were constructed in the Delta National Wildlife Refuge (DNWR) (Fig. 6). Several organizations, including oil and gas companies, the US Army Corps of Engineers, and the US Fish and Wildlife Service, were involved in their construction[39].

The Mississippi River has a thalweg that is below sea level for more than 560 km upstream from the Gulf of Mexico. Salt water intrusion into the channel is a common occurrence during low stages of the river. In 1988, drought in the midwest lowered discharge sufficiently so the saltwater wedge advanced upstream at rates of more than 3 km/day. It appeared the wedge would reach at least 186 km above Head of Passes (therefore, above New Orleans), if it was not stopped. A saltwater barrier was built at a location 102.5 km above Head of Passes from the river bottom to the navigation depth of 13.7 m. The saltwater wedge did not reach the New Orleans water intake. However, at locations downstream from New Orleans, fresh water had to be imported by barge because of high salinity. Although it has been estimated that a sill will be needed every five years, since 1988 one has not been needed[40].

12. References

1. Roberts, H.H. 1998. Delta switching: early impacts of the Atchafalaya River diversion. *Journal of Coastal Research*, 14(3), 882-899.
2. Kolb, C.R. and Van Lopik, J.R. 1966. Depositional environments of the Mississippi River deltaic plain, southeastern Louisiana: In: Shirley, M.L. Ed. *Deltas in their geologic framework*, Houston Geological Society, 17-61.
3. Fisk, H.N. 1952. *Geologic investigation of the Atchafalaya Basin and the problem of Mississippi River diversion,* U.S. Corps of Engineers, Mississippi River Comm. Vicksburg, MS, V. 1, 145 p. v. 2, 36 p.
4. USACE. 1992. *1992 flood control and navigation maps of the Mississippi River. 59th edition.* Mississippi River Commission, New Orleans.
5. Kniffen, F.B. 1935. Bayou Manchac: a physiographic interpretation. *The Geographical Review*, 25, 462-466.
6. Smith, M.L. 1856. *Report of the Secretary of War.* 34th Congress, 1st session, Ex. Doc. No. 67, 3.
7. Pugh, W.W. 1856. *Obstructions in the Bayou Lafourche.* 34th Congress, 1st session, House of Representatives, Mis. Doc. No. 80, 1-2.
8. House of Representatives, 1887-8. *Internal Commerce of the United States.* 50th Congress, Executive Document 2552.
9. Pearson, C.E. and D.W. Davis. 1995. *Cultural adaptation to landlords in the Mississippi River delta plan.* In Geological Society of American Annual Meeting , New Orleans, Louisiana, Field Trip Guide Book.

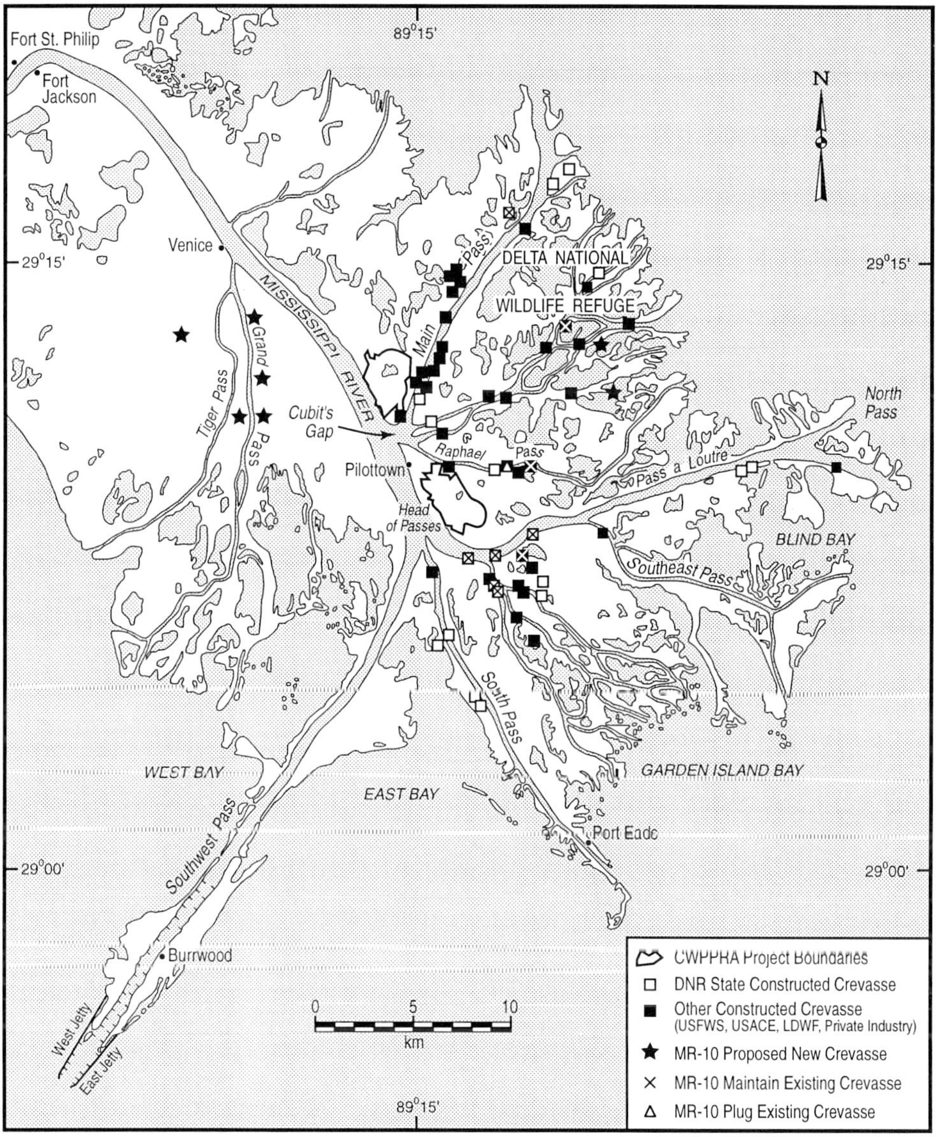

Fig. 16. Map showing the location of the recently constructed freshwater diversions (Map redrafted from Database Analysis Section, Coastal Restoration Division, Louisiana Department of Natural Resources. Map Data Source: 1994 Satellite Imagery).

10. Kniffen, F.B. 1936. Preliminary report on the Indian mounds and middens of Plaquemines and St. Bernard Parishes. In R.J. Russell, H.V. Howe, J.H. McGuirt, C.R. Dohm, W. Hadley, Jr., F.B. Kniffen and C.A. Brown. *Lower Mississippi River delta: report on the geology of Plaquemines and St. Bernard Parishes.* Louisiana Geological Survey Geological Bulletin 8, 407-422..

11. McIntire, W.G. 1958. *Prehistoric Indian settlements of the changing Mississippi River delta.* Coastal Studies Series No. 1. Baton Rouge: Louisiana State University Press.
12. Walker, H.J. and Grabau, W.G. 1992. Mudlumps. In: Janelle, D. Ed. *Geographical Snapshots of North America.* Guilford Press, London, 211-214.
13. McWilliams, R.G. 1969. Iberville at the birdfoot subdelta. In: McDermott, J.F., Ed. *Frenchmen and French ways in the Mississippi Valley.* Urbana: Univ. of Illinois Press, 127-140.
14. Lyell, C. 1889. *Principles of Geology.* 11th edition. Appleton, New York.
15. Martin, F-X. 1882. *The History of Louisiana from the earliest period.* James A. Gresham, Publisher, New Orleans. Republished in 1963 by Pelican Publishing Co., New Orleans.
16. Du Pratz, M. 1774. *The history of Louisiana.* Translated from the French by Tregle, Jr. J. and published by Louisiana State University Press. 1975.
17. Kesel, R. H. 1989. The role of the Mississippi River in wetland loss in southeastern Louisiana, U.S.A.: *Environmental Geology and Water Science*, v. 13, no. 3, 183-193.
18. Gagliano, S. M. and J. L. Van Beek, 1970. *Geologic and geomorphic aspects of deltaic processes, Mississippi delta system,* Hydrologic and Geologic Studies of Coastal Louisiana, Report pt.1, v. 1: Baton Rouge, Coastal Studies Institute, Louisiana State University, 140 p.
19. Russell, R.J., H.V. Howe, J.H. McGuirt, C.F. Hohm, W. Hadley, Jr., F.B. Kniffen, and C.A. Brown, 1936, *Lower Mississippi river delta: reports on the geology of Plaquemines and St. Bernard parishes.* Geological Bulletin No. 8. New Orleans: Louisiana Geological Survey. 454 p.
20. Mitchell, H. 1883. *Review of surveys and gauging of Cubits Gap, made in 1868, 1875, and 1876.* House Document, pt. 2, Report of the Secretary of War, vol. 2, pt. 3, Appendix U of Appendix SS (Mississippi River Commission) 48th Congress, 1st Session: Washington, D.C., U.S. Government Printing Office, 2302-2304.
21. Welder, F.A. 1959. *Processes of deltaic sedimentation in the lower Mississippi.* Baton Rouge: Coastal Studies Institute, Louisiana State University. 90 p.
22. Benson, H.A. and Boland, Jr., R.A. 1986. *Mississippi River Passes Physical Model Study, Report 2.* USACE, New Orleans.
23. Corthell, E.L. 1880. *A history of the Jetties at the mouth of the Mississippi River.* John Wiley & Sons, New York.
24. *Frank Leslie's Illustrated Newspaper.* April 14, 1877.
25. *Harper's Weekly.* 1893. New York.
26. Cowdrey, A.E., 1977. *Land's End.* USACE, New Orleans.
27. Dabney, T.E. Undated.*The man who tamed the Mississippi.* Ms in Hill Memorial Library, Louisiana State University.
28. Trowbridge, A.C. 1921. *Preliminary geological report on the Mississippi delta.* 105 pp. unpublished ms.
29. Gormin, K. 1949. *Jetties from Georgia.* Times Picayune, New Orleans.
30. Saucier, N.H. 1995. *Water resources development in Louisiana.* USACE, New Orleans District, New Orleans, 143 pp.
31. Anon 1937. *Eads daring plan saves New Orleans Commerce.* Times Picayune 1912.
32. USACE. 1997. *Project fact sheet, Mississippi River Gulf Outlet. Water Resources in Louisiana,* http://www.mvn.usace.army.mil/ops.
33. Humphreys, A.A. and Abbott, H.L. 1861. *Report upon the physics and hydraulics of the Mississippi River...*USA Engineers, Prof. Paper No. 4.
34. Ellet, C., Jr. 1852. *Report on the overflows of the delta of the Mississippi*, Senate Executive Document No. 26, 22nd Congress, 1st Session: Washington, D. C., United State Government Printing Office, 1-106.
35. Elliott, D. O. 1932. *The improvement of the Lower Mississippi River for flood control and navigation,* v. 1: Vicksburg, Mississippi, U.S. Army Corps of Engineers, Waterways Experiment Station, 1-158.
36. Gunter, G. 1950. *The relationship of the Bonnet Carré spillway to oyster beds in Mississippi Sound and the "Louisiana Marsh," with a report on the 1950 opening and study of beds in the vicinity of the Bohemia Spillway and Baptiste Collette Gap.* Report prepared for The District Engineer, New Orleans District, Corps of Engineers, 13 p.
37. Warfield, A. G. 1876. *Mississippi River levees,* House Report No. 494, 44th Congress, 1st Session: Washington, D. C., U.S. Government Printing Office, 1-26.

38. Cowan, J.H., Jr., R. E. Turner, and D.R. Cahoon. 1989. Marsh management plans in practice: do they work in coastal Louisiana, USA? *Environmental Management* 12(1):37-53.
39. Boyer, M.E. 1996. *Constricted crevasses as a technique for land gain in the Mississippi River delta.* Unpublished MS Thesis, Louisiana State University, Baton Rouge, LA.
40. Laurent, A.C. 1997. *An overview of the Mississippi River's saltwater wedge.* http://www.mvn, usace.mil/eng/saltwater/wedge.

TOKYO BAY REFORMATION

KENJI HOTTA
Nihon University
7-24-1 Narashinodoi Funabashi-shi
Chiba 274, Japan

1. Overview of Tokyo Bay

Tokyo Bay is a long, oval-shaped body of water surrounded by Tokyo, Chiba Prefecture, and Kanagawa Prefecture (Fig. 1). It is connected to the Pacific Ocean by Uraga Channel (90 m deep, 7 km wide). Tokyo Bay, from its mouth to its most distant point, has a length of approximately 90 km; it has a width of about 20 km. The bay encompasses a water area of about 1,200 km^2. It is relatively shallow, 18 m on the average. The total length of the coastline inside the bay is approximately 180 km. Tokyo Bay has 31 ports, of which Yokohama, Yokosuka, Kawasaki, Tokyo, Chiba, and Kisarazu are the largest. As a result, many ships navigate the bay and the traffic is always heavy.

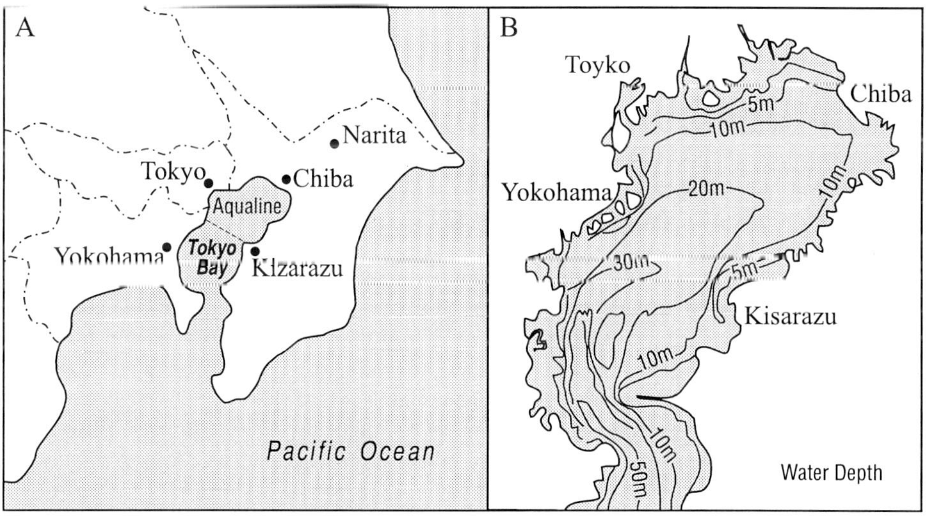

Figure 1. Tokyo Bay: A. Location; B. Water depth.

Water exchange in the bay is via Uraga Channel. It is estimated that two billion tons of seawater go in and out of the bay mouth with every tide. This volume is 11 to 12% of

J. Chen et al. (eds.), Engineered Coasts, 85–102.
© 2002 *Kluwer Academic Publishers. Printed in the Netherlands.*

the entire volume of Tokyo Bay (18 billion tons). The tide current at the point farthest from the bay mouth is extremely slow (no more than 0.4 to 0.5 knots), which means that the bay is slow to recover from pollution. A total of 23 rivers flow into Tokyo Bay, of these Edogawa River, Arakawa River, and Tamagawa River are the largest. Total annual inflow reaches approximately eight billion to 10 billion tons.[1]

The inflow from these rivers has aggravated the bay's pollution to such an extent that the bay is in a constant state of eutrophy. Red tides and blue tides have occurred repeatedly. Although the rivers carry 300 tons of nitrogen and 200 tons of phosphorus into the bay every day, the volume of these materials are decreasing every year. The load of COD reached 1,500 tons per day in the 1970's; today it is less than 500 tons (Fig. 2). Although the bay is rich in fish and other organisms, the continuing loss of tideland and sea jungles to landfill is weakening the bay's ability to support reproduction of sea life[2].

2. Reclamation and Use of Tokyo Bay

Because Tokyo Bay is adjacent to massive plains, it has been used as a port since ancient times. Land reclamation has been carried out at the Port since at least the day of the triumphal entry of Ieyasu Tokugawa into Edo Castle in the late 16th century. The land in the vicinity of present-day Hibiys and Nihonbashi was reclaimed during the Edo Period. Land reclamation began to be carried out in earnest after the Meiji Restoration in 1886, the year that marks the beginning of modern Japan. From 1906, when the project to improve the Sumida River estuary was initiated, to the present, approximately 4,396 ha of land, or an area equivalent to that of Chiyoda, Chuo and Minato Wards of Tokyo, have been reclaimed. Reclamation projects continue to be implemented at the Outer Central Breakwater, reclaiming approximately 314 ha of land through the disposal of waste and other matter. Thus, hand in hand with construction works, the reclamation of land has been an important activity in the Port for several centuries.[3]

Since 1955, in particular, as economic growth shifted into high gear under the leadership of the Japanese heavy chemical industry, people and industries began to concentrate in the city. As a result, Tokyo grew into a giant metropolis within a short period of time. As consumption soared in the city with the sudden improvement in living standards, it became necessary to transport massive volumes of goods. Therefore, the harbor was forced to increase service. The development of Tokyo Bay had been concentrated in the Keihin area until that time, but then the emphasis shifted to Chiba Prefecture, where a coastal industrial zone was developed by reclaiming a large area of land. This development triggered the rapid reclamation of the bay. To deal with this situation, the government devised a major harbor plan for Tokyo Bay in 1966. One reason that landfill increased in Tokyo Bay is that industrial land was scarce due to the intensive use of the cities real estate. Another reason is that land formed by landfill is less expensive than land in the city and landfill could be used to create a large property convenient for industrial use.

As the development of Tokyo Bay proceeded in about 1970, pollution problems became apparent and people's perception of the environment changed. After the Oil

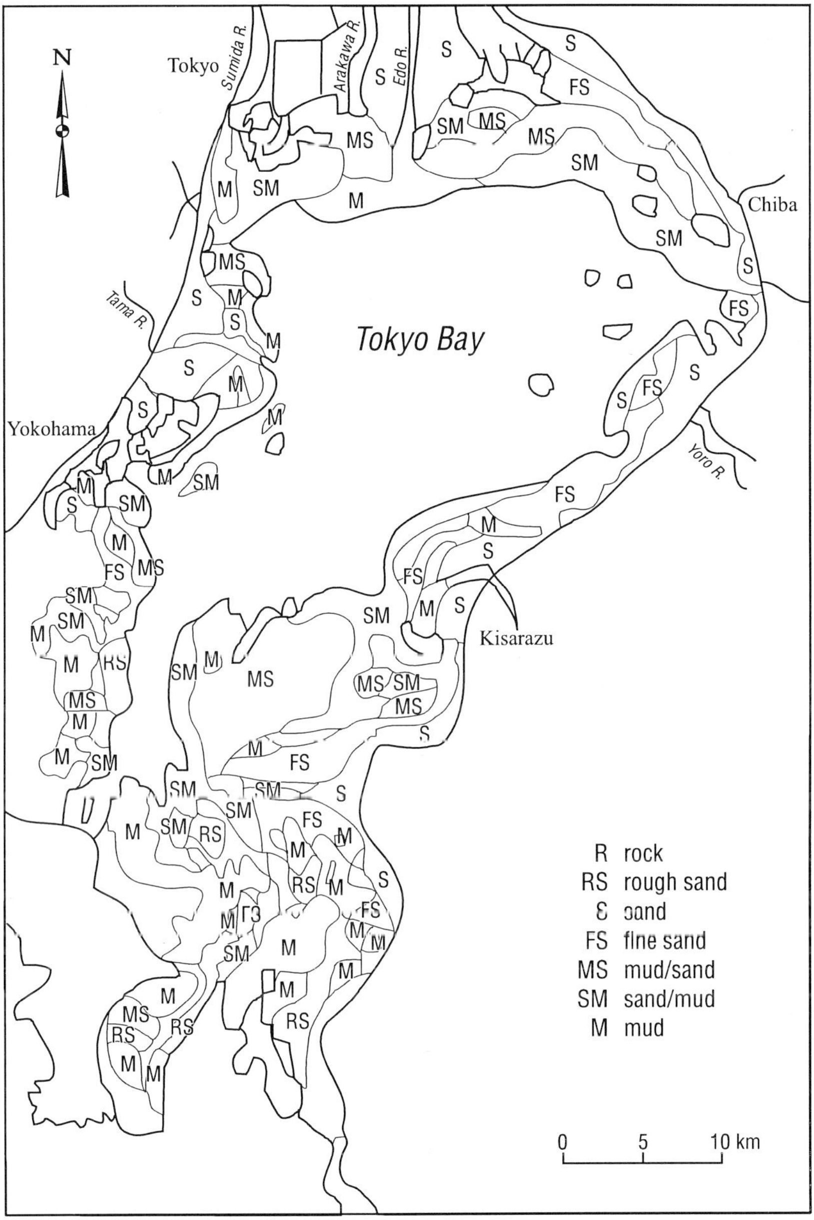

Figure 2. Tokyo Bay geology and sediment type.

Crisis, the government restricted landfill in Tokyo Bay, thus putting a brake on the expansion of the coastal industry.

Since 1968 landfill in Tokyo Bay totaled 2,200 ha (Fig. 3). The industrial development in the 1970's was accompanied by increased urban activities. As a result, various urban functions were enhanced and pollution followed. Amidst such a situation, the reclaimed land has come to be used as parts of urban infrastructure including public facilities, homes, and parks. Since the 1980's, in particular, emphasis has shifted to the restoration of lost environment, such as the restoration of tideland and wildlife

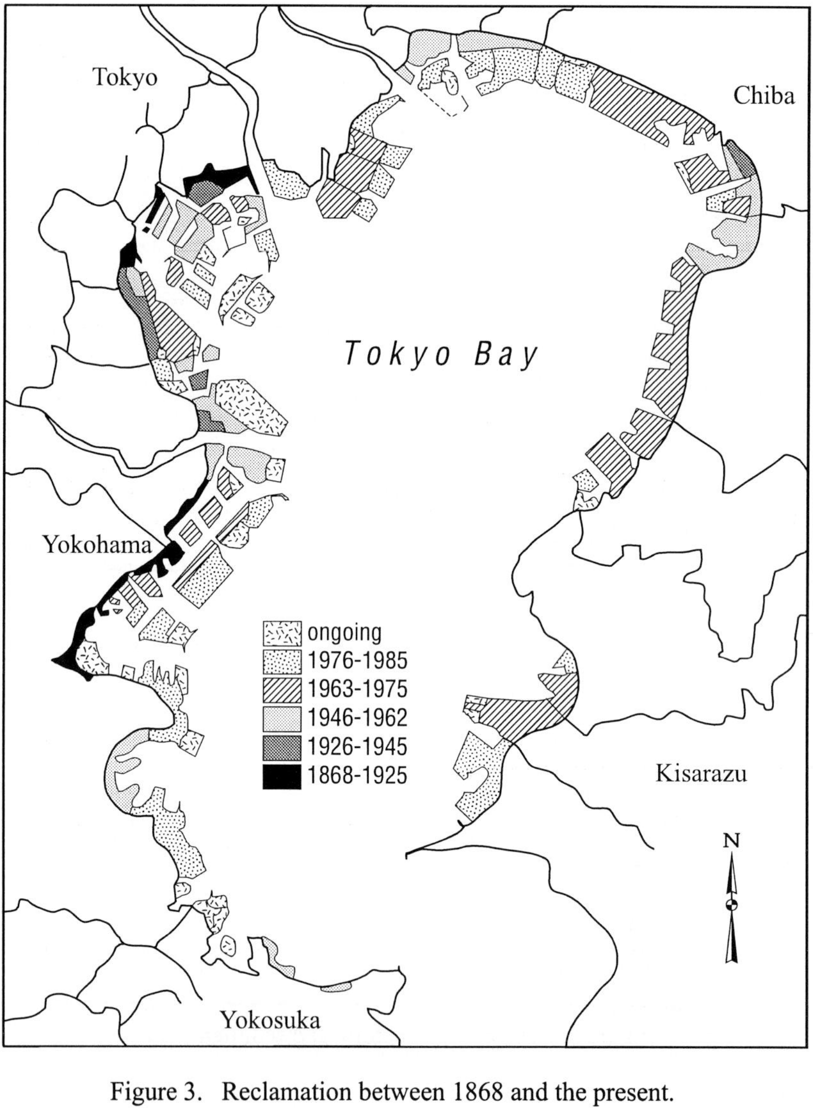

Figure 3. Reclamation between 1868 and the present.

preserves that were lost to coastal use and landfill and improvement of public access. At the same time, effort has been made to use the coastline and develop the waterfront in ways that will allow citizens to enjoy the area for leisure and recreation.

2.1 LAND RECLAMATION DEVELOPMENT PROJECTS INSIDE AND OUTSIDE THE CENTRAL BREAKWATER OF TOKYO BAY

The full-scale implementation of the Tokyo Bay Reclamation Plan began in 1961 with the formation of reclaimed land in the ocean area of southeast Tokyo. The objective of the reclamation was to enhance Tokyo's capital functions. In particular, reclaimed land was situated to accommodate Tokyo's industries and general population growth. Land use was not limited to just the expansion of harbor functions, but also included diverse uses such as housing, waste disposal facilities, redevelopment, parks and green areas[4] (Figs. 4 & 5).

In particular, the reclaimed land sites inside and outside of the central breakwater and the reclaimed land off Haneda's shore, planned since 1965, were formed as places to bury Tokyo's solid waste.

2.2 DESCRIPTION OF THE OVERALL PROJECT

Reclaimed land totaling 190 ha was formed inside the central breakwater by building a bulkhead around the area originally used to dispose of dredged spoils from navigation channels. Of this area, 78 ha was devoted to the disposal of solid waste. About 12.3 million tons of solid waste was disposed of at this site between 1973 and 1987.

In addition, two man-made islands created through land reclamation were built outside the breakwaters and planned as waste disposal sites (Fig. 6). The total disposal site area of the two islands is 314 ha. The total outer circumference of the bulkhead is 10,121 m and the facilities can handle 59.39 million m^3 of waste (Table 1).

Table 1. Construction in Tokyo Bay (1974-1998)

Location	Tokyo
Land Use	Waste Disposal Area*
Reclamation Area	313.9 ha
Construction Period	1974 to 1998
Construction Cost	2,300 billion yen
Amount of Soil	5.940×10^8 m^3
Water Depth	-6 m
Total Length of Seawall	10,121 m
Distance from Coast	200 m
Access	Bridge 1

*After filled, site will be used for other purposes.

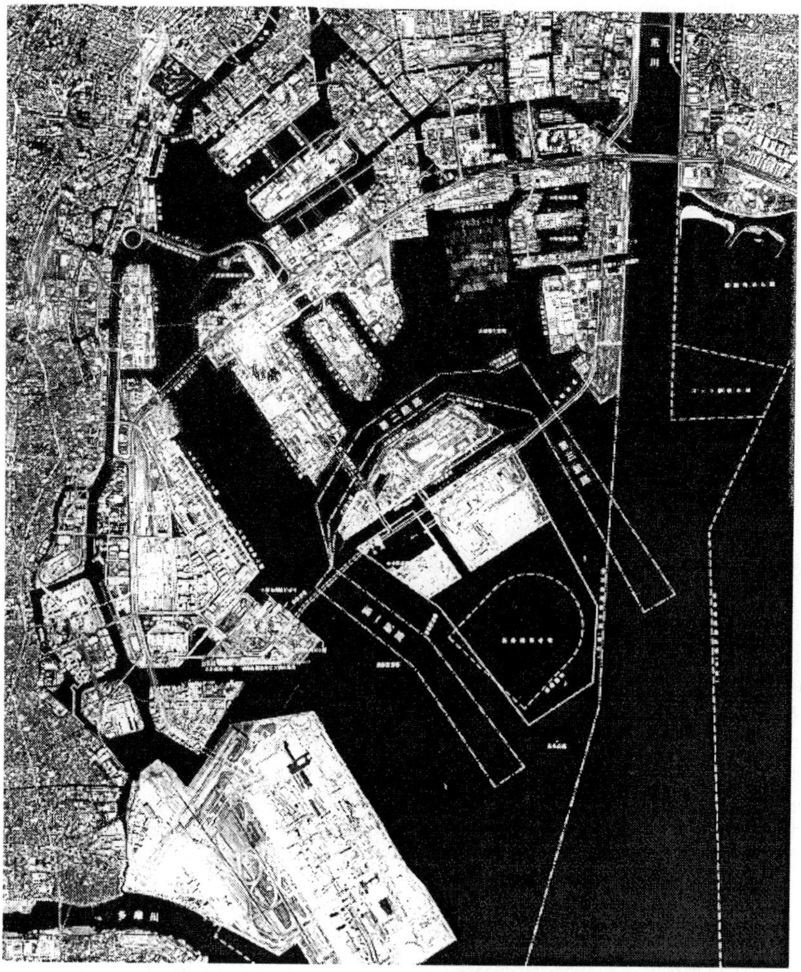

Figure 4. Tokyo Port, view from 2,300 m.

2.3 BULKHEAD STRUCTURE AND CONSTRUCTION

The sea floor at the reclaimed land outside of the central breakwaters is 10 m below mean sea level (MSL). On the offshore side of the reclaimed land, the sea floor is soft up to 40 m below MSL. Considering that the submerged land is soft and considering that the reclaimed land was being planned as a waste disposal site, the following points had to be taken into consideration:

1) All possible measures had to be taken to preserve the environment and prevent pollution. In particular, the leaching of sewage had to be prevented by physically containing the effluent,

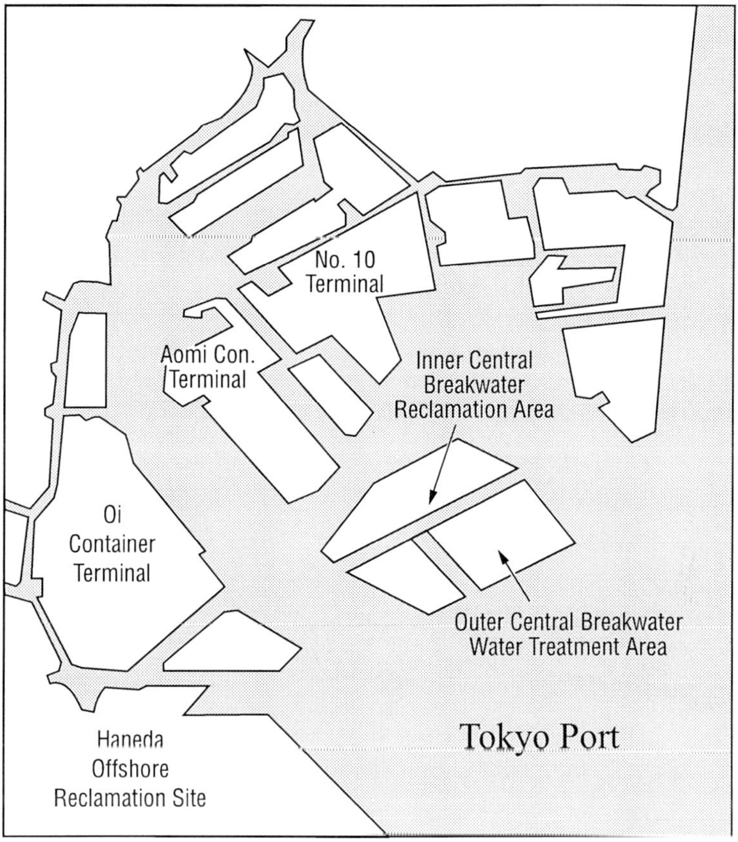

Figure 5. Overall view of Tokyo Port construction.

2) Since the sites are in rough waters outside the breakwater, the reclaimed land had to be designed to withstand the impact of waves both during and after construction,
3) Because settlement of the reclaimed ground is unavoidable, simple bulkheads which can be repaired were desirable,
4) Roadways had to be built to accommodate waste disposal operations, and
5) The construction period had to be minimized.

As a result of the above criteria the selected bulkhead design was a double steel pipe, sheet-pile for the outside bulkhead (Figure 7).

2.4 ENVIRONMENTAL PRESERVATION MEASURES

The numerous measures undertaken to help preserve the environment included:

1) The development of environmental impact assessment procedures which had not been done before the plan was implemented in 1973. Thus, a water quality

simulation for the entire Tokyo Bay was conducted for the first time. The simulation which forecast water quality considered the impact of solid waste disposal practices as well as the tidal characteristics of Tokyo Bay,

2) Channels were built throughout the project and in the middle portion of the disposal site to improve water circulation as well as to provide for the migration of marine life,
3) An impervious double steel pipe type bulkhead was employed to prevent off site leaching of sewage,
4) A sewage treatment system was constructed which collects and partially treats sewage then transports the treated effluent to a public sewer system for further treatment and disposal, and
5) In order to minimize pollution from the land filling operations, 50 cm of clay was placed over each 3 m of compacted refuse. The final clay cap was 1 m in depth. Such practices minimized airborne litter, odor and vectors.

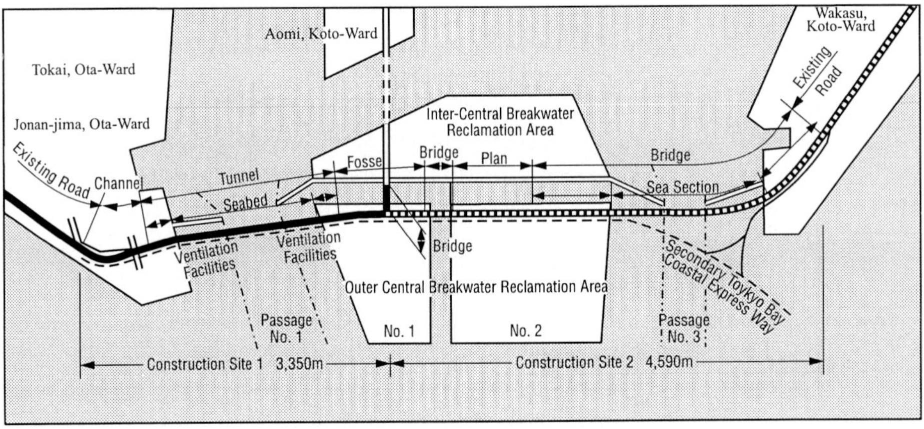

Figure 6. Plan view of constructed island.

2.5 LAND USE PLAN

The use of the reclaimed land inside the central breakwater was based on the type of fill material used. Those land uses that needed soil, such as parks, were placed in the area filled with soil. Those uses which did not need soil were placed in areas where the fill material was solid waste. Such land uses include harbor functions (42 ha), waste handling areas (19 ha) and land for accommodating redeveloped facilities (18 ha). Development is currently in progress.

On the other hand, the reclaimed land outside the breakwater is reserved as public space for future growth.

3. Kasai Marine Park

The restoration and preservation of the natural environment is an important aspect of the Port's functions. With 39 parks already providing ideal opportunities to engage in sports, fishing, bird watching or other activities, on-going development throughout the entire Port area will continue to bring into being recreational seashores where local residents can take in the sights of the harbor and become familiar with its waterway and lush greenery. An example is Sakai Marine Park which is newly created for living marine creatures[3] (Fig. 8).

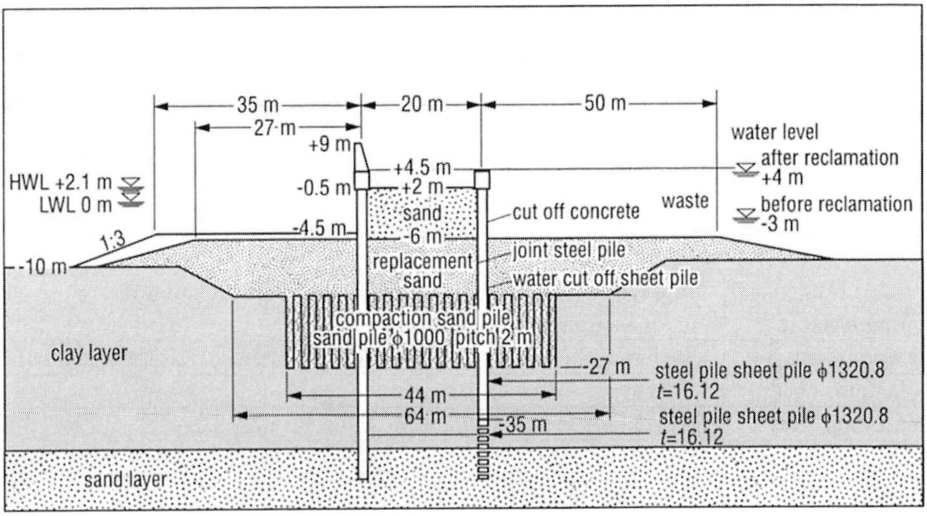

Figure 7. Cross-section of bulkhead island.

Figure 8. Sakai Marine Park

4. Tokyo Bay Aqualine

Tokyo Bay Trans-Highway (Tokyo Wan Aqualine) is a 15 km long toll way that runs across the central portion of Tokyo Bay, connecting Kawasaki (Kanagawa Pref.) and Kisarazu (Chiba Pref.). The Tokyo megalopolis, located along the coast, consists of Tokyo, Yokohama, Kawasaki, Chiba, Keaau and other cities.

In order to meet the yearly increasing traffic demand in the greater Tokyo area, it was decided to build a bay-crossing highway (Table 2). The highway consists of an undersea tunnel, two man-made islands and a bridge. About 10 km from the Kawasaki side is a dual tunnel that eliminates disturbance to the heavy traffic of surface shipping. The balance of the 5 km section of Kisarazu, where shipping traffic is less heavy, is crossed by a bridge.

Table 2. Scope of construction license of Tokyo Wan Aqualine[6]

Road Name	Trans-Tokyo Highway (renamed Tokyo Wan Aqualine)
Route Name	National Highway No. 409
Work Area	Ukishima, Kawasaki (Kanagawa) to Nakashima, Kisarazu (Chiba)
Length	15.1 km
Carriage Way	Dual, 2 lane at first stage, triple 2-lane at final stage
Design Speed	80 km/hr
Traffic Forecast	25,000 veh/day in the first year, 53,000 veh/day after 20 yrs
Work Schedule	1987 to 1997
Project Cost	1,440.9 Billion Yen

The underwater tunnel of the highway is ranked as the fourth longest in the world. Three of the world's top five underwater tunnels are in Japan.

Tokyo Bay Highway is able to provide a shorter connection between the east side (Tokyo-Kawasaki-Yokohama) and west side (the Boso area which still offers much room for development). As a matter of fact, the driving distance from Kawasaki to Kisarazu is only 30 km whereas it is 100 km via the Tokyo Bay Ring Road. The travel time is one-third that needed to use the conventional route along the semi-circular coastal road.

The first feasibility study, initiated in 1966, concluded with a total cost of approximately 1.48 trillion yen. The project was completed at the end of 1997. After completion and being opened to traffic, this highway was renamed Tokyo Wan Aqualine.

The man-made island Umihotaru (meaning marine firefly) with a five-story observation deck commanding a 360-degree panoramic view of Tokyo Bay has become one of the most popular scenic points for tourist. The parking lot on this man-made island is always full, particularly on weekends and holidays. Another cylindrical-shaped man-made island, that houses a ventilation station for the tunnel, is named Kaze-no-tou (means tower of wind).

In the past, sea-bottom tunnels were designed and constructed as sunken and buried tunnels. However, shield tunnel engineering technology, which was developed in recent years, represents remarkable technological progress in the construction of urban tunnels and even for underwater large-diameter and long distance tunnels. This method is able to minimize the impact on surface shipping during the construction period so that it is ideal for use in congested sea-lane areas. Moreover, this method also provides advantages when coping with the soft soil of the sea bottom [5].

The bridge from Umihotaru to Kisarazu pier section has a superstructure of box girders with a steel deck and forms a continuous multi-span structure. The ventilation station on the Kawasaki man-made island, has a unique configuration; it was constructed using the results of a series of wind tunnel experiments and numerical analyses that were based on the distinctive features of the wind characteristics in Tokyo Bay[6]. The location is 5 km offshore from Ukishima Access, Kawasaki side, where the water depth is about 28 m.

The seabed is so soft that the sand compaction method and deep mixing method were used to stabilize the ground. The structure includes a diaphragm wall that extends into the sea bottom between the inner and outer jackets. The concrete structure of this man-made island itself was constructed inside of the jacket structure. Precise control and monitoring of distortion had been carried out to ensure that the cylindrical structure could resist wave forces and water pressure. During tunnel work, this island was used as bases for the shield tunnel (Fig. 9).

Environmental preservation was also intensively carried out to insure harmony with the natural and social environment. For example, underwater curtains were deployed to prevent turbidity-related pollution, and, of course, environmental assessments were done. Water quality, air quality, noise, vibration and other items are targets of periodic surveys on site. Long-term assessments covering land flora and fauna, marine life and other factors are being made.

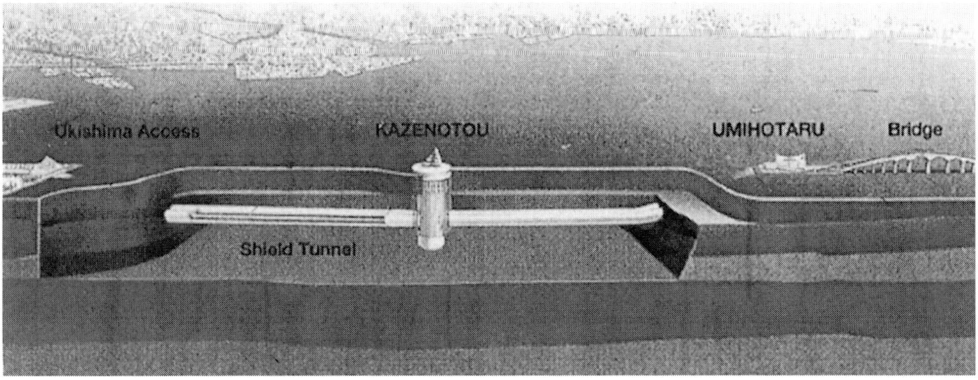

Figure 9. Cross-section of the terminal area.

Around Umihotaru Island, an artificial seaweed forest has been successfully established. It is a kind of new concept of mitigation, compensating for the reclaimed and lost ocean space by creating a favorable environment for the marine life around the man-made island.

5. Yokohama Port

The city of Yokohama lies about 30 km to the southwest of central Tokyo. Its population of about 3.3 million makes it the second largest city in Japan with 130 years of service as an international port. Because Yokohama is so close to Tokyo, many of the two cities' economic and other activities are linked.

To the southwest, Yokohama has a long waterfront on Tokyo Bay. The most important feature of Yokohama is its role as a port city, which has led to bustling activity in the field of marine transportation, shipbuilding, trade and commerce. The industrial belt covering the eastern part of the city, the western part of Tokyo (Keihin Industrial area), and the area in between, is one of the best-known in the world. It is home to a variety of manufacturing business, created on electrical appliances, machinery, and metal products, that engage in everything from the supply of basic materials to the processing, assembly and production of household goods.

The Minato Mirai 21 (MM21) Project, located on an extensive coastal district adjoining central Yokohama, is one of the largest development projects in Japan. It aims to create a new city center with a population of 200,000 and offers office buildings, cultural and commercial facilities, international organizations, and other amenities on 186 ha of land[7] (Fig. 10).

6. Yokohama Bayside Marina

Yokohama is developing the Yokohama Bayside Marina. Located in Kanazawa Ward on the south part of Yokohama, when completed, it will be the largest marina in Japan. Situated midway between Tokyo Bay and the open sea, Yokohama Bayside Marina is a prime spot for boating and sailing as well as for fishing.

The marina facilities opened in 1996 (Fig. 11). The Yokohama Bayside Marina Development Project is a marine leisure activity center, bringing together marina-related businesses such as commercial facilities, hotels, marine leisure showrooms, research facilities, club houses, repair shops, refueling stations as well as parking lots, roads and green belts. The marina has been developed on 13.9 ha of reclaimed land encompassing 27.9 ha of water with a capacity for 1038 ships[8].

7. Development of Coastal Industrial Zone in Tokyo Bay: A less successful project

After World War II, by making full use of the coastal area, Japan's economy has achieved tremendous growth. Until the late 1950's, Tokyo Bay was a major recreation area for city dwellers, who bathed in its water in summer. They also gathered shellfish and seaweed from its tidal flats and fished in the waters of the bay all year around. In other words, most of Tokyo Bay was opened to people for recreational purposes and as a fishing area for fishermen. In the 1960s, most of the bay's coasts was occupied by industry and port facilities as a result of reclamation, a condition that continued until the middle of the 1980's. As a result, bay waters were severely polluted. Subsequently, environmentally oriented projects including environmental preservation and restoration

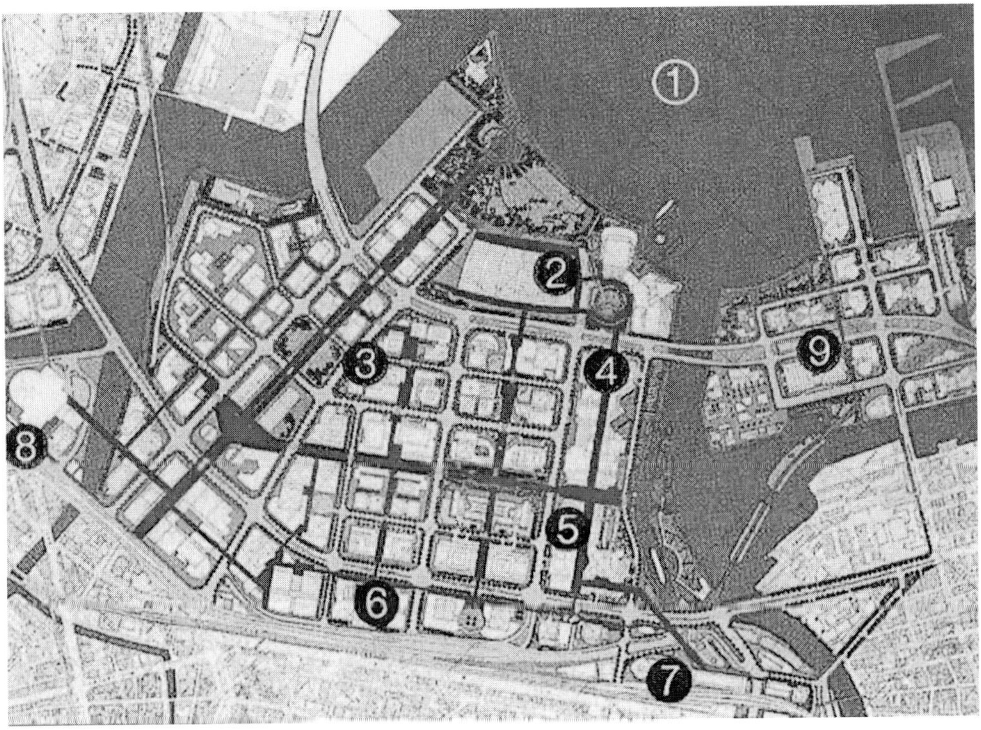

Figure 10. MM21 Land Use Plan.

1) Port of Yokohama
2) Pacifico Yokohama
3) Minato Mirai 21
4) Queen's square 21 Line
5) The Landmark Tower
6) Minato Mirai Ramp
7) Sakuragicho St.
8) Esst Exit District
9) Yokohama Import Mart

as well as technology development have attracted attention. At the same time, interest in the amenity and other values of coastal environments have been revived and a coastal access rights movement has grown out of the changing life style of Japanese citizens. Under these circumstances, city dwellers around Tokyo Bay (and people of the other coastal urban areas as well) have also recognized that the water front area is essential for enjoying recreational activities. People desire to have places where they can be in close contact with nature. Currently, because of Japan's depressed economy, the development and use of Tokyo Bay are also depressed.

Taking the Keiyo coastal industrial zone (which is located in the eastern part of Tokyo Bay) and its surrounding inland area in Chiba prefecture as an example, the influence of coastal area utilization is examined below from the viewpoint of the local communities and their economic environment. The coastal industrial region in Chiba

Figure 11. Yokohama Bayside Marina.

prefecture was developed from a 12,035 ha land reclamation area along the region's 76 km long shoreline. The main industrial facilities in this region were two iron and steel industries, four oil refineries, three petrochemical complexes and electric power, gas, ship building, and aluminum and glass plants. The Keiyo coastal industrial zone become one of the largest energy supply areas in Japan. The construction of the industrial zone aimed at expanding employment, upgrading the industrial infrastructure, and improving the lives of those living in its Chiba prefecture.

As a result, Chiba prefecture, which had traditional primary industries in fisheries and agriculture, underwent a complete change and became one of the leading industrial regions in Japan. The value of shipping from the manufacturing industry in Chiba was US$ 7.95 million in 1981, ranking sixth in Japan. The coastal industrial area accounted for 62% of the entire shipping value of the prefecture. Unfortunately, however, the increase in industrial shipping did not increase the employment opportunities of the area, nor did it improve people's lives. Chiba's industrial development has rather caused adverse effects on the local economies and the lives of the local residents. A more detailed analysis of these adverse affects is given below.

7.1 DEFORMATION OF THE INDUSTRIAL STRUCTURE

First, fishing declined because of land reclamation. Agriculture also declined; the number of farming households decreased from 104,094 in 1960 to 21,898 in 1981 as a result of imports of agricultural products, the use of farmland for other purposes, and the outflow of the agricultural labor force to various other fields. Moreover, the local industries were not given favorable administrative treatment because they were considered a "low productivity sector".

Looking at the industrial structure of Chiba prefecture, the heavy and material-producing chemical industry (iron and steel, petroleum, chemical) accounts for 60 %. This is the largest proportion of all prefectures and far greater than the national average of 26%. This brought about an unbalanced and skewed industrial structure. Such a skewed industrial structure is vulnerable to stagnation and other economic changes.

Looking at the employment figures and tax revenue of the prefecture (Fig. 12) it is clear that the coastal industrial zone makes only a minor contribution to the whole prefectural economy. In other words, the financial resources of the prefecture were based on the taxes paid by individuals, rather than those paid by large corporations, in spite of the fact that Chiba has one of the largest industrial zones, with the 6th or 7th largest figure in shipping.

Table 3 is a comparison between the coastal and inland industries. Coastal industries occupy 69% of the total industrial area, use 74% of industrial water consumption, consume more than 70 % of the fuel, and require massive quantities of resources and energy. The same region consumed 90% of the sulfuric compounds used, and is the largest pollution generator. Nevertheless, the direct employment (number of employees) in the coastal region amounts to only 24% of the entire work force in the prefecture.

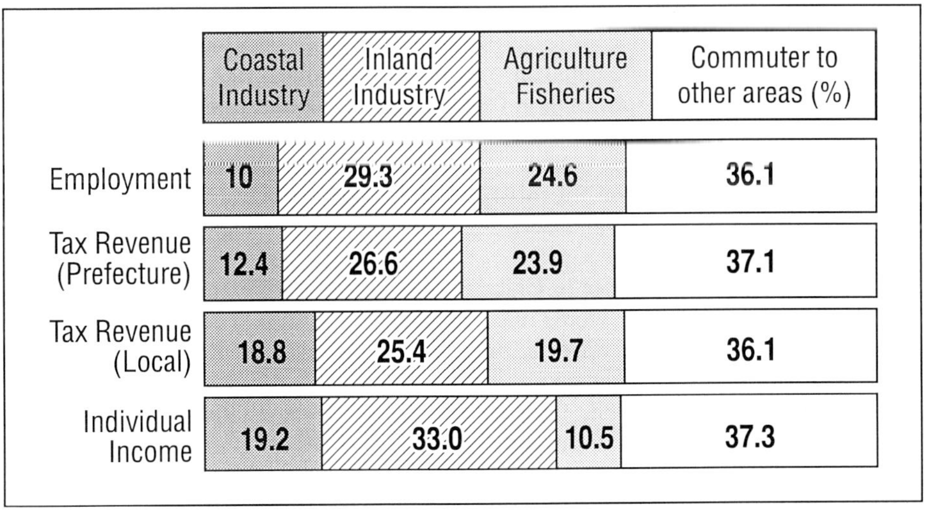

Figure 12. Tax revenue from each industrial sector.

The positive effects of the coastal industries are also smaller than that of the inland industries. In other words, small and medium industries are far greater contributors to the development of local economies than the huge coastal industries. The question is, why have the economic transmission effects of the coastal industrial zone remained minimal?

Table 3. Comparison between Coastal and Inland Industries

	Coastal (%)	**Inland (%)**
Number of Industry	6.0	94.0
Amount of Products	62.1	37.9
Area of Industrial Site	68.8	31.2
Amount of Industrial Water Use	73.8	26.2
Amount of Fuel Consumption	70.0	30.0
Amount of Sulfur Oxides Discharge	90.0	10.0
Amount of Direct Employment	42.4	57.6
Prefecture Tax Payment	36.5	63.5
Local Tax Payment	41.4	58.6
3rd Industrial Sector Inducement	42.6	57.4

7.2 ECONOMIC EFFECT DISPERSAL

The primary reason for the minimal economic effect is that the coastal industrial zone serves as production space for large corporations that have their headquarters outside Chiba prefecture. These headquarters are mostly located in Tokyo. That is, trade sales and the settlement of accounts are not conducted in Chiba but rather elsewhere. In particular, the profits of the tertiary industries which have recently been aggressively introduced in Chiba prefecture are being removed to the prefectures where their head office are located.

Second, products manufactured in the coastal industrial zone have nothing to do with the local industries. For instance, Chiba prefecture is the largest producer of steel and ethylene. Nevertheless, there are few factories which process and use these materials in Chiba prefecture. In this regard, the Keihin industrial zone, which is located on the western side of Tokyo Bay, has much greater advantages. It purchases raw materials and produces value-added products which have higher economic values in their own region.

Third, the coastal industries in Chiba prefecture are chiefly composed of facility-intensive industries. In other words, the manufacturing system is largely automated to serve manpower costs and labor-saving, large-scale equipment and facilities are employed. These heavily equipped and full automated factories do not create large employment. As a result of the three reasons given above, a large proportion of the economic potential flows out of the area.

7.3 THE LOSS OF LIVING ENVIRONMENTS AND THE INCREASED FINANCIAL BURDEN TO THE LOCAL AUTHORITIES.

In the area of social life, people have lost valuable recreational space where they can enjoy direct contact with nature, including beaches for sun bathing and fishing. The industrial zone created a barrier which hinders people from going near the beach. This is an immeasurable loss for local residents. They have lost the cultural, historical and spiritual benefits they used to gain from access to the beach.

The second loss is that of environmental destruction. The tidelands have declined from 7,757 ha in 1945 to about 985 ha at present. Crowded ocean traffic and oil discharged from ships and boats have resulted in deteriorating water quality. Industrial pollution, in the form of air, noise and odor pollution, have caused serious damage to the health of the local residents. In addition, drawing underground water for industrial use has caused wells to dry up and soil to subside in some areas.

A third problem is an increase in the financial burden on the local governments. It is true that the establishment of the coastal zone expanded tax revenues and increased financial resources. However, local administrations were forced to spend huge amounts of public money on projects related to the development of the coastal industrial zone; reclamation, improvements for plants and factories, ports and harbors, the building of industrial-water-supply facilities, industrial roads, housing, and the provision of educational and welfare facilities. In addition, local governments have had to pay the cost of pollution-control measures and also for disaster prevention measures required by plants.

As stated above, the industrial activities in the Keiyo industrial zone did not always produce the desired results, and often put an extra financial burden on the local government. However, such negative effects did not become salient, because towns in Chiba prefecture have become bedroom communities for workers commuting to Tokyo, and as a result, a huge population immigrated into Chiba prefecture. Because of this, the tertiary industries began to make rapid progress and could contribute to sustaining the increasing financial burden of the prefecture, and helped offset the negative effects of the coastal industrial zone described above.

8. References

1. Research Institute for Ocean Economics Tokyo Bay Utilization Planning in 21st Century Research Report, 1988.
2. Research Institute for Ocean Economics Research on Water Quality Investigation Research Report, 1990.
3. Tokyo Metropolitan Government, Port of Tokyo 1996, 1996.
4. Masaomi Koike, Man-Made Islands, Vol.78-12, Society of Civil Engineer.
5. Hiro Nakahara, Tokyo Wan Aqualine and MEGA-Float; Two Major Projects on Ocean Science and Technology in Japan, 1998.
6. Trans-Tokyo Bay Highway Corporation, TOKYO WAN AQUA-LINE, 1998.
7. City of Yokohama, Yokohama: Your Gateway to Japan and Asia, 1996.
8. City of Yokohama, Yokohama Bayside Marina, 1997.

9. Bibliography

1. Hotta, Kenji 1993. Software Technology for Coastal Development and Utilization *Proceedings of Engineering for Coastal Development*, Kozai Club.
2. Hotta, Kenji 1996. *Coastal Management in the Asia-Pacific Region*, Japan International Marine Science and Technology Federation.
3. Koike, Kazuyuki. Artificial Beach Construction on the Shore of Tokyo Bay, *Journal of Coastal Research*, Summer 1996
4. Walker, H. J. 1977. Tetrapods, Igloos, and the Coast of Japan, *Scientific Bulletin, Dept. of the Navy Office of Naval Research,* Tokyo.
5. Walker, H. Jesse and Joann Mossa 1986. Human Modification of the Shoreline of Japan, *Physical Geography.*

OFFSHORE CONSTRUCTION AND OCEAN SPACE UTILIZATION IN JAPAN

KENJI HOTTA
Nihon University
7-24-1 Narashinodoi Funabashi-shi
Chiba 274, Japan

1. Introduction

Japan is a small country with a dense population. It consists of four major islands and more than 4000 smaller islands. Combined, the length of the shoreline is more than 34,000 km or about equal to that of the contiguous United States which has an area more than 20 times larger. Because Japan rests on the western part of the Pacific plate it is mountainous and subject to frequent seismic activity. The bulk of the population and most of the industry and commercial activities are concentrated on the coastal plain which is small in area. Because of the pressure this concentration places on the landscape, there is little natural room for expansion. As a result, the Japanese have been very aggressive in protecting the shoreline from erosion, in ocean space utilization and in reclamation[1].

Reclamation for the creation of rice paddies and the growing of other crops and for solar-salt production has a long history in Japan. However, it was only recently—especially since World War II—that reclamation for industrial, commercial, transportational and residential purposes has been undertaken in a major way in Japan. It is this non-agricultural type of reclamation that is the subject matter of this chapter.

2. The Utilization of Coastal and Open-Ocean Space

The utilization of coastal and open-ocean space in Japan is of three types: 1) shoreline reclamation, 2) nearshore artificial-island construction and 3) offshore artificial-island construction (Fig. 1).

Shoreline or seaside reclamation (the process of extending land out from the coast) is usually centered around a bay or inland sea. Bays, such as those of Tokyo and Osaka, and inland seas, such as Seto Inland Sea, are usually shallow bodies of calm water. They are reclaimed by utilizing land fill and can be used for a variety of purposes including industry, agriculture, transportation, urbanization and recreation.

Artificial islands constructed close to shore are designed to make good use of existing coastal regions. They are usually closely associated with cities and/or industrial complexes. These near-shore islands are created by reclamation for the purposes of protecting the coastal environment and utilizing the separating ocean space as harbors

J. Chen et al. (eds.), Engineered Coasts, 103–119.
© 2002 *Kluwer Academic Publishers. Printed in the Netherlands.*

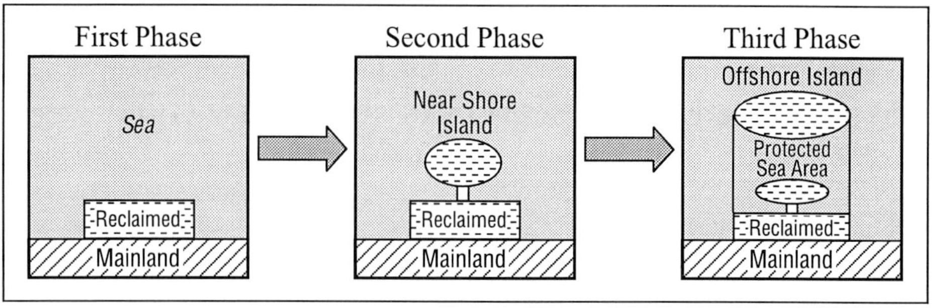

Figure 1. Phases of space utilization.

and fairways. The industrial region in the Keihin area (between Tokyo and Yokohama) is an example. It was constructed to create a channel for vessel navigation along the coast. Kobe's Port Island, Rokko Island and Osaka South Port were all built for use as harbors, for urbanization and for the treatment of wastes.

Artificial islands constructed offshore at some distance from the shoreline, are designed for efficient utilization of the surrounding ocean as well as to provide for development on the islands themselves[2].

2.1. ARTIFICIAL ISLAND STRUCTURAL TYPES

Artificial Islands can be made by utilizing a number of different construction techniques. Three of the most common construction procedures used are: 1) with fill, 2) supporting with piles, and 3) by floatation (Fig. 2). The specific method used is closely related to the proposed function and size of the structure and to the natural conditions present at the construction site. Natural conditions include those of both sea and land such as waves, tides and earthquake potential.

Reclamation is accomplished by filling the ocean space to be utilized by land that is bounded by some kind of shore protective structure such as bulkheads. They may be of sheet-piling, cell type or gravity-type. This method of construction has been widely used. Its technical reliability is high and there is no restriction on its size. Further, planning for it can be flexible.

The pile-supported structure grounds the island to the sea floor. Supporting the structure's weight piling provides protection against external forces such as ocean waves. The main methods used include 1) bottom fixing, 2) piling and 3) jacking-up (Fig. 2). The bottom-fixing method is accomplished by floating the structure's base to its location and then sinking it to the seafloor before its super-structure is built. In the pile method, piles are driven into the seafloor upon which the platform that will hold the superstructure is built. In the jack-up method, the structure, which has expandable legs attached, is floated to the desired location. Once on site, the legs are dropped to the seafloor and the floating body is secured by jacking.

In the floating type artificial island the structural weight is supported by buoyancy and stabilized by mooring equipment that is capable of resisting the forces of waves.

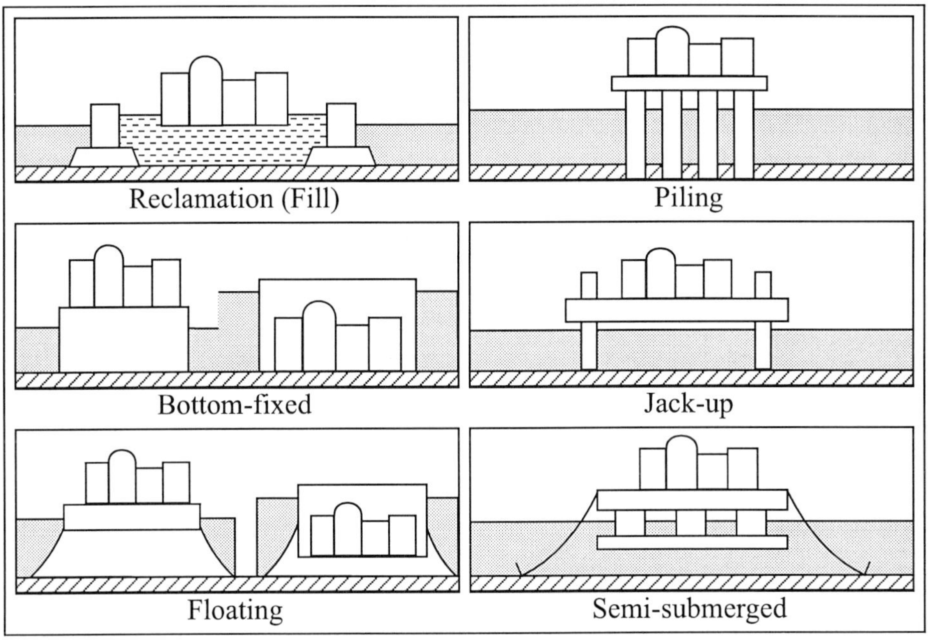

Figure 2. Types of offshore island construction.

This type may be completely floating or semi-submerged (Fig. 2). Semi-submergence reduces the movement of the floating body[3].

There are still some unsolved problems with these methods including wave resistance, stability and draft variation control with respect to changes in the structural mass. Therefore, this type of offshore island has been used only for relatively small projects and for temporary facilities. Experiments are being conducted in order to determine the feasibility of using the methods for large-scale floating structures. Presently, the major method used for creating large offshore areas is reclamation, the technique emphasized in the rest of this article.

2.2. OFFSHORE ISLAND CHARACTERISTICS AND ADVANTAGES

Offshore islands, natural as well as artificial, can be used for a variety of purposes and their utilization can, in most cases, be easily accomplished. In the case of man-made islands, the total marine, seashore and hinterland combinations can be planned in advance of construction. A major disadvantage, however, is the length of time construction requires.

2.2.1. *Planning Flexibility*

The rapid change occurring in the economic characteristics of Japanese society is creating new, diverse needs for coastal and marine space. Among these new conditions is the requirement that the load on the coastal environment be reduced and that a clear distinction between marine recreational space and present-day sea-area usage be developed. When compared to mainland reclamation and attached-island construction, offshore-island construction is flexible in regard to location, size and shape. It is also relatively easy to utilize the offshore islands in a variety of ways (Table 1). Hakata Island City, for example, is providing space for a number of functions including transportation and urbanization (Table 2).

Table 1. Characteristics of Artificial Islands

Step	First Phase	Second Phase	Third Phase
Type	Coastal reclamation	Attached Island	Offshore Island
Environmental Preservation Effect	Small	Medium	Large
Distance from the coast	Small	Medium (200 to 500m)	Large (2 to 5m)
Possibility of deep water area use	Small	Medium (0 to 20m)	Large (20 to 50m)
Possibility of hinterland use	Possible	Small (waterway, etc.)	Medium to Large
Use of existing coastal facility	Possible	Possible	Possible
Accessibility	Easy	Relatively easy	Relatively difficult
Use	Harbor facility, Industrial use, Urban use, etc.	Harbor facility, Industrial use, Urban use, etc.	Offshore harbor, Industry, Ocean city, Airport, Power Sta.
Examples	Many areas	Rokko Island, Port Island, and others	Kansai International Airport, Shimonoseki Offshore Island

Table 2. Use of Offshore Artificial Islands

Name	Use
Kobe Port Island (Phase 1 and 2)	Harbor facility, Urban function
Rokko Island	Harbor facility, Urban function
Hakata Island City	Harbor facility, Urban function
Osaka Bay, North Harbor – North	Urban function, Recreational facility, Waste disposal facility
Osaka Bay, North Harbor – South	Harbor facility, Urban function
Osaka Bay, South Harbor	Harbor facility, Urban function

2.2.2. *Length of Construction Time*

Man-made islands are, in general, large scale projects, requiring from several years to decades from the time of conception and planning to completion. Especially time-consuming is the period of time devoted to the part of planning that is concerned with adjusting the project to the natural environment.

2.2.3. *Advantages of Offshore Islands*

Compared to onshore reclaimed and attached island areas, offshore islands have a number of advantages, including: 1) deeper water around them and thus, greater accessibility to large draft vessels and more openness to navigation routes, 2) being offshore their shape is more flexible making planning easier, 3) the calm water created landward of the island provides a calmer sea for navigation, pleasure boating and the development of aqua-culture projects, 4) development on such islands away from shore allows more freedom in preserving the existing natural coastline, 5) being structurally disconnected, they can function independently in times of disaster (e.g. earthquakes, typhoons, fires, river flooding) on the main land, 6) they are compatible with large scale developments such as airports with their noise problems, power stations with safety risks and waste treatment centers[3,4,5].

2.2.4. *Evaluating Offshore Islands*

In an evaluation of offshore islands, due consideration must be given to both their economic and environmental impact. Evaluation economically must include not only profitability but also: 1) the economic and social effects due to construction, 2) the economic and social effects due to functions introduced to the offshore island and 3) the economic and social effects accompanying the creation of a calm-water area between the island and the mainland.

Environmental evaluation involves mainly the impact construction has on the natural (physical and biological) environment. This includes the alteration and elimination of prior conditions as well as the creation of new parameters.

3. **Examples of Man-made Islands in Japan**

Artificial islands have been constructed in Japan for a variety of purposes. Included are those islands used as airports, industrial sites, residential areas, commercial harbors and fishing harbors.

3.1. KOBE: PORT ISLAND AND ROKKO ISLAND

Kobe is a city located in northwest Osaka Bay. It has an area of 545 km^2 and a population of $1.5x10^6$ people. Kobe, backed by mountains, has been built on a narrow band of land that is only about 2 km wide and 4 km long. Thus, the city is surrounded by mountains and the sea. As population pressure mounted the city has had to turn either to the mountains or the sea for additional space. Eighty years after the old port was opened (1886) Kobe began construction on Port Island in the sea (1966). The objective of this construction was not only to integrate harbor and city activities but also to create a new city on the ocean capable of serving as a base for international exchanges in commerce and culture. In 1972 construction was begun on a second island (Rokko Island) in the sea to the east of Port Island.

3.1.1. Port Island

The first phase in the construction of Port Island lasted from 1966 to 1992 (Table 3). It was a land reclamation type project with fill being brought from the mountains that backed Kobe. The objective of this construction was to build a container wharf capable of handling deep draft vessels and to supply a community with housing, commercial facilities, hotels, convention facilities as well as other city function (Fig. 3).

Beginning in 1987, the second phase of construction on Port Island was begun (Table 3). The land use in this second phase involved five functional areas: 1) advanced visitor facilities including a multi-function convention center with international conference and exhibition hall, 2) a factory square for the transfer of manufacturing companies from downtown as well as new facilities, 3) an area for entrepreneurial corporations including research and development and personnel training facilities, 4) an international exchange zone and 5) a major entertainment square[4].

3.1.2. *Rokko Island*

The main objective for the construction of Rokko Island (Fig. 4) was to establish an international information city with two basic elements. The first element was the preparation of a modern harbor that could handle increasingly larger vessels and cope with the diversification of distribution systems. The second element was the construction of a multifunctional, multipurpose city that possesses residential, commercial, educational and recreational facilities[5].

Rokko Island construction began in 1972 and the reclamation portion of the project was completed in 1992 (Table 4). Because the fill portion of the project was

Table 3. Port Island Statistics

Reclamation area	Phase 1 436 ha Phase 2 390 ha
Construction period	Phase 1 1966-1992 Phase 2 1987-1998
Construction dost (Island)	Phase 1 2.3×10^{12} Yen Phase 2 5.2×10^{12} Yen
Amount of soil	Phase 1 $8 \times 10^7 m^3$ Phase 2 $9.2 \times 10^7 m^3$
Total length of seawall	Phase 1 10,879 m Phase 2 8,304 m
Distance from shore	400 m
Access from coast	Bridge 1 Underwater tunnel 1
Water depth	-4 m to –14 m

Table 4. Rokko Island Statistics

Location	Kobe
Reclamation area	580 ha
Construction	Completion of reclamation 1992
Construction cost	Total cost: 12.4×10^{12} Yen Reclamation: 5.4×10^{12} Yen
Amount of soil	$12 \times 10^7 m^3$
Planned population	30,000
Planned no. of housing	8,000
Construction body	Japanese Government, Kobe City, among others

done in stages, many facilities were constructed before reclamation was completed. Land use is broadly divided into zones that support three different functions: 1) those related to a harbor, 2) those that are mainly industrial and 3) those relevant to city activities (Fig. 5).

The harbor zone includes container berths with water depths of 13-14 m, a quay that is 350 m long and a container yard that has an area of 12.5 ha. The industrial zone is configured to promote the development of unique local industries. The city portion of Rokko Island is designed to serve business, residential, cultural and recreational

Figure 3. Low oblique photograph of Port Island.

Figure 4. Low oblique photograph of Rokko Island.

demands. In the residential portion of the area housing ranges from high-rise condominiums to single-family dwellings. Because of the desire to become as modern as possible, the business and commercial zone is designed to attract new and diverse industries such as those related to fashions and information sciences. An overall objective is to energize the economy of Kobe and its surrounding area.

3.1.3. *Other Aspects of Kobe's Offshore Islands*

Along with the development of Port Island and Rokko Island, Kobe City has seen a resurgence of activity. The private sector has become heavily involved in the generation

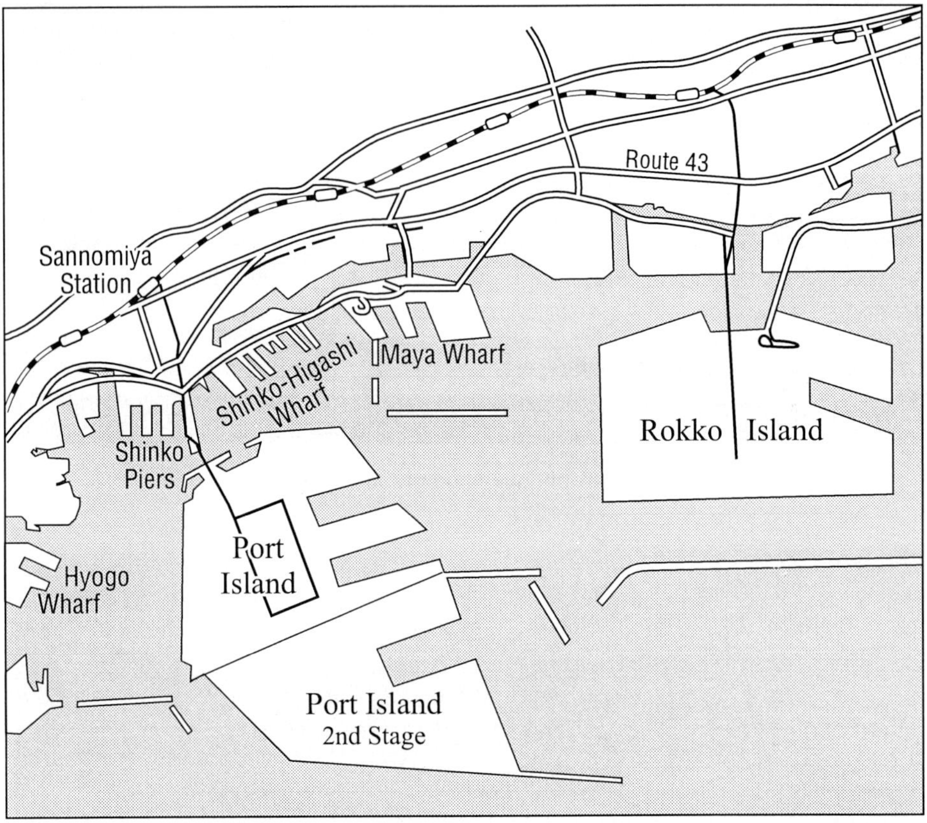

Figure 5. Map of construction site.

of concepts, project execution and sales in the area. The economic impacts of the two islands have been great. Port Island with its 12 deep draft berths and Rokko Island with seven have made the two among the largest container ports in the world. Some idea of the economic impact can be seen from the statistics for the construction of Rokko Island. The investment for the project totaled 1.24×10^{12} yen, an investment that led to a production totaling 69×10^{10} yen with 63,000 employees.

An additional benefit from the construction of these offshore islands was the creation of building sites at the locations where the fill was excavated from the surrounding mountains. These sites now contain universities and athletic parks as well as other public facilities.

3.1.4. *Kobe's Offshore Islands and the Kobe Earthquake*

Before dawn on 17 January 1996 an earthquake with a magnitude of 7.2 on the Richter scale struck Kobe. The city area along the coast was seriously damaged and the wharf section with its 239 berths became almost completely unusable[7]. Both Port Island and

Rokko Island were damaged but in a somewhat different fashion than Kobe City itself. The major damage to the islands resulted from liquifaction of the fill. Many parts of the islands became oceans of mud. In addition, parts of the islands subsided as much as 3 m. Some of the residential areas subsided between 30 and 50 cm. Surprisingly, no buildings collapsed; damage to them was limited to cracks in the walls.

The surrounding harbor facilities suffered greatly; bulkheads collapsed and cranes were dislodged. The piers supporting the bridge connecting the island to Kobe were dislocated which caused a temporary suspension of traffic. The loss of the bridge connection led to a serious shortage of products between the islands and the mainland. As a result of the earthquake a tunnel has been added so that there are now two connections between the islands and the mainland.

The building foundations on the islands were strong enough so that they suffered little damage compared to those in the city proper. The greatest problem stemmed from the loss of vehicular contact, a loss that was partly compensated for by boat transport. It has been concluded that being able to use the sea after such a disaster is indeed one of the advantages of man-made islands[6].

3.2. ISLAND-TYPE FISHING PORTS

Ordinarily fishing ports are found on naturally rocky coasts. However, some island-type fishing ports have been constructed off sandy shorelines. Such offshore ports reduce the impact on the nearby seashore, especially in those locations which are sandy and where the shoreline gradient offshore is gentle[7].

Generally the land behind sandy beaches is level which has been an enticement to industry that often relegates fishing to a lesser status. In addition, the construction of fishing facilities has a low priority and the beaches near them are subject to longshore drifting. Many of the fishing ports are small and often protected by breakwaters. These shoreline structures are impacted by beach drift which often results in filling the port, making it unusable. Because of these problems, the construction of a fishing port necessitates that it not only serve the fishing industry but also that it does not impact the surrounding shoreline. One way to avoid the problems associated with shoreline drift is to establish the fishing port on an artificial island.

3.2.1. *Kuninui Fishing Port*

Construction of Kuninui Fishing Port was begun in 1988. The area is one of fine sand with diameters of 0.16-0.20 mm and has a slope of 1/100. Maximum wave heights are 2.1-2.5 m with an interval between waves of 4-5 s[8]. After a study of wave and drift conditions along the shore it was calculated that the best depth for the creation of the fishing port would be a water depth of 6 m. It was further determined that the best shape for the port would be approximately circular (Fig. 6). The strand line shows some accumulation but it is not considered to be of major importance. Further, there seems to be no impact of the structure in a downdrift direction.

Figure 6. The fishing port.

3.3. FLOATING PLATFORMS

About half of the shallow water around Japan has already been reclaimed and beforetoo many years have passed the rest of the sea area will be saturated. There are many problems associated with utilizing the deeper areas off the coast. One is the high cost, another is the lengthy time it takes to complete a project and yet another is concerned with environmental impacts, a concern that has become increasingly important in Japan as it has elsewhere in the world.

In 1993, the Transport Technological Council made a recommendation to the Minister of Transport to promote research into the development of ultra-large floating structures. Then, in 1995, the Technological Research Association of Mega-floating Structures was established. Its charge was to implement research and development of floating structures that would have a span of several kilometers and a lifetime of 100 years[9].

3.3.1. *Advantages of Large Floating Platforms*

The many advantages of the floating structures being designed by the TRAM[10] include: 1) the possibility of utilizing most areas of the sea regardless of depth and bottom conditions, 2) the minimum impact of earthquakes, 3) the minimal impact they have on the natural environment and ecosystem, 4) the possibility of moving them to other sites and 5) the availability of the areas below deck for a multitude of uses.

3.3.2. *The Structural Characteristics of a Large Floating Platform*

Between 1995 and 1997, research was conducted into the development of a large floating platform (Fig. 7). The objective was to establish technological systems that will be verified through experiments with a large floating model at sea. The location selected for the experiment is at Oppama in Tokyo Bay. The 5 major parts of the

selected for the experiment is at Oppama in Tokyo Bay. The 5 major parts of the research deal with: 1) design, 2) construction technology, 3) durability, 4) operational functions and 5) environmental assessment.

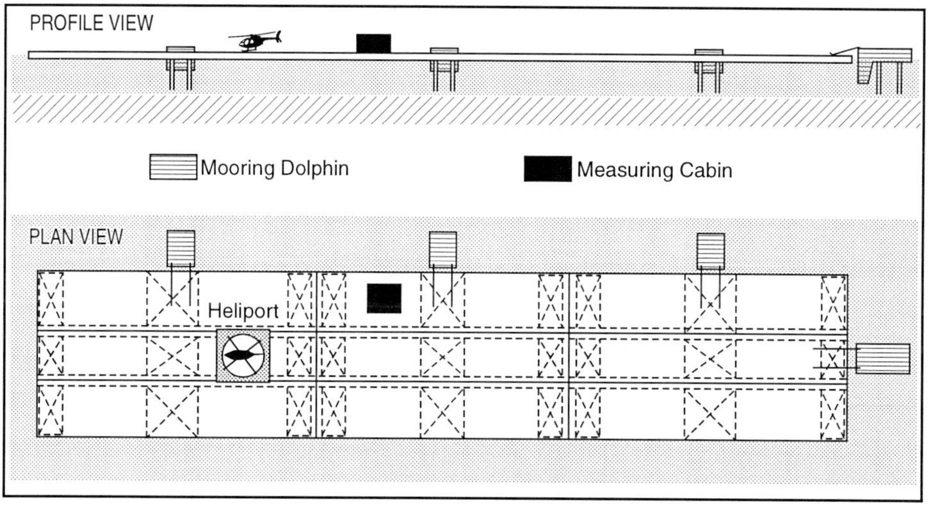

Figure 7. Structural system and plan of offshore island.

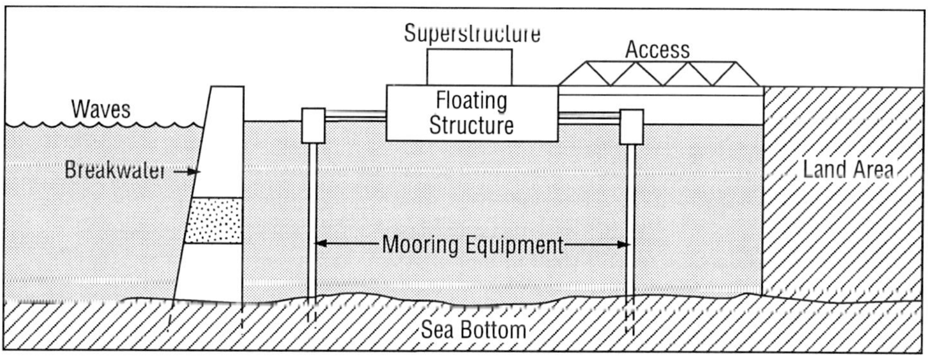

Figure 8. Cross-section of floating island showing mooring system.

Research into Design Technology. Ultra-large floating structures will be thinner than conventional structures in relation of depth to length. Therefore, the objective is to develop a structural design that adapts to elastic movements (Fig. 8). The research involves precisely determining the external and mooring forces acting on the floating structure, analyzing elasticity by empirical methods and with wave tanks and verifying the results at sea.

Research into On-Site Construction. Because ultra-large floating structures are too large to be built on land, they are constructed by joining floating units at sea. The objective of at-sea construction is to complete a 500 ha scale floating structure in three years. The research involves how to maintain precise dimensions for each unit, to reliably join the units which will be acted on by waves and other forces and develop underwater welding techniques.

Research on Durability. Various methods are conceivable in the construction of floating structures durable enough to last more than 100 years. In preparing for such durability new anti-erosive materials (e.g. titanium plates) must be developed. Further, advanced monitoring systems capable of detecting cracks and corrosion of underwater structural parts will be needed as will systems capable of repairing underwater components.

Operational research. In order for mega-floating systems to be effective they must function at high levels of performance whether they be used as airports, harbors or factories. Research must be focused on such things as vibration, noise, temperature and magnetism all of which can impact adversely on effectiveness. Some of these problems have been tested in the model by using helicopter takeoffs and landings since 1996.

Environmental assessment research. It is generally considered that floating structures have relatively little impact on the environment. However, few data substantiating this belief have been collected. Pertinent research showing the characteristics of water flow around such structures is needed as is that about associated ecosystems.

Stage 2 in Mega-Floating Structure Research. Toward the end of the construction of the 9-part, 300 m x 60 m x 2 m structure in 1996 there was an onsite demonstration for the purpose of verifying the technology involved in an at-sea connection of floating units (Fig. 9). It was then decided to proceed with Phase 2 of the mega-floating structure research. This involved the construction of a large-scale model capable of accommodating large aircraft. However, because of monetary constraints, the program was downsized. Instead, the airport to be constructed is of a size that could handle small commuter aircraft (Fig. 10). The 1000 m long platform will cover about 8.4 ha and have a standard width of 60 m with an expansion to 121 m in its center. Construction began in 1998 with plans calling for actual takeoffs and landings commencing in 2000 and research to be completed in 2001.

4. Recent Oceanic Environmental Improvement Technology in Japan.

The recent disappearance of shoals and beaches, which are vital as habitats to marine organisms, has had a drastic impact on the ecosystem in a number of locations along Japanese coasts. The amelioration of such disasters and the improvement of marine habitats are now considered high priority tasks. To this end detailed programs are being developed in coastal and offshore environmental protection, preservation and enhancement (Table 5)[11]

Table 5. Matrix of environmental preservation and technology.

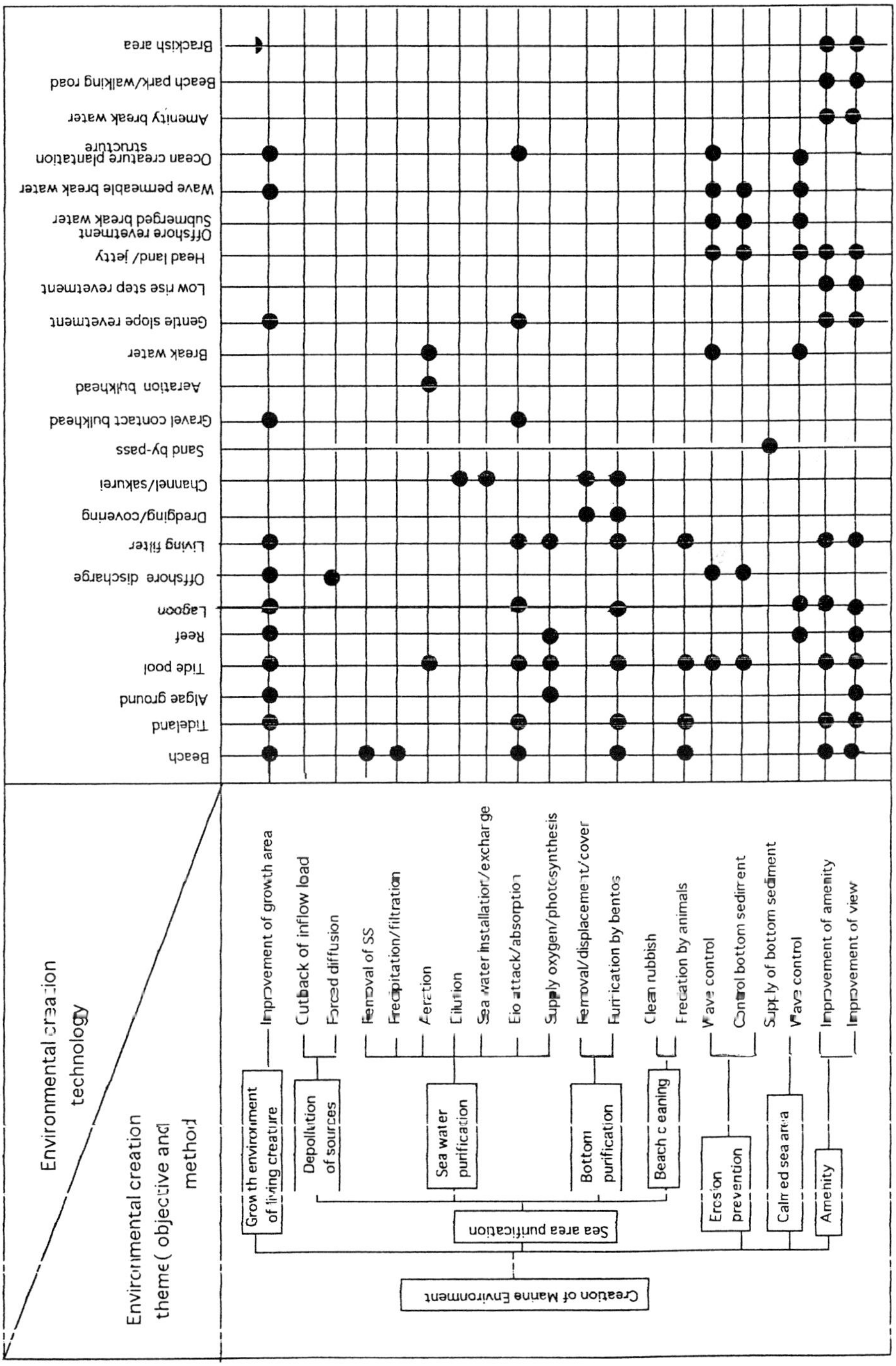

Figure 9. Floating island construction.

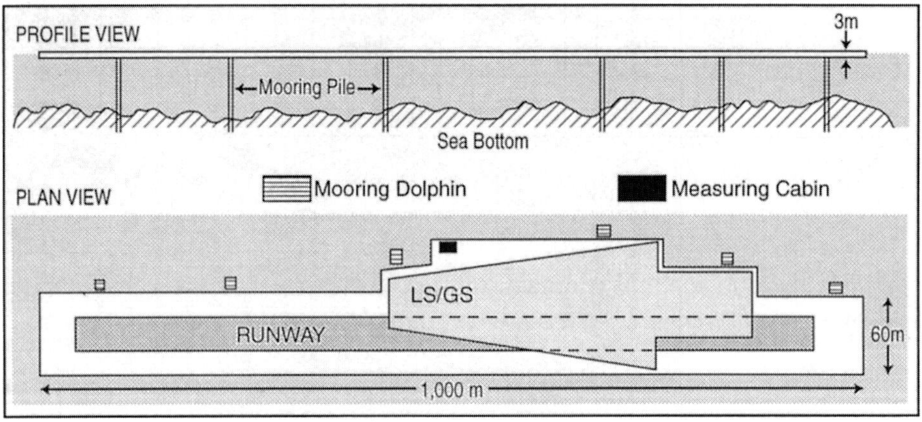

Figure 10. Profile and plan views of second stage offshore island.

4.1. EXAMPLES OF NEWLY APPLIED TECHNOLOGY

4.1.1. Gravel Contact Purification Wavebreaker

The gravel contact purification method is a system that utilizes a gravel-filled wave breaker (Fig. 11). Water passes between pieces of gravel that fill the breakwater. Because there is a film of microbes living on the surface of the gravel (as they do in a normal shingle beach), organic pollution substances are decomposed. Experiments on this type breakwater are being conducted in the port of Amagasaki[12].

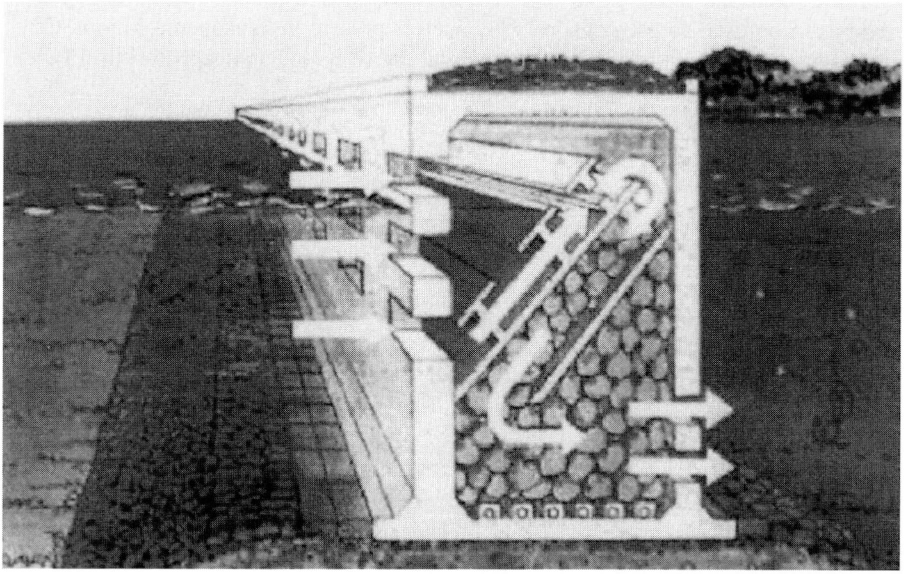

Figure 11. Breakwater with gravel for micro-organism growth.

4.1.2. *Water Quality Control and Algae Adhesion Structures.*
In the coastal areas of Japan, concrete has been used as a structural material for breakwaters and revetments as well as other offshore structures. In water, concrete produces alkali which can convert the surrounding water into a strong alkaline region with pH values often reaching values of 10 to 11 in seawater and 13 in freshwater. In order to eliminate this problem concrete blocks are coated with ferrous sulfates which not only reduces the release of carbon hydroxides from the concrete blocks but also serve to provide micro-nutrients to algae[13].

4.1.3. *Electrical Energy Generator*
Caisson breakwaters can be equipped with small turbines which are powered by water as it enters and exits the structure's chambers. Experiments are being presently conducted in the port of Sakata[14].

4.1.4. *Dual-cylinder Caisson*
The dual-cylinder caisson is a breakwater that is constructed in such a way that incident waves crash together dissipating their energy. This type breakwater is most suitable for shorelines that are deep or have a high tidal range. This type breakwater has been used in the ports of Sakaiminato and Shibayama[14].

4.1.5. *Other Types of Experimental Breakwaters*
A type in which the wave force acts vertically, known as the slit caisson breakwater, is suitable for high tide or soft ground shorelines. One is being tested in Miyagaki Port (Fig. 12).

Some locations today have breakwaters that have a level surface on top and therefore can be used as promenades. One such is present in Wakayama Marina City. This type has the advantage of being usable by those who are handicapped (Fig. 13).

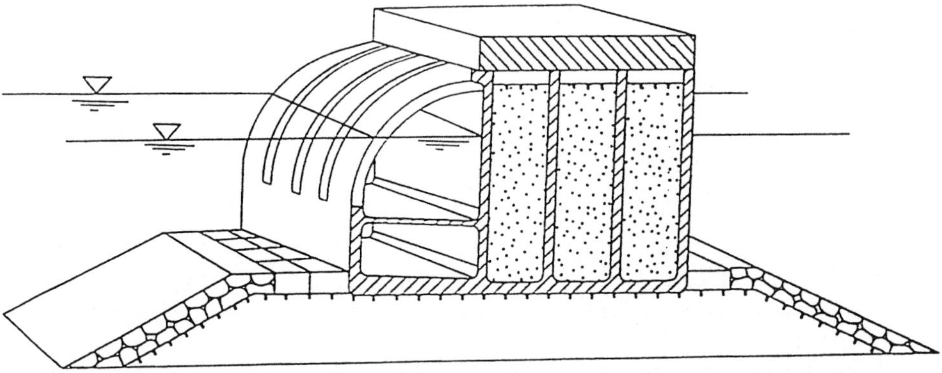

Figure 12. Cross-section of slit-type caisson.

Figure 13. Promenade breakwater.

5. References

1. Nagao, Y. and Fujii, T. 1991. Construction offshore man-made island and preservation of coastal zone. In: Nagas, Y. Ed. Coastlines of Japan. American Society of Civil Engineers, New York. 212-226.
2. Committee on Ocean Development, 1997. Offshore Artificial Islands in Japan. Keidannen and Kozai Club, Ministry of Transport, 17-19.
3. Committee on Ocean Development, 1992. Man-made Island, Kozai Club, Ministry of Transport, 22-25.
4. Committee on Offshore Man-made Island, 1996. Hand Book of Man-made Island, Keidannen and Kozai Club, Ministry of Transport, 18-20.
5. Josuke Akedo, 1993. Man-made Islands, Journal of Japan Society for Civil Engineers. 107-109.
6. Hanshin – Awaji Disaster Investigation Committee: Hyougoken Southern Area Disaster Investigation-Ocean area, Japan Architectural Institute, 1997. 146-150.
7. Akira Nagano, 1993. Man-Made Islands. Journal of Japan Society for Civil Engineer. 112-113.
8. Committee on Offshore Man-made Islands, 1996. Hand Book of Man-made Island, Ministry of Transport, Keidannen and Kozai Club, 46-47.
9. Technological Research Associates of MEGA-Float, 1994. MEGA-Float, 34-40.
10. Technological Research Associates of MEGA-Float, 1998. New Research Project of MEGA-Float, 5-8.
11. Ports and Harbors Bureau, Ministry of Transport, 1994. Ports that can coexist with the environment. Printing Bureau, Ministry of Finance, 5-6.
12. Nishiumi, Hiroshi, 1995. Ocean environmental creation technology. Japan Ocean Development and Construction Association. 31.
13. Kenji Hotta and Tetsuo Suzuki, 1995. Technology for creating an artificial seaweed community using ferrous sulfate, proceedings of ECOSET, Japan International Marine Science and Technology Federation, 180-191.
14. Nishiumi, Hiroshi, 1995. Ocean environmental creation technology. Japan Ocean Development and Construction Association. 34-35.

RECLAMATION AND RIVER TRAINING IN THE QIANTANG ESTUARY

HAN ZENGCUI
Qiantang River Administration
Hangzhou, China

DAI ZEHENG
Zhejiang Institute of Estuary and Coast
Hangzhou, China

1. Introduction

The Qiantang Estuary, with an upstream boundary at the tidal limit near the Fuchunjiang hydro-power station and a downstream boundary at the mouth of Hangzhou Bay, is a macro-tidal estuary that is world-renowned for the magnificent tidal bore in the estuarine stretch between Wenyan and Ganpu. It lies south of the Changjiang River Estuary and north of the hilly region of eastern Zhejiang. The rich sediment supply from the Changjiang River and the coastal regions of the East China Sea has formed a vast area of coastal plains on both banks. The plain at the north bank, known as the Taihu Lake Plain, is connected with the Changjiang delta; the plain at the south bank is the Xiaoshan-Shaoxing-Ningbo plain (Fig. 1). The total area of the plains stretching from both banks is about 50,000 km^2. The land is fertile, has well developed transportation systems, developed production, a prosperous economy, and a galaxy of talent in Chinese history. The evolution of the estuary has combined natural and cultural processes resulting in the construction of an artificial coastline. This artificial coastline reflects the development of the area's social economy and the progress of science and technology. In order to understand the historical development of the area it is helpful to review the interaction of man and nature.

2. Historical Changes in Shorelines of the Qiantang Estuary

About 5,000-6,000 years ago, the world climate became warmer and sea water reached the hillsides of Hangzhou and Huzhou to the north and Xiaoshan, Shaoxing, and Yuyao to the south of the Qiantang Estuary (Fig. 1). During the period between 4,000-6,000 BP, several lines of sand bars formed along the south coast of the Changjiang River Estuary in a direction perpendicular to strong wave action from the east. Two lines of bars can still be found near Caojing inside the north bank of Hangzhou Bay[1]. This is the oldest natural coastline that caused a wave dissipation effect on the wet lands behind the sand bar. The coast was rather stable during the following 3,000-4,000

J. Chen et al. (eds.), Engineered Coasts, 121–138.
© 2002 *Kluwer Academic Publishers. Printed in the Netherlands.*

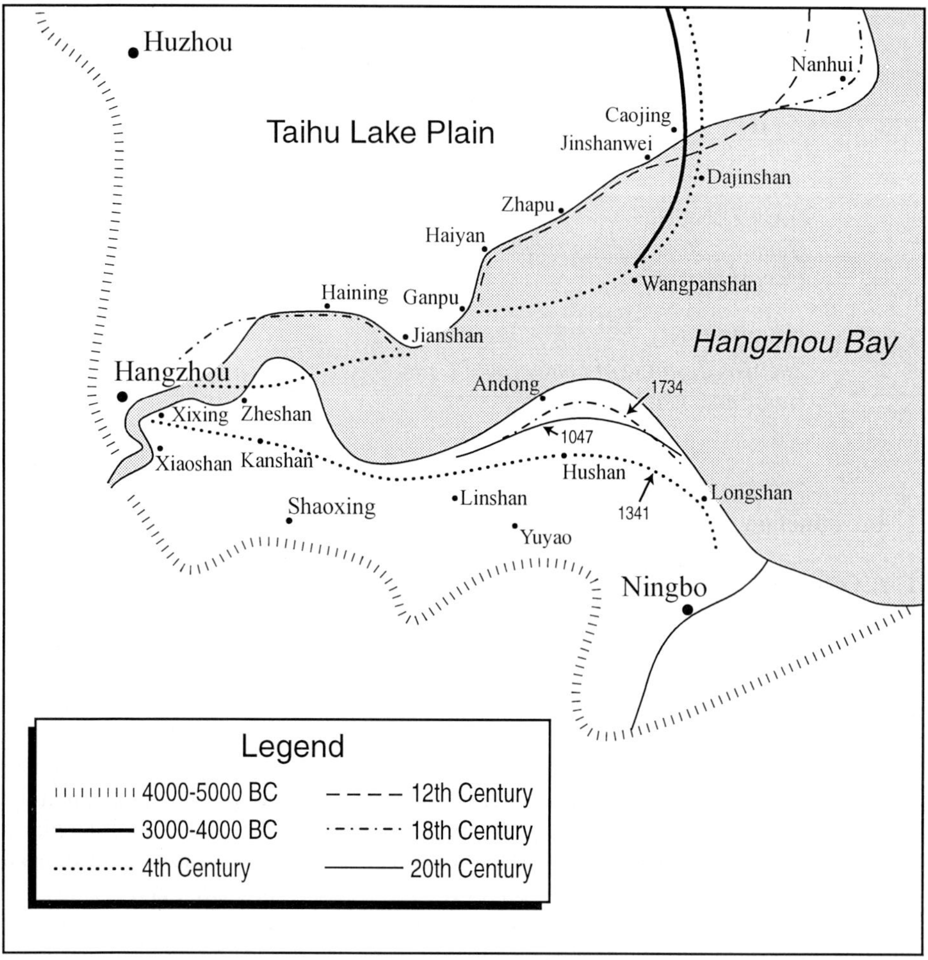

Fig. 1. Shoreline changes in the Qiantang Estuary

years. After the 4th century BC, as the people of the central plains of the Huanghe River migrated south, the hilly district in the Changjiang river basin developed and the sediment in the Changjiang river increased causing the estuarine spit of the Changjiang delta to grow rapidly eastward (more than 40 km in the last 1,600 years). Meanwhile, the southern part of the spit, i.e., the northern coast of Hangzhou Bay, was gradually eroded by tidal currents and wind waves from the SE. According to historical records, at Wangpanshan Island, which is now located in the bay at more than 20 km from the north bank, a military camp was located. The shoreline from Ganpu to Caojing receded progressively and did not become stable until the middle of the 16th century (the Ming Dynasty) when massive gravity stone seawalls were built as a result of the improvement of dike structures through the centuries.

To the west of Ganpu, before 1720 AD, the coastline ran through Jianshan and Zheshan to Hangzhou. There are three passages in the estuary between Kanshan in the south and Haining in the north; namely the Big South Passage, Small Middle Passage, and Big North Passage (Fig. 2). In the stretch from Hangzhou to Ganpu, the main channel frequently shifted northward/southward and the tidal flats underwent alternate erosion and siltation. During the period of the Tang-Song-Ming-Qing dynasties (7th-18th centuries) there were many recordings of the loss of large areas of land and salt fields, e.g., an area approximately 20 km x 15 km in the region south of Haining was lost. This gives a really vivid portraiture of the old Chinese saying "Seas change into mulberry fields and mulberry fields change into seas". Before the middle of the 17th century the main channel ran through Big South Passage. It then gradually shifted to Small Middle Passage and finally, in 1720, to Big North Passage. With a view to eliminating the threat to the northern dike, a diversion channel was successfully dug in 1747 through Small Middle Passage and the main channel shifted again to Middle Passage. This is an excellent example of 18th century river regulation engineering in China. Twelve years later (1759), the main channel shifted again to North Passage. The Qing Dynasty emperor decided to build a gravity stone seawall along the entire line from Hangzhou to Jianshan. According to the incomplete records, during the 268 year period ruled by the Qing Dynasty, about 120 km of stone seawall was built and 26 million ounces of

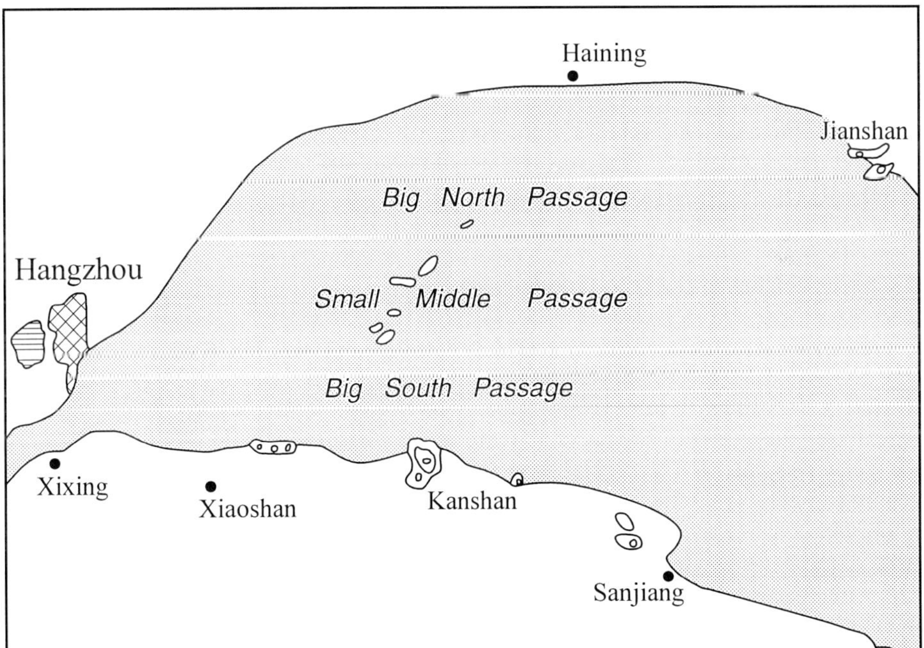

Fig. 2. Location of the three passages in the Qiantang Estuary

silver (equivalent to 1.3 million ounces of gold at that time) were spent in dike construction and reparations.

Along the south bank of Hangzhou Bay, the coastline in the 4th century AD ran through Longshan, Hushan and Linshan. North of this line, the tidal flat underwent siltation and then erosion; in the middle of the 14th century, the coastline returned nearly to the same location it had in the 4th century. Afterward, the southern tidal flat silted up and the coastline advanced northward gradually by successive reclamation projects. By 1950, a total of eight lines of sea dikes were built in different periods which formed a bow-shaped plain area of about 550 km^2, the apex of the bow being 16 km north of its 14th century base line.

To the west of Linshan, beyond the north bank of Xiaoshan and Shaoxing counties, the coastline in the 4th century ran through Kanshan to Xixing. Since 1759, as the main channel shifted to the North Passage, the South Passage silted up gradually and another bow-shaped plain formed. By 1950, the total area of the so-called South Sandy Plain was about 390 km^2, the apex of the bow was about 15 km north of its 18th century position. In summary, since the 4th century, the north bank with a stable stretch of hills from Jianshan to Ganpu in the middle, was eroded on both sides and receded. In the eastern part, the coast receded 20 km from Wangpanshan to the present coast, in the western part, the coast receded 15 km from Zeshan to the present coastline. Along the south bank, two bow-shaped plains silted up and were gradually reclaimed. In the eastern part of this area, the coast advanced to a maximum distance of 16 km north of Hushan, while in the western part the coast advanced to a maximum distance of 15 km north of Kanshan. The original coastline on both banks was comparatively straight, but gradually became curved.

Since 1950, large-scale reclamation and extensive river training works have been completed The artificial shoreline from Hangzhou to Ganpu was advanced rapidly, and a total of 700 km^2 of new land was reclaimed during the past 40 years (see Section 4, below).

3. Characteristics of the Qiantang Estuary[2]

The Qiantang Estuary is a funnel-shaped macro-tidal estuary. The mouth of Hangzhou Bay at Nanhui is nearly 100 km wide, at the inner end of the bay at Ganpu, 85 km westward, it is only 20 km, and at Hangzhou, 104 km farther upstream, the width is only 1 km. According to an analysis by the late Professor Qian Ning (Ning Chien) and his colleagues, the river flow/tidal flow ratio at Ganpu is only 0.01[3], and the sediment supply is mainly from the sea. Thus a big longitudinal sand bar forms inside the mouth rather than outside as it does in many other estuaries. The big sand bar that stretches from Zhapu to Wenyan is 130 km long with an apex 10 m above its base line (Fig. 3)[4]. The width/depth ratio is as much as 40-60; the water depth during low tide is only 1-2 m. This special geomorphologic feature impacts tides and river regime in a number of ways.

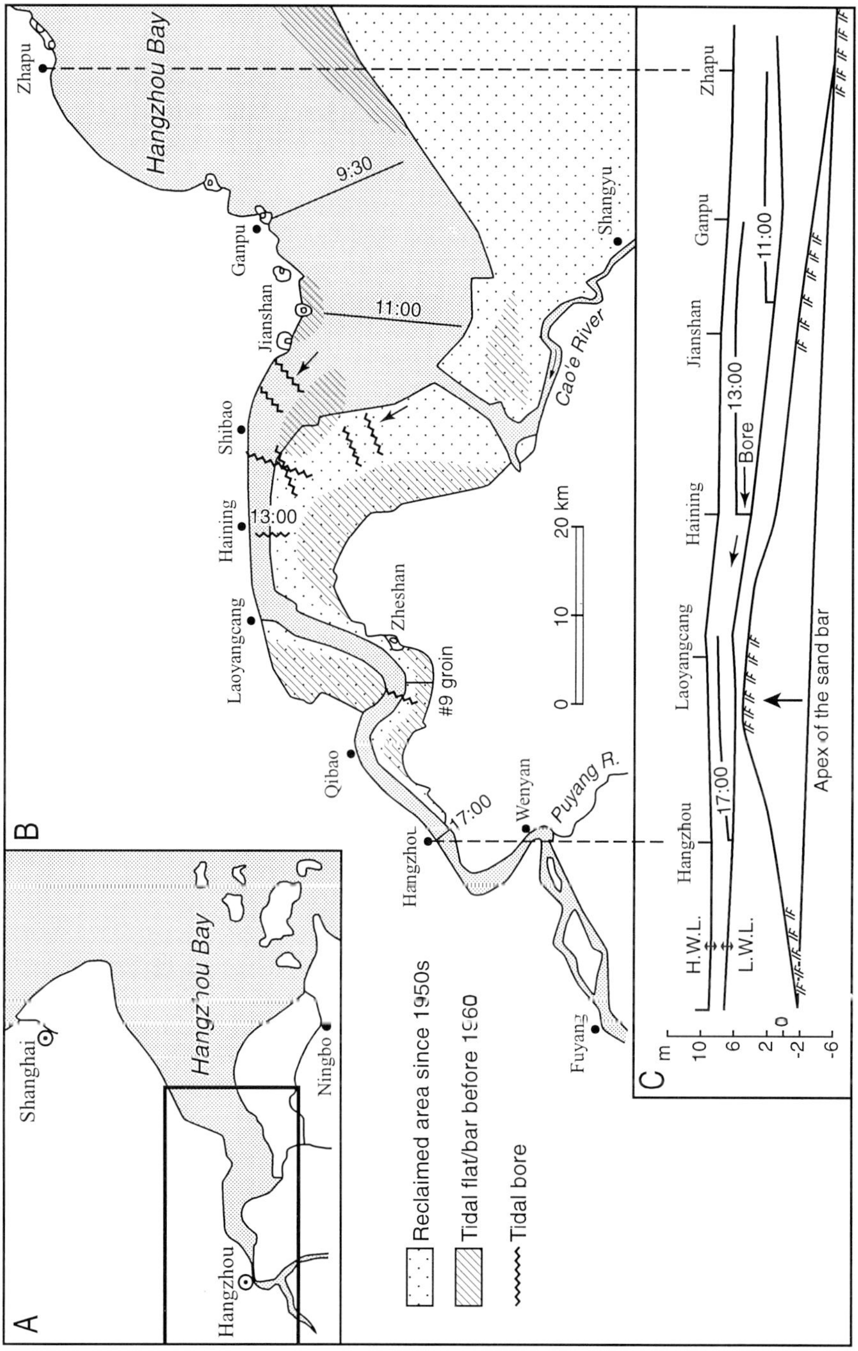

Fig. 3. Qiantang Estuary: A) general section, B) the estuary, C) profile and advance of tidal bore.

3.1 SHARP DISTORTION OF THE TIDAL WAVE

In the course of tidal wave propagation, the rate of distortion depends mainly on the ratio of tidal range and the water depth below low tide level. Upstream from Ganpu, the depth decreases rapidly, thus the ratio increases rapidly, being 1.16 at Ganpu and 2.44 at Jianshan where a magnificent tidal bore begins to form. The bore attains its maximum height at Haining, an amount that is generally 1.5-2 m with a maximum of 3 m (Fig. 4). The propagation velocity of the bore is usually 5-7 ms^{-1}, and the measured maximum pressure is 7 ton m^{-2}[5]. Meeting a protruding bank or a spur dyke, the bore may run to a height of more than 10 m. The destructive force of the bore is enormous and must be dealt with in reclamation and river training projects. In one instance, a large block of stone weighing 30 tons, which had been cemented to the top of a groin, was dislodged and moved 8 m away and finally lodged against the gravity seawall.

With a large tidal range and shallow depth, the current is inevitably swift: maximum point velocities of 4-5 ms^{-1} have often been measured and the flood velocity is about twice the ebb velocity upstream of Haining.

Fig. 4. The tidal bore approaching a groin in Hangchow Bay. Photo by J. Walker.

3.2 FREQUENT SHIFTING OF THE CHANNEL AND THE ACCOMPANYING EROSION AND/OR SILTATION OF TIDAL FLATS

The sand bar is composed of silt and very fine sand with a medium diameter of 0.02-0.04 mm, a sorting coefficient of 1.2, a non-uniformity coefficient of 1.1 and with little scour-resisting ability. The estuary is rather wide and shallow, and both the flood and ebb flow usually diverge. As the relative strength between river flow and tidal flow varies, the main channel frequently migrates between both banks; the distance of migration can exceed 20 km.

As the main channel shifts, it has been observed that the concave bank can erode at a rate of 245 m per day along a stretch several kilometers in length, forming a cliff 3-4 m high; while along the convex bank, the tidal flats can be silted up to mean tidal high level within several months. Although the tidal flats are not stable, being frequently eroded or silted, high flats are often present along the convex bank.

3.3 CONTINUOUS SEDIMENT INTERCHANGE BETWEEN UPPER AND LOWER STRETCHES AND INTENSE VARIATION OF THE LONGITUDINAL PROFILE

Because the silt content of river runoff is low, river flow possesses great scouring ability. During the wet season, the river flow is high while the bed material will be scoured from the upper stretch and transported to the lower stretch: hence the apex of the sand bar will be lowered and shifted downstream. When the wet season is over, the river flow is low and tides strengthen: aggradation occurs in the upper stretch, because the tidal flood velocity is higher than ebb velocity, causing much more sediment to be carried by flood flow than by ebb flow, resulting in a net sediment transport upstream. In the lower stretch, erosion occurs because the silted cross-section is too small for the larger tidal flow; this is caused by the enlargement of the cross-section in its upper stretch during the wet season. For example, from April to August, 1974, 67×10^6 m^3 of sediment was scoured in the upper stretch between Hangzhou and Haining, and 230×10^6 m^3 of sediment was deposited in the lower stretch between Haining and Ganpu. From September, 1974 to May, 1975, 98×10^6 m^3 of sediment was deposited in the upper stretch and 218×10^6 m^3 was scoured in the lower stretch. During successive dry years, heavy siltation can occur along nearly all of the estuarine reach: e.g., from July, 1977 to July, 1979, 413×10^6 m^3 was deposited from Hangzhou to Haining and from Haining to Jianshan respectively, the thickness of the deposits were between 1.4-4.8 m, averaging 2.78 m.

3.4 AMOUNT OF THE FLOOD PASSAGE CAPACITY GREATLY AFFECTED BY THE DEGRADATION AND AGGRADATION OF THE RIVER BED

Because of the greater depth and the small variation of the sea bed in Hangzhou Bay downstream from Ganpu, the intensity of the tides is almost unaffected by the variation of the sea bed. However, upstream from Ganpu, because of the smaller depth and larger variation in the riverbed, the tidal effect is different. Low water will be lower and the tidal range larger whenever the channel is straighter and deeper. For example, in 1973, the low water at Haining was at 1.91 m and the mean tidal range was 4.08 m. On the contrary, the low water will be higher and the tidal range smaller whenever the river is relatively curved and shallow; e.g., in 1969, the mean low water was at 4.04 m and mean tidal range was1.96 m.

The elevation of the apex of the sand bar can be chosen as the parameter characterizing the profile: the variation of high water at Hangzhou city will be 0.5-0.7 m corresponding to every 1 m variation of the apex elevation under the same fresh water discharge and the same high tide level at Ganpu which is the downstream boundary of the estuarine reach. Because of the great difference of the riverbed and cross-sectional area during different periods, the difference in fresh water discharge can be 17,700 m^3 s^{-1} under the same high water level at both Hangzhou and Ganpu (Table 1).

TABLE 1. Discharge and water level

Date	Ganpu	Hangzhou (m)	Fresh water discharge (m^3s^{-1})
June 22, 1955	6.45	9.07	29,000
June 10, 1968	6.46	9.01	11,300

4. Reclamation and River Training since the 1950s

4.1 PLANNING

In the estuarine reach (Wenyan to Ganpu) of the Qiantang Estuary, the river is very wide and shallow with a vast area of tidal flats, while the main channel frequently shifts. It was decided in the 1930s and 1940s to aim at contraction in the river, to stabilize the channel, and to reclaim the tidal flat. Only a few groins were built and the effect was not obvious at that time. Since the 1950s, systematic hydrometric observations, channel surveying, physical and mathematical modeling and the investigation of regulation schemes for the estuary were constantly conducted.

Based on the understanding of the characteristics of the estuary, the regulation principle was determined as to decrease the tidal volume as much as possible so that the harmful effects of the strong tide, such as the safety of the seawall and the heavy sedimentation problems, could be largely eliminated with beneficial effects on the reclamation of new land, navigation, etc. Also considered were the building of a tidal barrier near Jianshan to exclude the tide and its sediment load and a submerged sill to lessen tidal intrusion. Both of these plans were discarded because of the very heavy sedimentation problems. Finally it was decided to make the river width smaller by river training and reclamation as a first stage of regulation. This scheme enabled the stabilization of the channel and increasing deposition over a large land area that is very valuable to the densely populated Zhejiang province. Moreover, this scheme could be implemented by mobilizing the riparian people without too much investment.

In planning the river training project, the first problem was how to design the constricted width along the river. A rough estimation made early in 1962 indicated that even if the width at Ganpu was narrowed from 20 km to 4 km, the lowering of the apex of the sand bar would not be more than 1 m, while the navigable depth could not be increased very much, and the bore would still be present. At the initial stage of river training, the main purpose was to narrow the very wide river course step by step and thus to stabilize the shifting channel gradually, while at the same time, increase land area by reclamation which would also improve the navigation conditions to some extent. Only after the wide river was constricted could the regulation of the navigation channel begin.

The widths along the river course were preliminarily designed in the 1960s: 1 km at Zhakou, 1.5 km at Cangqian, 4 km at Haining, 7 km at Jianshan, and 12 km at Ganpu. The dike alignment was designed with reference to the contours of equal probability of development of a high tidal flat, and was drawn based on the statistical analysis of the survey charts completed by 1961 (Fig. 5). There is a close resemblance in the shape of the flats shown on Figure 5 and Figure 6 which is the actual plan view after the implementation of the reclamation up to 1997. An important principle was also set forth in drafting the dike alignment: the existing mountains (e.g., near Cangqian and Jianshan), spur dikes (e.g., the 3 km long #9 spur dike in Zheshan Bend) and the old massive seawall (e.g., along the northern bank from Laoyancang to Jianshan) should be fully utilized so as to lessen the construction costs by a large amount. The reasoning behind this principle is straightforward because the building and maintenance of new polder dikes would be quite difficult and very expensive, especially in the stretch of the estuary that is under the attack of a strong tidal bore.

In planning the river training project, cross-sectional area A, width B, and mean water depth H referring to mid-tidal level were calculated by the following empirical formulae:

$$A = 4.7Q_e^{0.9}\,S^{-0.22} \qquad (1)$$
$$B = 7.5Q_e^{0.62}\,S^{-0.12} \qquad (2)$$
$$H = 0.63Q_e^{0.28}\,S^{-0.34} \qquad (3)$$

Q_e is the mean ebb discharge in m^3s^{-1} (including river flow) and S is the mean sediment concentration in kg m^{-3} during ebb flow. The above formulae were obtained by the analysis of more than ten estuaries in China and abroad, including three sections at Zhakou, Haining, and Jianshan of the Qiantang Estuary. Tidal calculations were then carried out, especially for the case of high river flow confronted with high tides in order to check whether the high water near Hangzhou City would be raised or not. If raised, would it be within a tolerable range?

In the 1980s, having developed various 1-D and 2-D mathematical models in the Qiantang Estuary for numerical simulation of tide, flow, sediment and pollutant transportation, and bed deformation, a detailed investigation of the training project and prediction of its impact was carried out. Meanwhile, the construction of deep-water harbors and coastal industries in Hangzhou Bay was started and speeded up following the execution of the open-to-outside-world policy of China. The river training project had to be reviewed taking into account such new demands. The large-scale reclamation

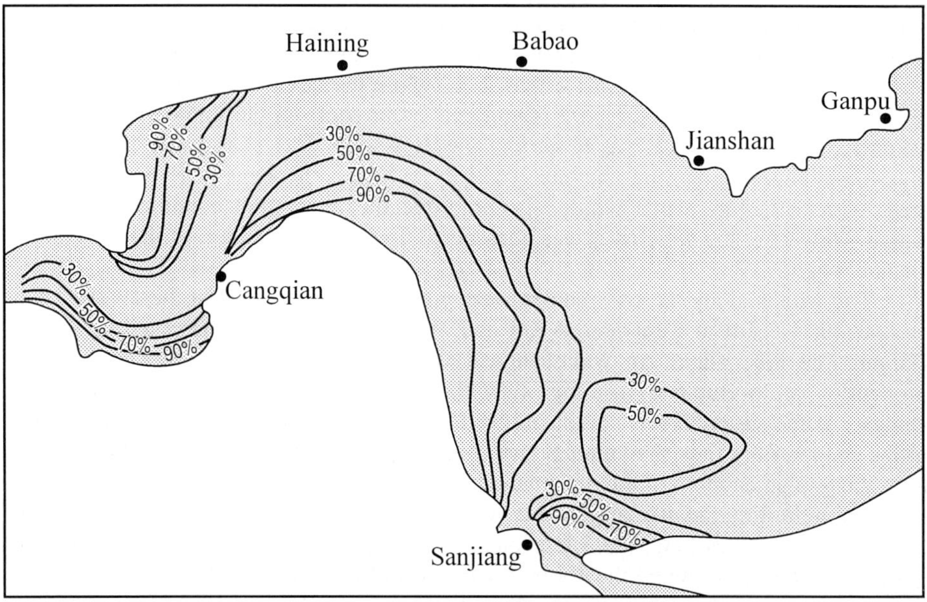

Fig. 5. Contours of the probability of tidal flat development

above Ganpu would bring about a reduction of tidal volume and sedimentation downstream which might cause siltation of the existing navigation channel and deep-water harbors. After a forecasting calculation of the bed deformation for the different training projects, it was decided to keep the narrowed width at Ganpu to 18 km instead of the formerly designed 12 km and correspondingly the polder area was to be limited to about 20,000 ha. For the present stage, along the southern convex bank of Hangzhou Bay, an additional polder area of about 27,000 ha was planned. All these studies were mainly carried out by means of mathematical modeling supplemented by physical modeling and both approaches were based on the analysis of field data.

4.2 IMPLEMENTATION OF RIVER TRAINING AND RECLAMATION WORKS

Owing to the strong tidal bore, the ever shifting channel, shallow water and rapid scour and deposition, the construction of groins and embankment dikes is confronted with many difficulties. When the rules of erosion/siltation of the tidal flats in Qiantang Estuary are known, river contraction work can be achieved gradually by building polder dikes on the high tidal flat which is relatively easy. Of course, whenever the main channel shifts to the side of the polder dike, the dike foundation has to be protected from heavy scour. The protection measures are not easy to carry out, however, but are still much easier than to fight against the tidal bore through building of groins/polder dikes in the deep channel.

On the convex bank, large areas of high tidal flat are often silted up and completely exposed during neap tide and ordinary tide. They are submerged by less than 1 m during high spring tide. This provides an opportunity to build up an encircling polder earth dike during neap tide and then to gradually carry out slope protection and base protection works. About 70% of the reclaimed land has been reclaimed in this manner by mobilizing tens or hundreds of thousands of peasants to complete dike building within 7-10 days. From 1968-1971, along the southern bank from Cangqian to Toupeng and Xinwan, five polders with a total area of 14,000 ha were successively reclaimed from upstream to downstream. The polder dikes prevented a possible shifting of the channel to the south and each upper polder promoted siltation on its downstream side, thus making follow-up reclamation possible. Since 1990, textile bags filled with earth by hydraulic pumping have been mostly used in building the polder dikes; large amounts of labor were saved and the construction schedule was speeded up.

Along the concave bank, there is little opportunity for silting a high tidal flat. Groins and longitudinal dikes have to be constructed to divert the current and to promote siltation. Only when the tidal flat has silted up to a high level is the building of polder dikes possible.

The progress of reclamation at different phases is shown in Figure 6, in which the impoldering year is indicated for each polder. The figure shows that reclamation progressed generally from upstream to downstream. Nevertheless, some of the polders have been reclaimed alternately at upstream and downstream stretches. By the end of 1997, about 70,000 ha had been reclaimed upstream of Ganpu. Together with those that are reclaimed below Ganpu, the total polder area amounted to 90,000 ha.

5. Environmental Changes after Reclamation and River Training

The large-scale and difficult training of the Qiantang Estuary is a rare example in the world, its width having been narrowed to 1-4 km from the original of 1.5-20 km in the 64 km stretch from Hangzhou to Shibao, Haining county. A new equilibrium of the river has been established since the regulation of the past forty years.

5.1 MAIN CHANGES OF THE RIVER CHANNEL

5.1.1 *Shifting Range of the Main Channel Greatly Decreased*

When the original width of the training reach was 1-20 km, the main channel of the river could shift across the entire area between both banks; after contraction, however, the shift of the main channel became limited; the river is now relatively stable (Fig. 7).

5.1.2 *Cubature Decreased and Tidal Flat Area Increased*

The cubature under HW and the area of the high tidal flats of the reach between Zhakou and Ganpu in the year of 1953 to 1998 and the related annual mean river discharge at the stream-gauging station named Lucibu (embodying three fourths of the drainage area upstream of Hangzhou) are shown in Figure 8. The average cubature was 6.6 billion m^3 before the river contraction and reclamation. After that, it was only 4.5 billion m^3; a cubature of 2.1 billion m^3 was silted up. The area of the tidal flat was about 40,000 ha

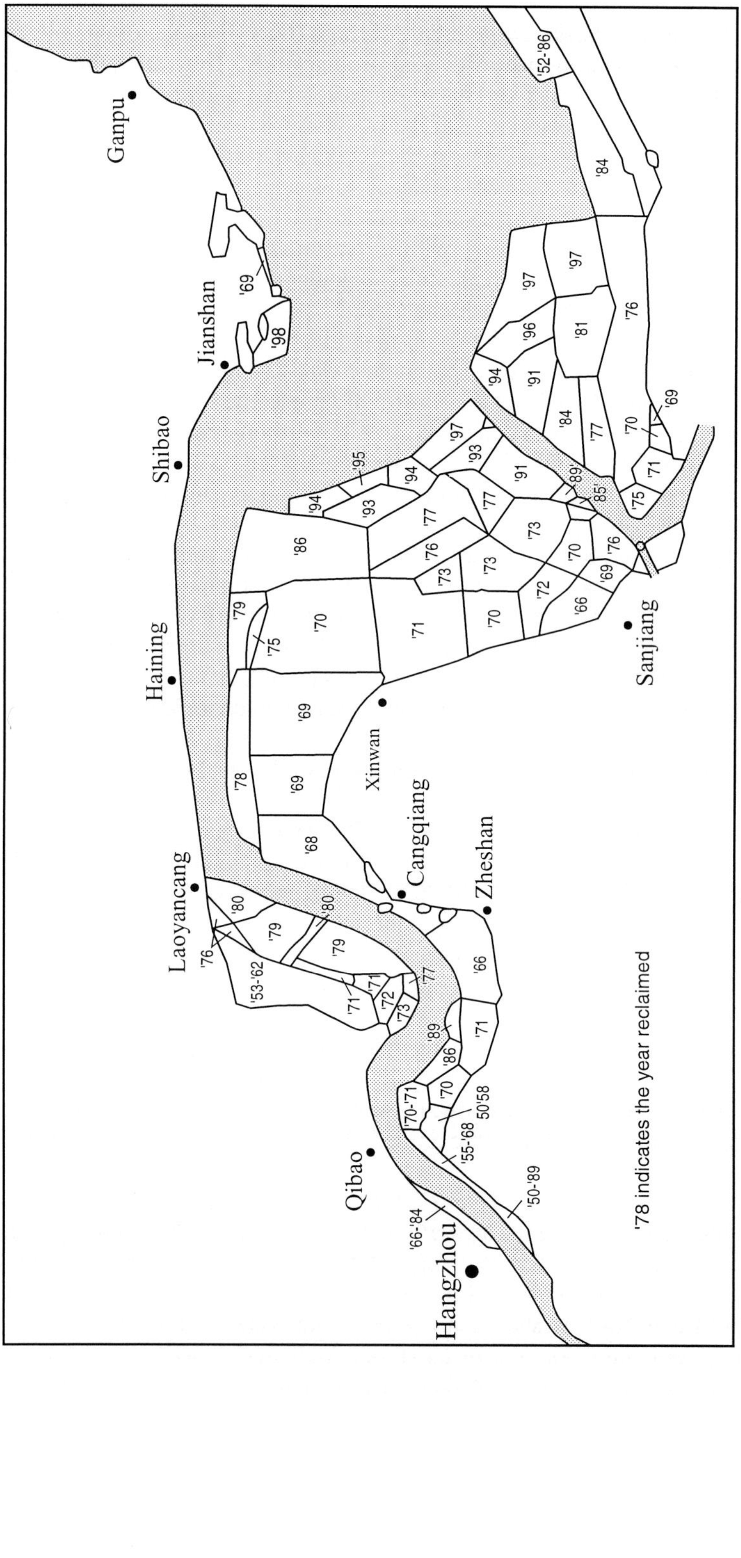

Fig. 6. Reclamation phases of the Qiantang Estuary.

before reclaiming. By 1998, 70,000 ha of flat had been reclaimed. The tidal flat area that remained totaled 13,300 ha. This shows that new tidal flats had formed after the polder works were completed.

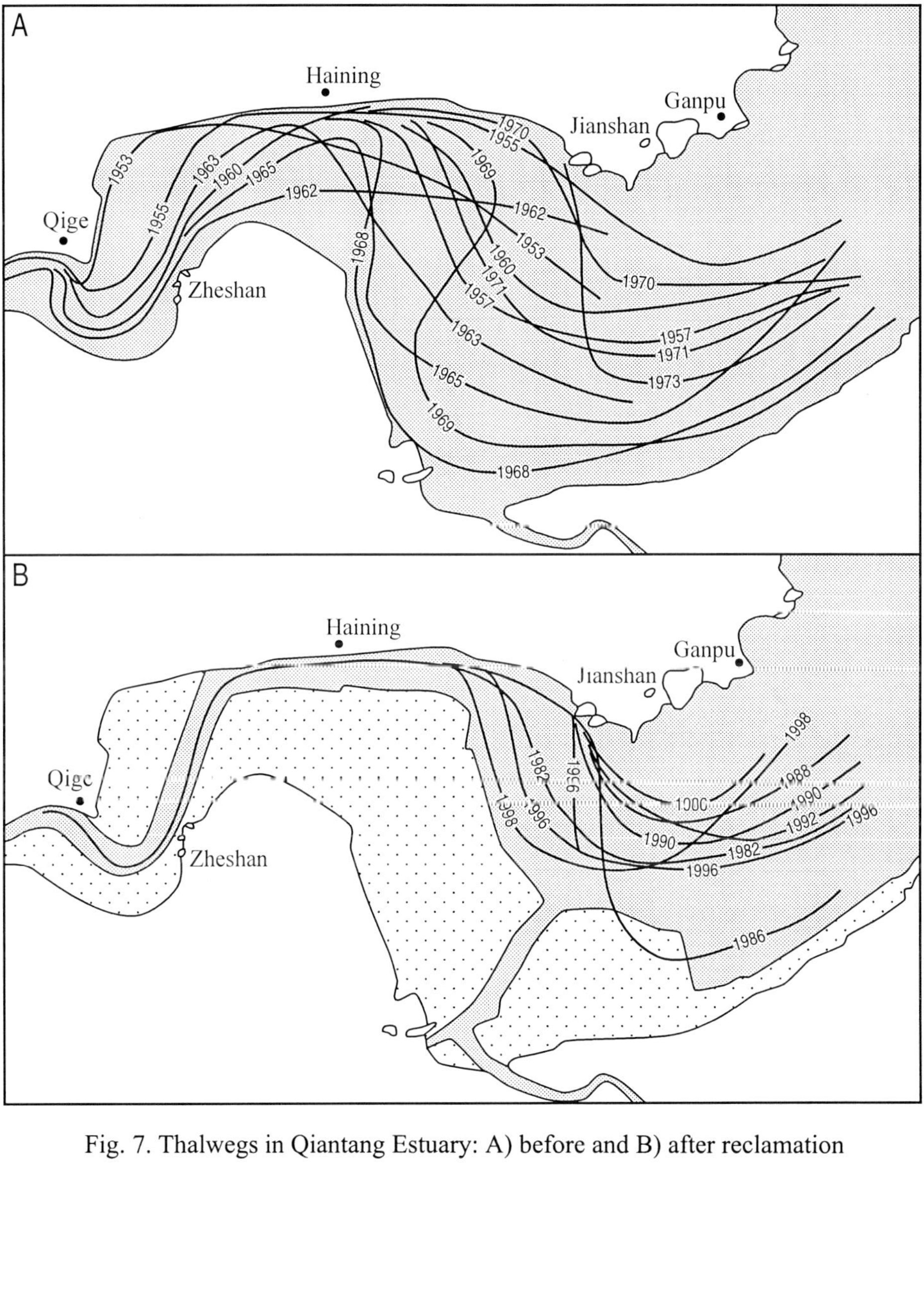

Fig. 7. Thalwegs in Qiantang Estuary: A) before and B) after reclamation

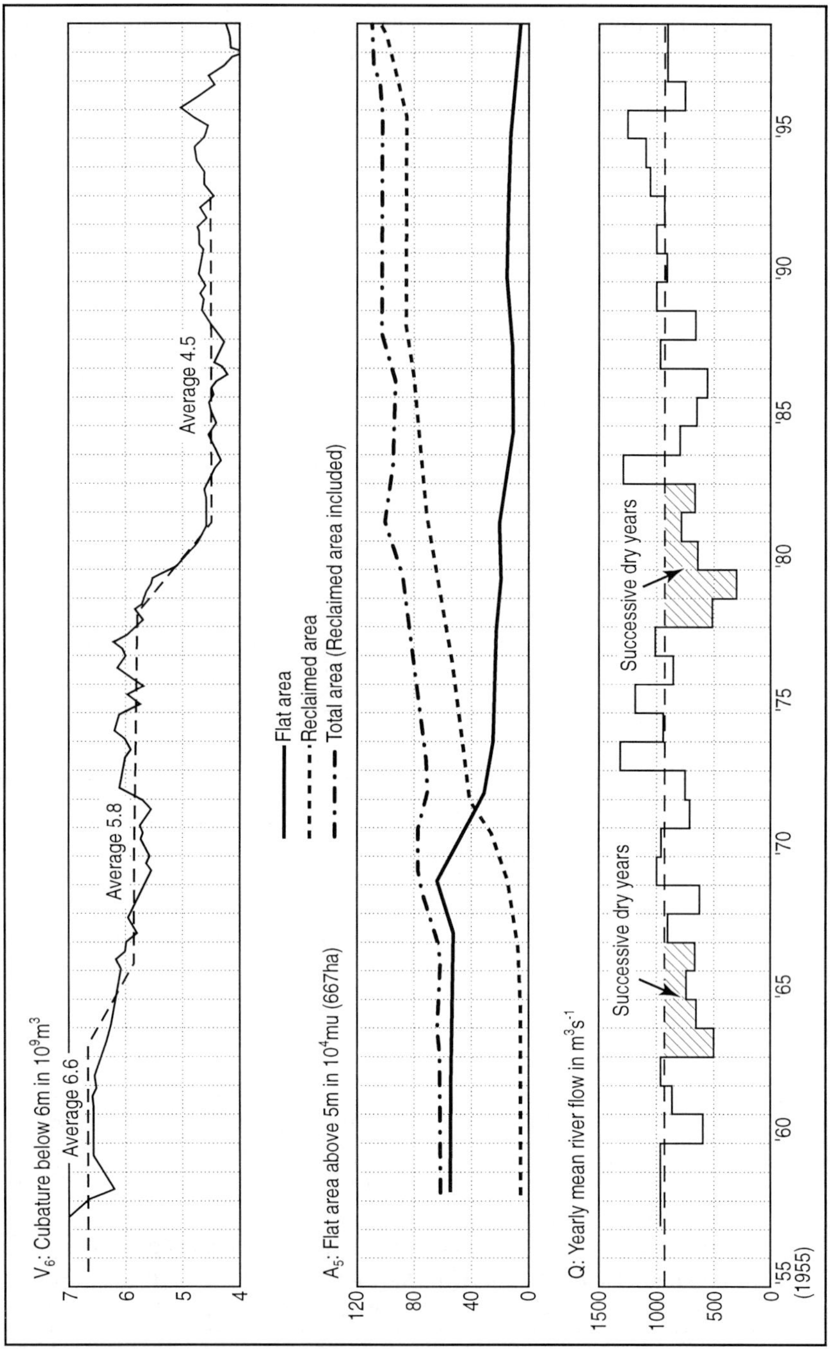

Fig. 8. Annual variation of cubature and flat area upstream of Ganpu vs river flow.

5.1.3 *Changes in the Longitudinal Profile of Average Bed Levels—Upper Reach Eroded and Lower Reach Deposited, while the Apex of the Sand Bar Moved Downstream*

The data for the years that had similar amounts of runoff and similar flow paths (i.e., along the downstream Jianshan stretch which was relatively straight) were selected to compare the longitudinal profile of the river bed. After reclamation, the reach upstream of Cangqian was eroded, the reach from Cangqian to Haining had slight deposition, and the lower reach below Haining had heavy deposition (Fig. 9). The apex of the sand bar was lowered about 0.5 m and moved downstream.

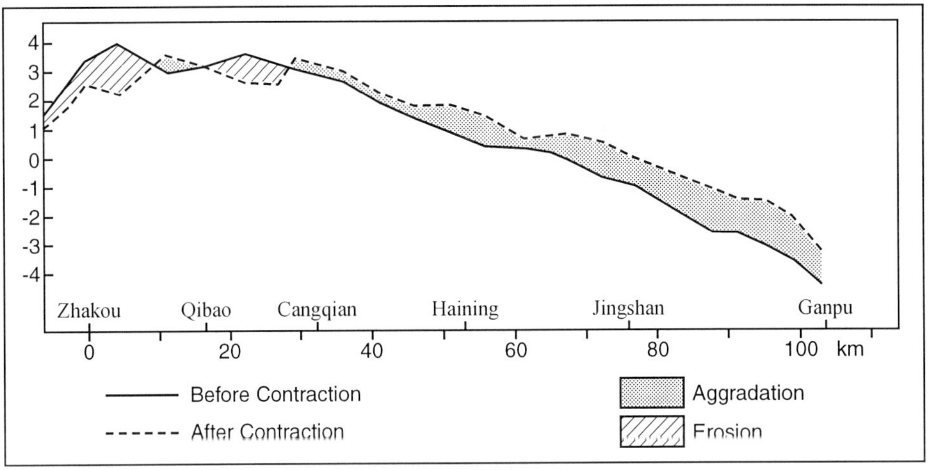

Fig. 9. Variation of average bed elevation before and after contraction

5.2 CHANGES IN TIDES

The variation of HW and tidal range of normal tides and spring tides (percentage of occurrence p=50% and 10% respectively) before and after reclamation was calculated for the condition that the Jianshan stretch was relatively straight (Table 2). The HW during normal tides and spring tides increased after reclamation, while the amount of increment increased upstream. It was about two times larger in spring tides than during normal tides that resulted from the blocking of a larger velocity by the contraction works during spring tides. Flow energy was transferred to potential energy. Whether the tidal range after reclamation increased or decreased depended on the variation of riverbed and the corresponding low water level. If the Jianshan stretch is rather curved, the increase of the tidal range will be lessened.

As there were no essential contraction works in the river reach upstream of Qibao, the tidal prism at Qibao increased because of the increase of tidal range. Downstream of Qibao, the tidal prism decreased because of the great contraction of the river width which was different at different locations. The flood tide duration was shortened by 30-90 minutes at most locations after reclamation.

TABLE 2. Increase of HW and tidal range after reclamation

Station	Normal tide (p=50%)		Spring tide (p=10%)	
	HW	Tidal range	HW	Tidal range
Zhakou	0.40	0.08	0.81	0.60
Qibao	0.44	0.10	0.99	0.87
Cangqian	0.38	-0.26	0.75	0.14
Haining	0.22	-0.34	0.77	-0.53
Ganpu	0.32	0.28	0.54	0.24
Zhapu	0.21	0.23	0.40	0.20

5.3 CHANGES IN THE SALT-WATER INTRUSION

The water supply for Hangzhou city and the irrigation of the farm land in the Xiaoshan-Shaoxing plain depend on the diversion of water from the Qiantang River. Every year, during the months from July to November, the fresh water intakes from the river usually can not be used continuously because of salt-water intrusion. The salinity in the river stretch near Hangzhou depends on the fresh-water discharge and the tidal range at Qibao. After the Xing'anjiang Hydro-power Station became operational, the river flow during dry seasons of the Qiantang River increased by about 200 m^3s^{-1}. Meanwhile, the tidal range and tidal prism at Qibao also increased as a result of river training. The data from field observations and physical-mathematical modeling showed that under the condition that the Jianshan Bend is rather straight and the tidal range at Qibao increased by 0.8 m, the duration that the water exceeded the industrial standards for water supply (chlorinity>250 mg l^{-1}) at Shanhusha (3 km upstream of Zhakou, a regulating reservoir) increased by 24 hours. When the Jianshan Bend was relatively curved, and the tidal range of Qibao increased by 0.4 m, the duration increased by 11 hours. It can be remedied by increasing the discharge flow from the Xing'anjiang Reservoir, or increasing the impounding capacity of the Shanhusha Reservoir or by building intakes further upstream.

5.4 CHANGE IN HIGH WATER LEVEL DURING FLOODING AT HANGZHOU

The origin of the highest water levels in Qiantang Estuary is different in two of its reaches: upstream of Qibao, it is the result of river water, whereas in its lower reach, it is caused by storm surge tides. After river training and reclamation, the high water level once in 100 years is expected to increase by 0.25 m at Zhakou, while it decreases upstream. It could be remedied by raising the top elevation of the seawall. While in the downstream reach, the top elevation is mainly determined by wave run-up and is not very sensitive to the high water level.

5.5 CHANGES IN NAVIGATION CAPACITY

The river reach from Zhakou to Shibao, which is 64 km long, generally became stable after the river training works were implemented in the past 40 years, and the navigation channel in the estuary had been improved accordingly. The navigable depth increased by 0.5-1.0 m in the reach from Zhakou to Cangqian which is 32 km in length. The river bed was raised by 0.5-1.0 m in the reach downstream of Cangqian, and boats can travel there taking the advantage of the rather large tidal range. After river training, since the 1980s, the boats with a water displacement of 300-500 ton can navigate in the river and the yearly throughput reached 300-400 thousand tonnage instead of the 50-100 thousand tonnage before, which was effectuated with boats of only 100 ton. At present, shoals and a tidal bore still exist in the river, the navigable capacity of the estuary is still limited.

5.6 CHANGES IN SOCIAL ECONOMY

River training and reclamation in the Qiantang Estuary were carried out mainly by local peasants under the support and guidance of Provincial and local governments: nearly 90,000 ha of tidal flats was reclaimed. According to the statistics at the mid-1980s, the total investment from government was less than 500 million Yuan RMB, and the total labor days paid by peasants converted into a monetary amount totaled less than 2 billion Yuan RMB. As the yearly GNP reached more than 10 billion Yuan in this area in 1996, the investment returned a great amount of wealth to the area. The new reclaimed land is quite valuable to the densely populated Zhejiang Province: it has made up for the loss of the farmland used for the infrastructure constructions and forms the demonstration district of model management in large areas. Moreover, the safety of the main seawall protecting vast plains is raised by new dike construction, which makes the main seawall a rear line of defense against erosion.

6. Conclusion

The Qiantang Estuary is world renown for its magnificent tidal bore. It ran straight eastward from Hangzhou to the East China Sea before the 4th century. Since then, it gradually curved because of the combination of natural processes and artificial activities such as the building of seawalls. Since the 1950s, large scale river training and reclamation works have been carried out: more than 600 km of new polder dikes were built and 90,000 ha of reclaimed land formed which is about the same area reclaimed during the previous 600 years. This achievement is the result of a combination of experience and contemporary scientific knowledge. There was no modern construction equipment used and no complicated structures completed in the course of the project's implementation. Although most measures adopted were simple and feasible, they were nevertheless elaborately conceived to suit local conditions. In this project tremendous benefit has been gained in the regulation of an ever shifting river channel under the attack of a violent tidal bore. It is an example of a major advancement in the regulation of macro-tidal estuaries.

7. References

1. Chen, J. 1989. Historical shoreline changes in China. In: *Development and Evolution of China's Coast*, Shanghai Scientific and Technical Publishers, Shanghai, 7.
2. Li, G. and Dai, Z. 1986. Fluvial processes of the Qiantang Estuary, *International Journal of Sedimentation Research,* 1(1) 58-62.
3. Ning, C., Han-siang, S., Chih-the, C. and Quan-pin, L. 1964. The fluvial processes of the big sand bar inside the Chien Tang Chiang (Qingtang) Estuary, *Acta Geographica Sinica*, 30(2) 124-142.
4. Chen, J., Luo, Z., Chen, D., Xu, H. and Qiao, P. 1964. The formation and historical evolution of the sand bar in Qiantang Estuary, *Acta Geographica Sinica*, 30(2) 109-123.
5. Dai, Z. and Zhou, C. 1987. The Qiantang bore, *International Journal of Sedimentation Research*, 1, 130-131.
6. Dia, Z. and Jiang, W. 1989. Fluvial processes and regulation practise of the Qiantang Estuary, *International Research and Training Centre on Erosion and Sedimentation*, Beijing, China, Circular 5, 27-34.

THE SEAWALL IN QIANTANG ESTUARY

JIANG WEI
Zhejian Institute of Estuary and Coast
Hangzhou, China

TAO CUNHUAN
Qiantang River Administration
Hangzhou, China

1. Introduction

The seawall (sea dike) situated in the estuarine reach of the Qiantang Estuary is subjected to the strong, destructive action of the world-renowned Qiantang bore; in Hangzhou Bay, on the other hand, the dike is subjected to strong wave action. The seawall protects the vast lowlands of the Taihu Lake plain on the north bank and the Xiaoshan-Shaoxing-Ningbo plain on the south bank; these plains are prosperous and densely populated. This seawall attracted much attention from the supreme sovereigns during China's history. Emperor Qianlong of the Qing Dynasty inspected the seawall in Haining County four times during his six patrols to the southern area of the Changjiang River. After many attempts at construction, repairing and reconstruction, large amounts of money and labor were expended and, finally, a magnificent seawall, nearly 300 km long, was completed. Along with The Great Wall and Grand Canal, the Qiantang Seawall is one of the three most famous civil engineering works of ancient China.

Since the early part of this century, because of the progress of science and technology and the development of new construction materials, a cheaper sloping dike with masonry revetment has been developed and is replacing the old gravity type of sea dike. In addition, new types of seawalls are being used to protect important industrial districts.

2. Hydrodynamical Conditions and Soil Foundations

The Qiantang Estuary, which is funnel shaped, has a tidal range that increases from its mouth (Nanhui across to Zhenhai) toward its upper end at Ganpu and then decreases upstream due to the friction of the river bed (Fig. 1). The annual mean tidal range is 3.2 m at Nanhui and 1.8 m at Zhenhai, 3.9 m at Jinshan, 5.6 m at Ganpu, 3.2 m at Haining, and 0.5 m at Hangzhou. The maximum wave height recorded along the northern bank of the Hangzhou Bay is 4-6 m. During summer/autumn, the estuary is occasionally hit by typhoons when water levels are raised significantly. A maximum additional rise of 2.34 m because of a storm surge has been observed at Zhapu. High water combined

J. Chen et al. (eds.), Engineered Coasts, 139–150.
© 2002 *Kluwer Academic Publishers. Printed in the Netherlands.*

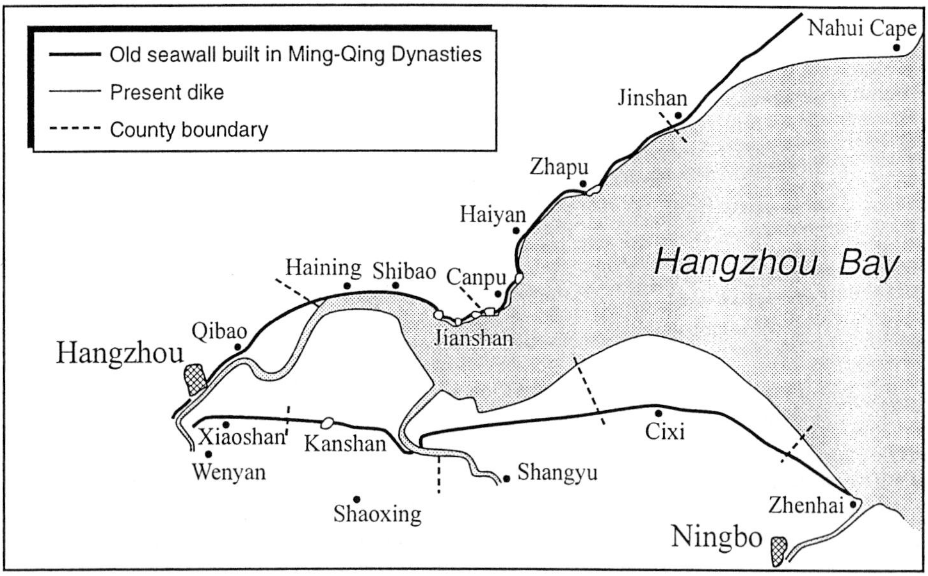

Figure 1. Location of seawalls in Qiantang Estuary.

with gigantic waves cause a violent destructive force to the seawall. In the estuarine reach above Jianshan, the front of the incoming tidal wave develops into a tidal bore (the Qiantang Bore). The height of the bore is generally 1.5-2 m, the propagation velocity is about 5-7 ms^{-1}, the current velocity following the bore may be as large as 12 ms^{-1} which impacts heavily on the seawall, especially its foundation. The river bed from Zhapu to Hangzhou is a giant 130 km long longitudinal sand bar which is composed of well-sorted fine silt with a mean diameter of 0.02-0.04 mm as the result of the constant to and fro movement by flood/ebb currents. A similar kind of fine silt is buried to about 10-20 m below the bed; it is of high bearing capacity and easily eroded. The bed downstream from Ganpu is mainly clayey silt, the content of silt particles being reduced to 50-60% with an increase of clayey particles, it has a higher scour-resisting quality.

3. The Historical Evolution Of Seawall Construction

3.1. THE EVOLUTION OF GUIDING PRINCIPLES.

About 4,000-6,000 years ago, sand bars naturally formed by wave action played the role of coastal defense structures. The remains of two such bars still exist near Caojing inside the north bank of Hangzhou Bay. Later, earth dikes were built to protect living space from the attack of tides and waves. Of course, the ancient dikes were very simple and were often destroyed and rebuilt. As recorded in historical literature, the original seawall of Haiyan County was far from the present city. Some eighteen sea dikes, built before the 13th century, were destroyed. In Haining County, a strip of land 20 km wide south of the present city was returned to sea during 1218-1219 AD when the main channel shifted northward. Facing the rapidly collapsing bank and dikes, the responsible officer (Liuwu) acknowledged that he was unable to protect the land from erosion and flooding: There was no alternative but to recede and guard the inner dike. At that time, it was not possible to build seawalls strong enough to guard against the force of the waves and the tidal bore, so that retreat was the only choice. Up to 1735-1762, the main channel shifted gradually from south to north along the north bank from Hangzhou to Jianshan. The Emperor of Qianlong in Qing Dynasty considered that when retreating and guarding the inner dike became a principle of "guarding the teeth and abandoning the lips", there would be no end[1]. Meanwhile, as experience accumulated, seawall construction technology greatly improved, and the principle of firmly guarding was adopted along with the decision to build gravity stone-laid seawalls along the entire coastline.

Of historical interest are the many ideological aspects that related to seawall construction in the past in China. Through centuries of hard struggle against the tempestuous tidal bore, seawalls were frequently destroyed and rebuilt, people became exhausted and puzzled. Assuming that the overpowering destructive force might be of magical origin, they prayed to god to suppress the tide. In the 10th century, King Qian Niu of the Yueguo nation prayed to heaven in Hangzhou to calm down the bore so that he could build a seawall. He also had three thousand arrows made and hired five hundred soldiers to shoot the bore (Fig. 2). Pagodas (Fig. 3) were built and an iron oxen was placed on the top of the seawall (Fig. 4). These activities reflect the anxious desire of the riparian people to calm the bore.

3.2. SEAWALL TYPES IN HISTORY

According to historical literature, a preliminary system of artificial coast had already been formed by 713 AD in the Qiantang Estuary (including Hangzhou Bay). However, most of the seawalls were built of earth or of earth with broken stone, which could not protect the land against the attack of waves and tides. In 910 AD, bamboo baskets filled with stone were used in seawall construction[3] (Fig. 5). Later faggots as well as wooden cribs were used for dike building. Those structures were better, but still could not sustain the impact of the bore and were frequently destroyed. Around the middle of the 14th century, seawalls with foundations of wooden piles with the outer face laid with ashlar and the back filled with broken stone and earth, first appeared. This was a major advance in the improvement of seawall construction.

At the beginning of the 18th century, a local scholar named Chen Yu wrote an article entitled "On the seawall construction in Haining and Haiyan counties"[4]. He made an incisive analysis of the different dynamic conditions prevailing, the different types of foundation required, and the corresponding requirements for seawall structures. In Haining county, he pointed out that the foundation soil was comparatively loose so that the swift current acting along the wall foot caused the foundation to be scoured, hence the wall footand its outside bed needed to be tightly protected against scouring

Figure 2. King Qian Niu organized soldiers to shoot the tidal bore.

Figure 3. Sketch of bore-suppressing pagoda built in the 17th century in Haining (after W.U. Moore[2]).

Figure 4. Sketch of bore-suppressing iron ox built in the 18th century.

while the dike body itself did not need to be very massive. In Haiyan county, on the other hand, the foundation soil was more cohesive, so that there was far less fear of scour. There the seawall was subjected to the impact of oncoming waves, hence the wall needed to be massive and laid carefully to withstand wave action.

The seawall in Haining county was of the type with single capping blocks with 2-3 apron steps, each with double rows of wooden piles acting as a sheet of piles to prevent scour (Fig. 6), while the seawall in Haiyan county was composed of massive double capping blocks that could withstand wave impact (Fig. 7). In both cases they were gravity structures, laid with ashlar stone. The outer face was stepped and named the "fish-scale-shaped ashlar seawall"[5]. Most walls of this type, built in the Ming-Qing dynasties, still exist in the front line now guarding against the bore and storms. The development of this type of seawall was a mile-stone in history.

Besides gravity type seawalls, an ashlar-laid slope dike was once tried in the middle of the 15th century. Because the facing of the revetment was too thin with no bedding materials between ashlars, the dike was soon ruined and this type of construction was never again used (Fig. 8).

In 1748, an important improvement was made in the structure of sloping dikes in Zhenhai county[6]. Besides strengthening the wall foot, the revetment facing was paved with two layers of stone plate, the top layer precisely chiseled and laid with keys and notches between transverse and longitudinal panel keels (Fig. 9). This type of sloping seawall still stands after 250 years with no essential damage. It was not used widely because it was extremely complicated to construct and expensive.

In the 18th century, 3-4 separate rows of piles filled with stone were used for wave dissipation and earth dike protection at Jinshan county. This type of seawall was largely used in Jinshan until the early 20th century (Fig. 10).

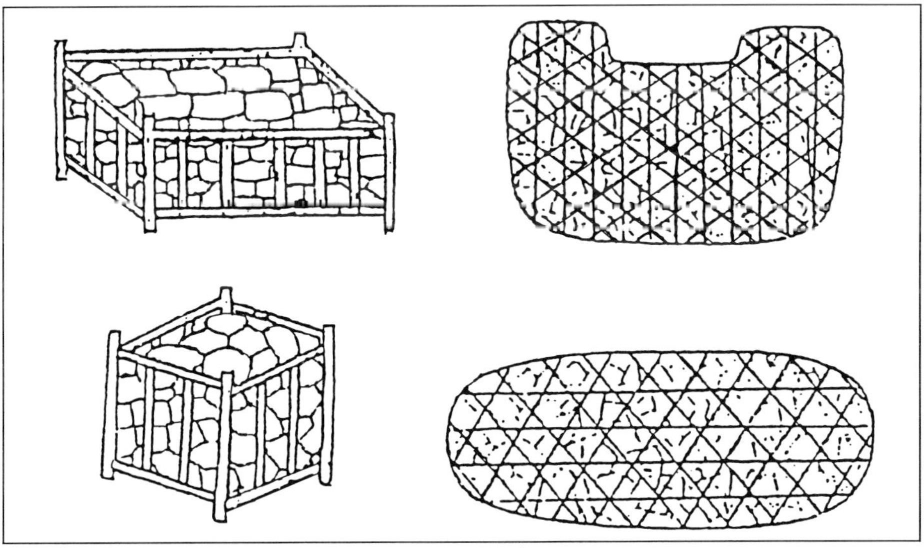

Figure 5. Bamboo baskets and wooden cribs filled with stone.

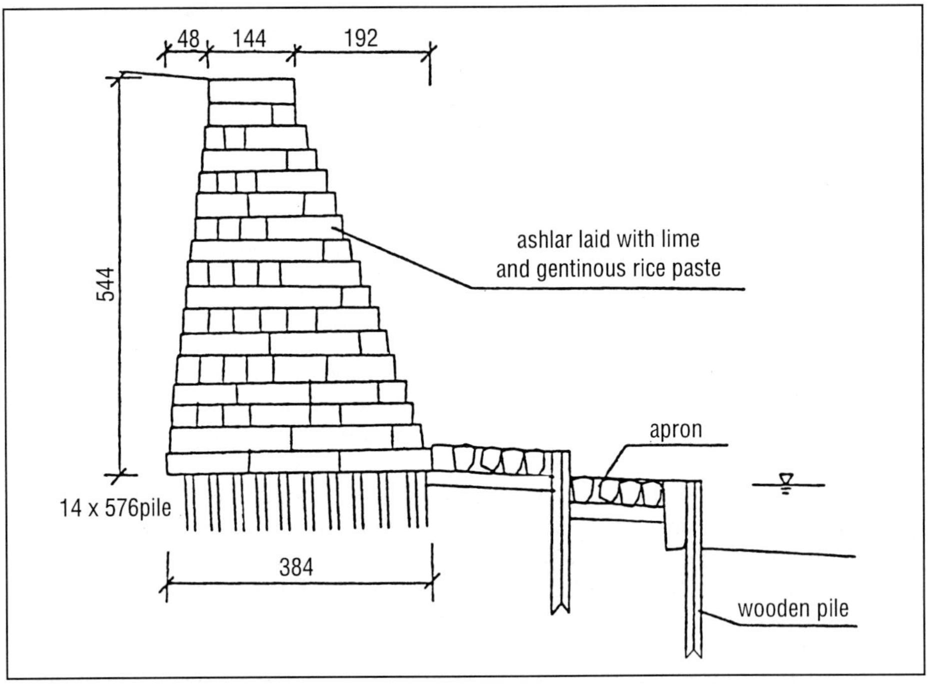

Figure 6. Fish-scale-shaped ashlar seawall with a single line of capping blocks (Haining County).

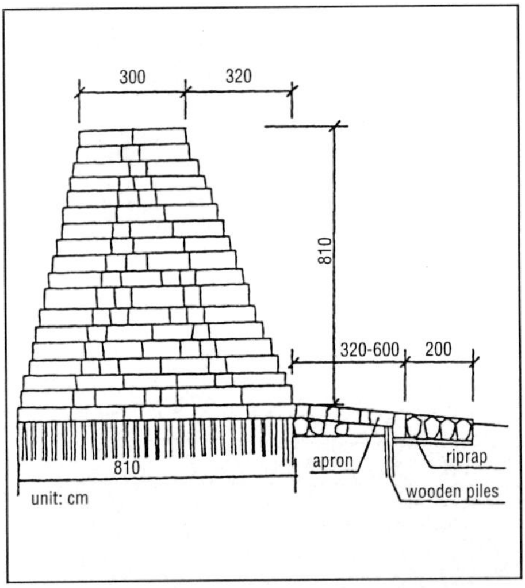

Figure 7. Fish-scale-shaped ashlar seawall with double capping blocks (Haiyan County).

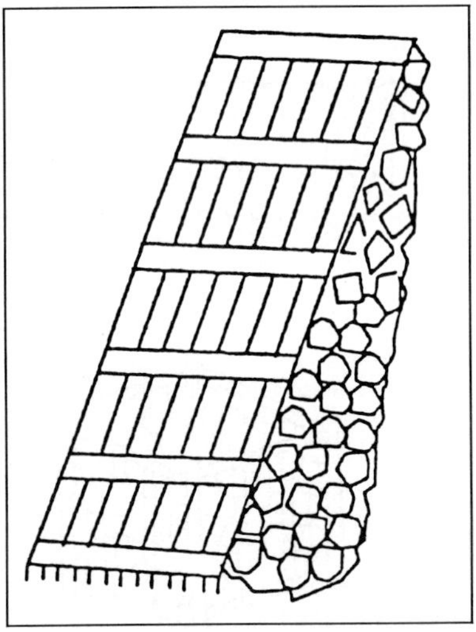

Figure 8. Ashlar laid slope dike built by Yang Xuan in the 15th century.

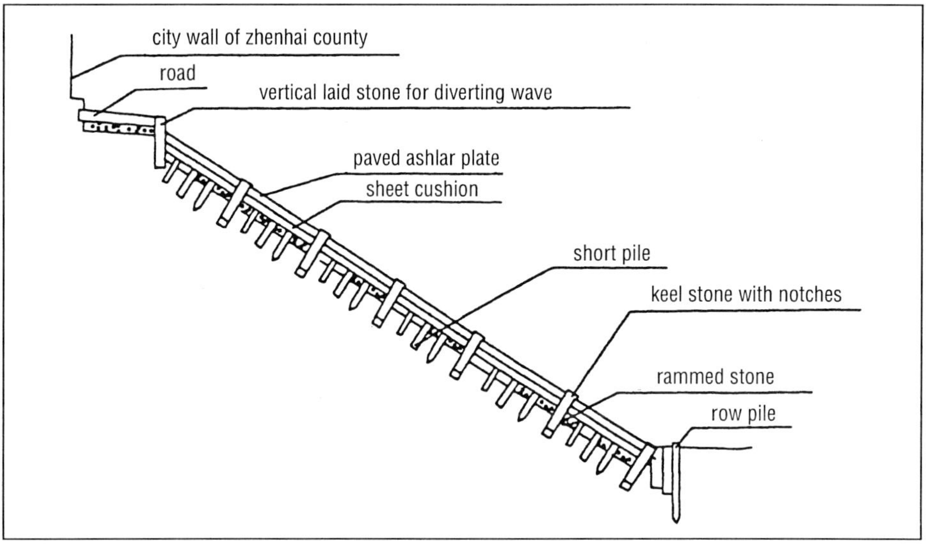

Figure 9. Improved ashlar paved slope dike built in 1748 at Zhenhai County.

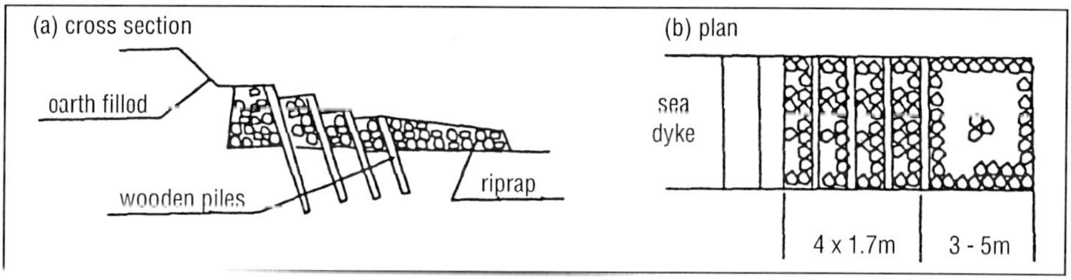

Figure 10. Sea dike of broken stone supported by rows of pile at Jinshan County in the 18th Century.

4. Contemporary Seawalls

4.1. CONCRETE GRAVITY SEAWALL

Since the beginning of the 20th century, concrete has been used instead of the expensive ashlar in building and repairing seawalls and reinforced concrete long piles replaced short wooden piles. New seawalls were mostly of the gravity type, although a few were of the buttress type (Fig. 11).

4.2. SLOPING SEAWALLS

Since the masonry work associated with construction of a sloping seawall is much less than that with gravity type walls, it costs much less. Moreover, the use of cement and concrete helps bind the broken stone/gravel into a unit mass reducing the impact waves. During 1925-1934, concrete blocks and masonry were first used in building sloping dikes to replace the old gravity seawalls that had been destroyed. This type has been gradually improved and is now constructed in various ways (Fig. 12). In river reaches with a weak tide and wave action, less expensive riprap revetments are still used.

4.3 SLOPING DIKES USED IN RECLAMATION WORKS

The present rate of growth of the population and the expansion of the economy bring an urgent need for new land. In the Qiantang Estuary, vast areas of tidal flat can be reclaimed in combination with training works to stabilize the ever-shifting main channel of the estuary. Since the 1950s, large-scale reclamation, in coordination with river contraction works, has been gradually carried out. About 600 km of new dikes have been built; part of them turned into inner dikes whenever new flats formed and were reclaimed with new dikes on the seaward side. At present, more than 200 km of new dikes form a new artificial coast, encircling 70,000 ha of new polder areas. Most of the new dikes, which also serve for river channel training, were built on high tidal flats within 7-10 days during neap tide. Formerly this work was done by organizing tens or hundreds of thousand of local peasants. In recent years, dikes have been built by using hundreds of hydraulic guns, suction pumps, and transporting and filling equipment. These new techniques have saved a large amount of human labor and money and could be used on low flats. As soon as the earth filling work was completed, riprap pavement was placed on the outer slopes to prevent erosion. When the main channel shifted toward the dike resulting in the high flat being scoured by 4-6 meters in depth, large amounts of riprap were added immediately for emergency protection. The masonry work was done later (Fig. 13).

Generally the dike foot below low water remains under the threat of scouring. Underwater works are particularly difficult to maintain in the reaches hit by the tidal bore. Recently, reinforced concrete wells were successfully placed, sunk down to the bottom and anchored to the apron to prevent scouring (Fig. 14).

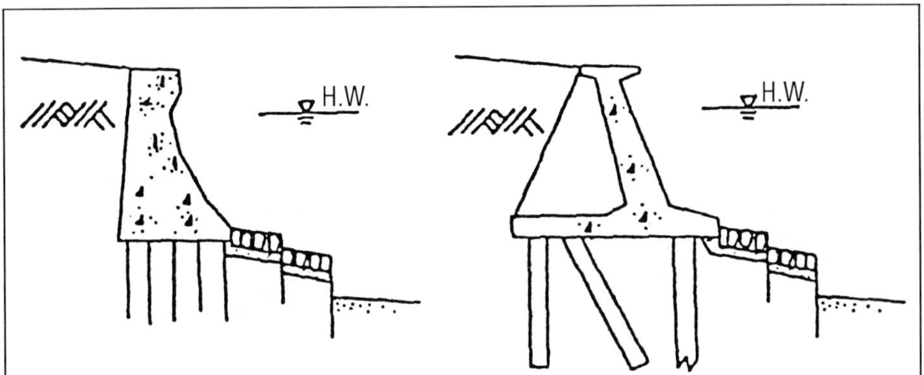

Figure 11. A. Concrete gravity seawall built in 1913, B. Buttress reinforced concrete seawall built in 1948.

4.4. HIGH STANDARD SEAWALLS BUILT FOR PROTECTING IMPORTANT INDUSTRIAL BASES

China's policy on reform and its opening to the outside world had greatly stimulated the economic development in coastal areas. Large scale industrial complexes such as nuclear/coal power stations, petroleum complexes, and harbors have been erected on both banks of Hangzhou Bay. Because of the lack of land in the coastal areas, the space to accommodate these industries had to be reclaimed from a narrow strip of the lower tidal flat alongside the deep navigation channel. Seawalls of high standards were constructed during reclamation as illustrated in Figure 15 which shows the dike constructed for the Qinshan Nuclear Power Station, Haiyan County.

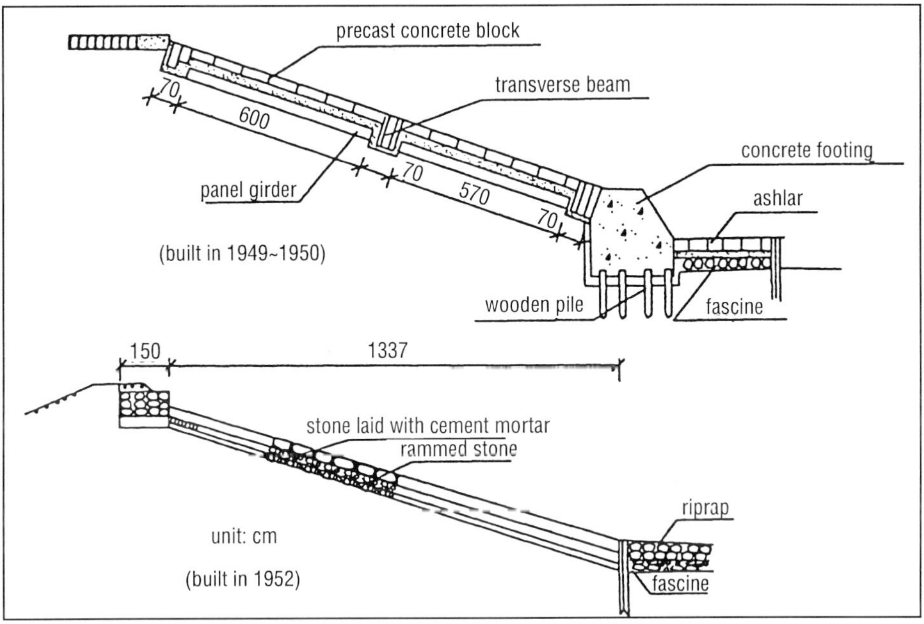

Figure 12. Sloping seawall paved with precast concrete blocks/masonry.

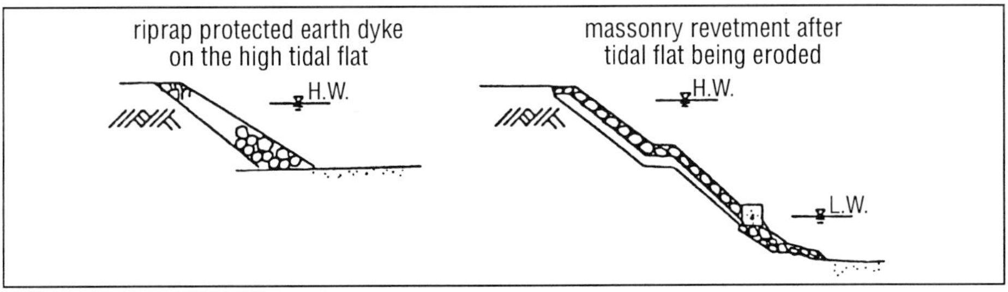

Figure 13. Reinforcing process of the earth dike.

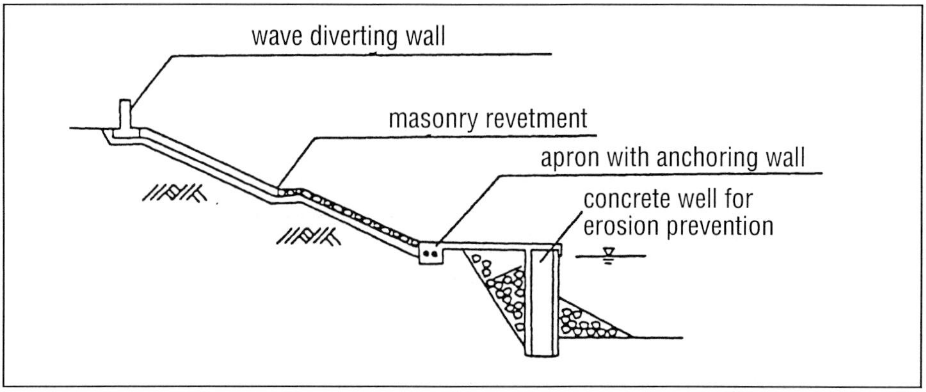

Figure 14. Finally completed dike with bottom erosion prevention structures.

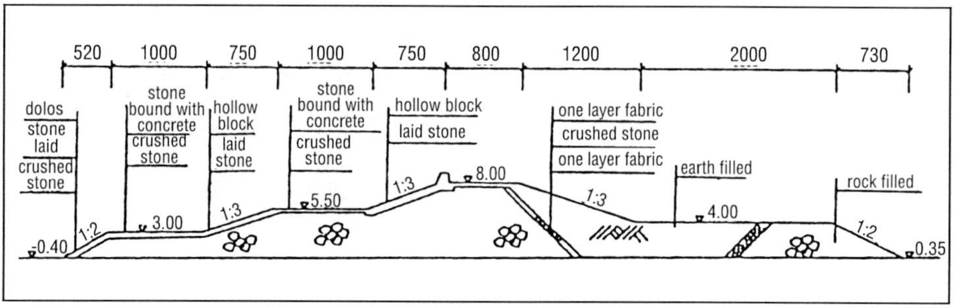

Figure 15. Dike section of the Qinshan Nuclear Power Station.

4.5. SCOUR PREVENTION STRUCTURES OF DIKE FOOT

Scour along the dike foot is the most important problem facing the stability of seawalls. This is especially true in the reach between Hangzhou and Jianshan which is hit by the tidal bore. Dike foot protection works are partly above low water level and partly below. In the region above low water, concrete/masonry work can be done only during 3-4 hours in each tidal cycle. In the region below low water, however, very difficult conditions are encountered. Because of the shallow water, swift currents and especially the attack of the dangerous tidal bore, construction work can never be done from boats. It is also impossible to use heavy cranes on top of a sea dike because that would require an arm about 20 m long. Therefore, underwater work can only be done at the foot during a few hours of low tide before the bore arrives. Generally two types of foot protection works have been adopted. One is direct protection, i. e., to build an ashlar laid/masonry apron with double row piles (Fig. 16). The other type is indirect protection, i.e., to build a group of short groins[7] to promote siltation along the dike foot (Fig. 17).

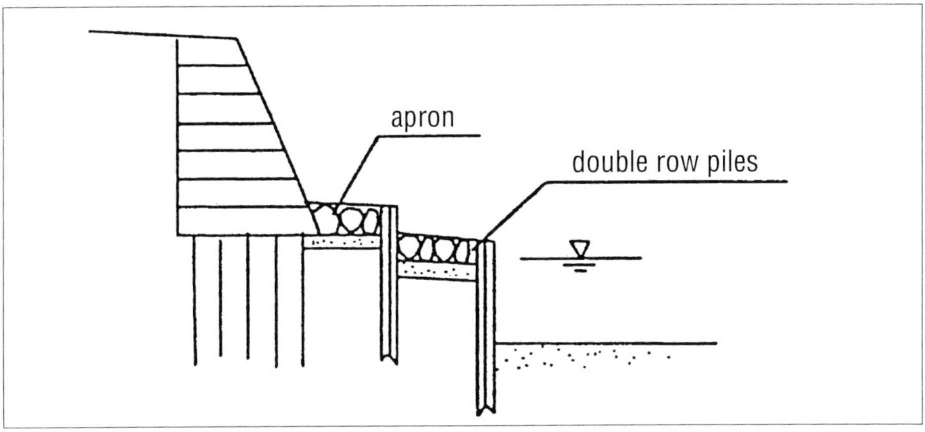

Figure 16. Apron and double row piles for scour prevention.

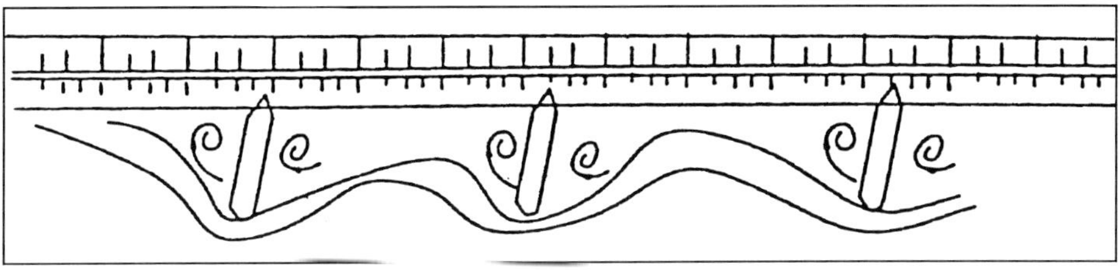

Figure 17. Group of short groins along dike foot.

5. Reserve Protection System

Inside the main outer dike at a distance of some tens to hundreds of meters, another earth dike is built to prevent inundation of the vast plain area in case the main dike breaks because of an unexpected high tide or poor maintenance. The top elevation of the inner dike does not need to be too high because the wave height will be largely reduced. Cross dikes are also built between the main dike and the inner dike to limit the inundated area. Canals and roads are arranged parallel to the sea dike for the transportation of construction materials and emergency uses. Reserve protection is a feasible and reliable system that enhances the safety of the coast.

6. Conclusion

The Qiantang seawall has experienced periods of unceasing destruction and reconstruction during its early history. With the constant struggle came improvements and renovation, and only at last, has it been developed into a comparatively complete system with an acceptable structure and with feasible measures of construction and maintenance that are suitable to local conditions. We believe that its history contributes valuable insights to the knowledge of artificial coast development.

7. References

1. Yang Rong, 1810. *Essentials of Seawall*, 1, The 15th year of Jiaqing period, Qing Dynasty, 22.
2. W.U. Moore, 1888. *The Bore of the Tsien-Tang Kiang (Hang-Chau Bay)*, J. China Branch, Royal Asiatic Society, 23.
3. Qian Wenham, 1796-1820. *Records of Seawall Protection, Jiaqing period*, Qing Dynasty.
4. Fang Guancheng, 1751. *Annals of Seawalls in Zhejiang Province*, 19. The 16th year of Qianlong period, Qing Dynasty, 23-29
5. Fan Weicheng, 1622. *Annals of Haiyan County*, 8. The 2nd year of Tianqi period, Ming Dynasty.
6. Sheng Hongtao, 1918. *Annals of Zhenhai County.*
7. Jiang Wei, 1990. Field Test of Floating Open Caisson in Qiantang Estuary, China. *Ocean Engineering,* 4(3).
8. C. Tao and Z. Dai, 1997. Qiantang Seawall in Ming-Qing dynasties, *Water Resource Planning*, 3, 73-78.

MARINAS, SEA-LEVEL RESERVOIRS, SOLAR SALT PANS AND OTHER ARTIFICIAL SHORELINES

H. JESSE WALKER
Department of Geography
Louisiana State University
Baton Rouge, LA 70803-4105

1. Introduction

Humans have modified coastlines in a variety of ways and for a variety of reasons throughout much of history. Some of the earliest modifications were nothing more than re-arranging boulders along a rocky coast to trap fish during tidal variations, building mounds in coastal swamps and marshes and creating middens along the shore (Fig. 1). For example, shell middens dating back to the last interglacial are present along the coast of South Africa[1] and underwater cameras have recently (2000) photographed remains of a dwelling site at 91 m deep on the bottom of the Black Sea[2].

Because much of the activity by early humans took place along shorelines that are now submerged, archaeological evidence is rare. Shoreline processes operating as sea level rose during much of the past 18,000 years destroyed most coastal sites or at least buried them beneath sea and sediment[3].

Along with the development of animal husbandry and agriculture came the development of cities, commerce and war. Such activities had their impact on coastlines. People living in the Mediterranean (e.g. the Minoans, Phoenicians, Greeks, Romans) and the coastal zone of China developed watercraft that needed good harbors. Breakwaters were built, swamps were drained, canals were constructed and solar salt pans established.

Not surprisingly, some of the early engineered structures failed to achieve the desired results. Strabo about 7 BC recorded just such a situation in his writings about the eastern Mediterranean Sea. About the Harbor of Ephesus, he wrote:

> The mouth of the harbor was made narrower by engineers, but they, along with the King who ordered it, were deceived as to the result . . . for he thought the entrance would be deep enough for merchant vessels . . . if a mole were thrown up at the mouth . . . But the result was the opposite, for the silt, thus hemmed in, made the whole of the harbor, as far as the mouth, more shallow. Before this time the ebb and flow of the tides would carry away the silt and draw it to the sea outside (Strabo as quoted in Russell[4]).

Nevertheless, there were early successes too. For example, the Phoenicians developed a 'continuous self-flushing' harbor at Tyre[5].

Dependence on the sea, especially around the Mediterranean, was so great that the results of their engineering endeavors are still much in evidence, some many kilometers inland (Fig. 2), others submerged because of tectonic activity (Fig. 3)[6].

J. Chen et al. (eds.), Engineered Coasts, 151–183.
© 2002 *Kluwer Academic Publishers. Printed in the Netherlands.*

Fig. 1. Shell midden on Red Sea coast. Note: All photographs in this chapter are by J. Walker unless otherwise credited.

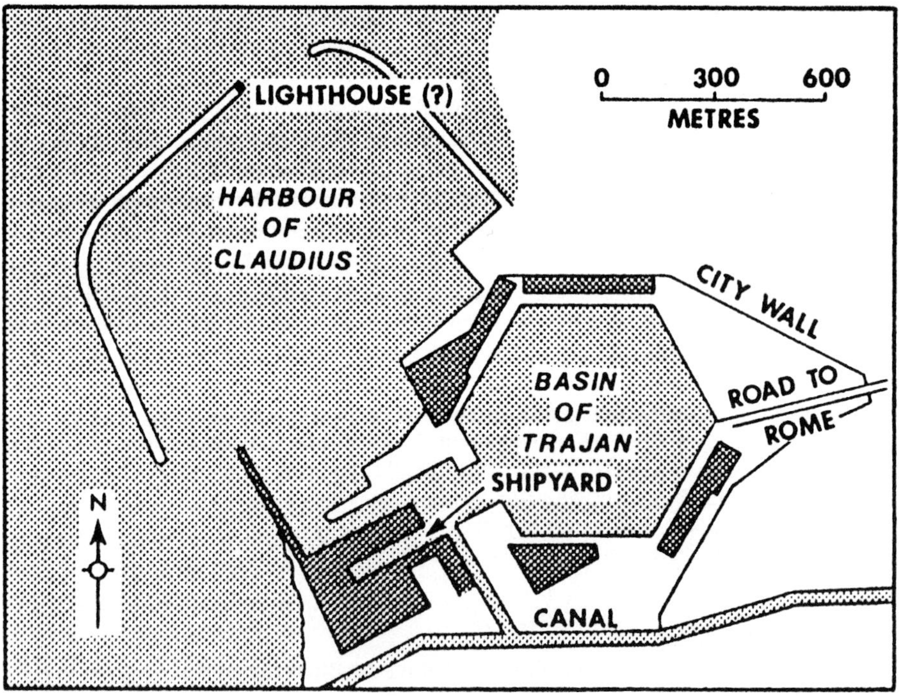

Fig. 2. The Port of Rome at Portus built in 46 A.D. It is now 8 km from the shoreline.

Although humans have long exercised their ingenuity in coastal modification, it has only been since the beginning of the Industrial Revolution that technology has advanced to the stage where lengthy sections of the coastline have been altered. These alterations reflect also the drastic changes in coastal perception that have accompanied increased urbanization, commercialization and industrialization of the shoreline.

All coastal structures are subject to a variety of natural and human "disasters". Some may occur slowly as, for example, because of sea level rise, others are almost instantaneous such as those by earthquakes and typhoons. Although all coastal structures are subject to destruction, coastal highways are especially susceptible. Storm surges along low-lying coastal areas (Figs. 4, 5) and landslides on cliffy coasts (Fig. 6) are but two examples.

Many aspects of coastal modification by humans are presented as separate chapters in this book. However, there are some that, while not deserving a full chapter on their own, will help illustrate the vast array of human activities vis-à-vis the shoreline. Some are relatively unique whereas others are more common and familiar to most humans. Included are such structures as coastal highways and railroads, solar salt pans, coastal freshwater reservoirs, marinas and fisherinas, coastal airports (Figs. 7, 8), tsunami and typhoon protective structures, offshore drilling rigs and artificial islands (Fig. 9), artificial crevasses, watergates (Fig. 10) and shoreline defense structures.

2. Sea-level Reservoirs: The Example of Hong Kong

An excellent example of the conversion of bays and straits into freshwater reservoirs is to be found in Hong Kong. Hong Kong (Fig. 11), a small area of slightly more than 1000 km^2, has rugged terrain, limited amounts of both ground and surface water, variable rainfall and a dense population. It does have a long and highly varied coastline relative to its area. Founded in 1841, Hong Kong began to experience water problems almost immediately. Less than 20 years later (1859) its first reservoir was built, a reservoir the dam of which collapsed six years later. By 1941, Hong Kong had a dozen reservoirs, a number that increased after WWII. Nonetheless, fresh water supply remained a problem. Hong Kong has utilized several other means of coping with water demands. It used sea water for flushing toilets, fighting fires and washing streets. In the late 1970s, about 16% of the water used in Hong Kong was from the sea. It also purchased water from China, some times even via barge and tankers from the Pearl River estuary. However, most of the water from China came by pipeline from the Sham Chun reservoir (Fig. 11).

Although salt water was used directly for some purposes, in 1975 Hong Kong began to use converted sea water. Whereas the first land-based desalination plant was constructed in 1912 in Egypt by the British, it was not until 1971 that Hong Kong decided to build a large desalter. The plant was built at Lok On Pai (Fig. 11), a location on the Pearl River side of the New Territory. During the rainy season the salinity at this location is reduced because of the flooding of the Pearl River.

When constructed, with a production of fresh water at a rate of 181,000 m^3/day, it became the world's largest such plant replacing the one in Kuwait which had a capacity only 63% as large. At the dedication the Governor of Hong Kong stated that ". . . it is

Fig. 3. Tectonic subsidence near Naples, Italy.

Fig. 4. Coastal erosion along the coast of southeast Italy.

Fig. 5. Armor unit protection for road in Japan.

Fig. 6. Highway across an unstable cliff on Pacific coast of California.

Fig. 7. Reef runway at Honolulu International Airport.

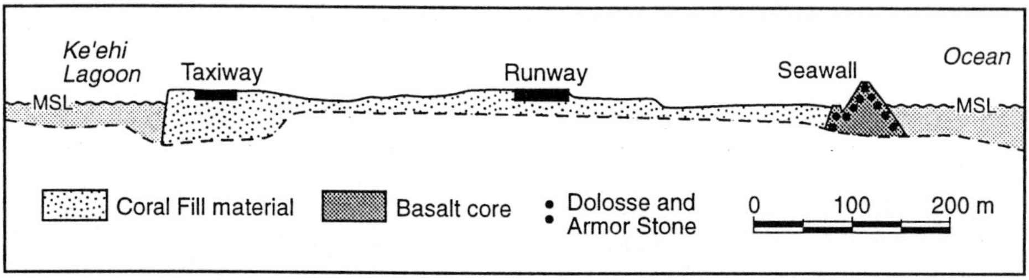

Fig. 8. Cross-section of reef runway, Honolulu: 3650 m long, 60 m wide, situated on a 4200 m x 650 m coral fill pad[8].

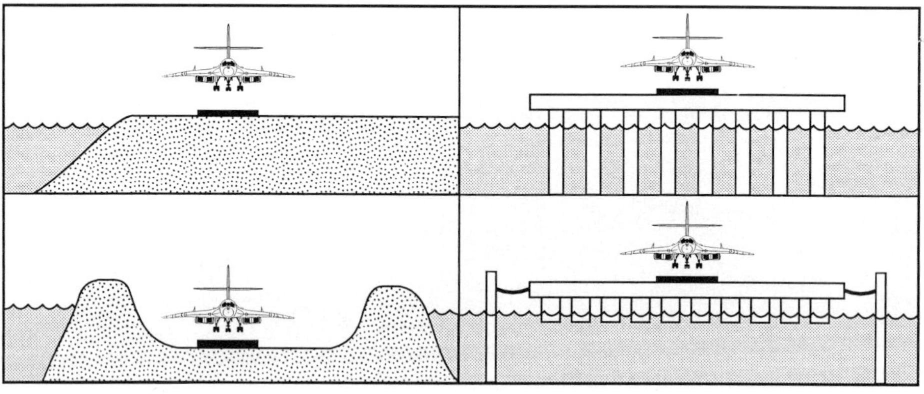

Fig. 9. Early plans for the artificial island airport in Osaka harbor.

Fig. 10. A watergate in a distributary of the Shijiang, China.

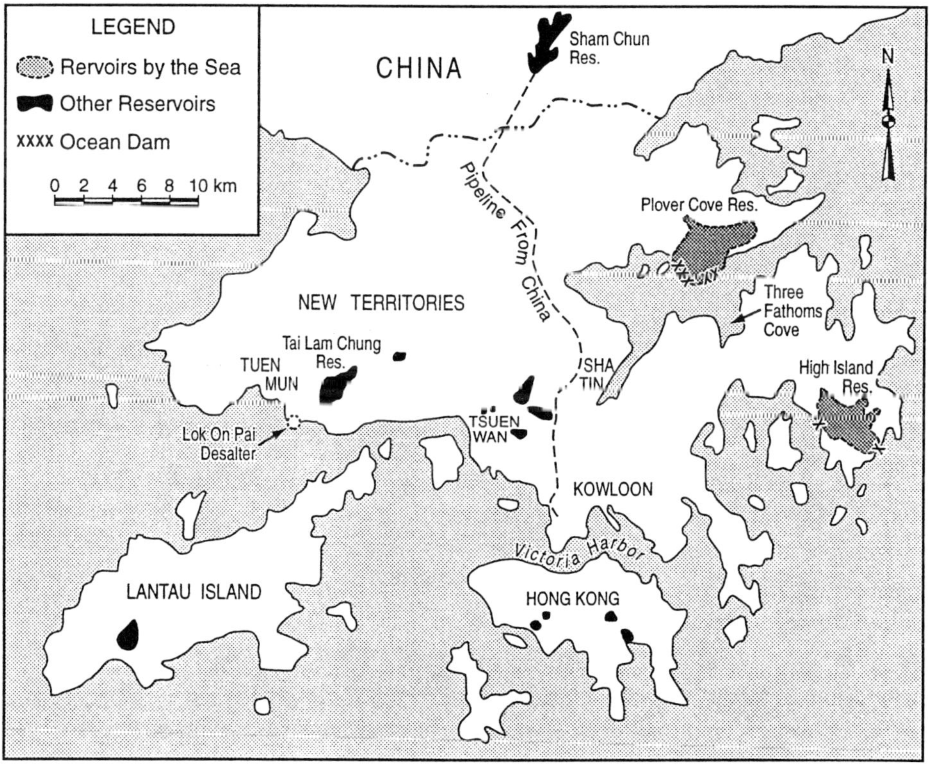

Fig. 11. Map of Hong Kong showing the major water systems[9].

an immense relief to have this new source of supply, equal to about one-fifth of full daily consumption, which is independent of rainfall." It only operated for four years, shutting down in June 1978[10].

The idea of converting coastal bays and inlets into fresh-water reservoirs was discussed by Hong Kong Public Work's personnel during the 1950s. Possible sites were examined and Plover Cove (Fig. 11) was selected not only to become the first reservoir from the sea in Hong Kong but also in the world. Engineering contracts during the eight year construction period were handled by French, Swedish, Japanese and local Hong Kong firms. Tests were conducted in England at the Wallingford Hydraulics Laboratory in the 1960s in order

> . . . to find a practical way of closing the dam, to determine the velocity patterns that would occur at various stages of closure, and to examine the dropshafts that would lead water from the small catchment basins scattered throughout the drainage area to the large reservoir[9].

Construction lasted from November, 1960 to December, 1968 although the first water to be furnished from the reservoir flowed in 1967. The structures involved in developing the Plover Cove reservoir included the main dam, two subsidiary dams and a spillway. Construction utilized sand and stone from a variety of locations in Hong Kong including the site that now houses the Chinese University of Hong Kong. By 1967 the dam was closed and the trapped sea water pumped out. Within four months $50x10^6$ m^3 of fresh water had drained into the reservoir. It is 5.5 km long, has an area of 12.1 km^2, and will hold $223x10^6$ m^3 of water an amount that was three times the storage capacity of all other reservoirs in Hong Kong at the time.

The success of Plover Cove led Hong Kong's engineers to construct an even larger reservoir. Instead of a cove or bay, they decided to dam off a strait, the strait that separated High Island from the mainland (Fig. 12).

A problem not faced in the construction of the Plover Cove reservoir was the removal of the 45 m of riverine and marine sediment that had accumulated in the strait. Its removal to bedrock, of course, enlarged the capacity of the final reservoir. In order to build the main dams at the ends of the strait two cofferdams were needed at each end—two outside ones to keep seawater out and two inside to keep channel water in.

Three of the cofferdams were considered temporary. The fourth, the eastern ocean-facing cofferdam (Fig. 13) was constructed to be permanent in order to protect the reservoir from the high waves often generated by the tropical cyclones that occasionally occur in the South China Sea. The sea-ward face of this dam consists of about 7000 dolos weighing 25 metric tons each. The inside of the dam is protected, in the case of wave overtopping, by concrete 'Svee' blocks.

The main dams, which required more than three times the rock fill of the east cofferdam, rise to a height of 66 m above sea level. They create a basin that is 5.6 km long and has an area of 6.9 km^2. The capacity of the High Island reservoir is $278x10^6$ m^3, or slightly more than that of the Plover Cove reservoir.

With the construction of these two reservoirs, 25 km of seashore were transformed into a fresh-water shoreline with major changes in the energy and biologic systems.

Fig. 12. High Island strait before dam constuction (photo courtesy of Hong Kong Government).

Fig. 13. The dam facing the Pacific Ocean showing the dolosse at its base.

Shoreline changes have been an integral part of Hong Kong's short (150 year) history. Because of its small size its engineers have frequently been involved with reclamation, the most recent example of which is the creation of the new airport.

3. Island Defense: Kojima

Japan has a shoreline that is about 32,000 km long. It is very irregular, subjected to extreme natural processes, intensively and densely populated, and highly modified. Today less than half of the coastline of the four major islands is in a natural state[3]. Some of the prefectures have engineered structures, such as seawalls, detached breakwaters and groins, along virtually their entire coastlines.

The extent to which the Japanese have modified their coastline is well illustrated on Kojima (which in Japanese means 'small island') located about 1 km off the coast of southeast Hokkaido. Kojima, only about 5 ha in area, is nearly rectangular with a narrow, elongated hill at its eastern edge (Figs. 14, 15). Composed mainly of tertiary shale, the ridge is about 30 m high and exposed to the open Pacific on its eastern flank. With an almost unlimited fetch, the windward side of the ridge is subject to intensive wave action generated by the SSE and S prevailing winds, occasionally typhoons and even tsunami.

The ridge, serving as a barrier, protects the lee side of the island where a circular shaped bar has developed. Because of the variability of refracting and diffracting waves around the ridge the shape and area of the bar have varied extensively through the years.

This island has served as a base for the harvesting of konbu, a seaweed important in Japanese culture. Early in the 20th century, herring fishing and iodine manufacture were important occupations. Iodine was produced in three factories on the island until 1917 when the industry was discontinued. Until the mid-1970s the island had full time residents with their well-maintained homes, konbu drying sheds, and connections to the mainland by underwater telephone, electricity and water. In the 1980s, the 10 families involved in konbu collecting and drying were only living on Kojima between June and September, the konbu season.

Because erosion was rapidly destroying the ridge and therefore the shingle bar, protective measures were begun in the 1960s. Because the major wave impact was on the northeastern side of the island, it was the first location to be protected. A short (about 85 m long) seawall was built in 1962-3 (Fig. 15). Soon thereafter seven groins were added from the western-most portion of the island toward the south. The groins, built between 1967-1970, consist mainly of hexalegs. They not only helped stabilize the shore but, in some locations, have also led to accretion. The inner sections of some of the groins are buried in shingle. Part of this may be the result of the residents long-time practice of annually dragging gravel up from the shoreline. It seems this practice has a dual purpose—it raises the level of the island but also serves to flatten the surface upon which the konbu is spread out to dry[3].

Whereas the structures discussed above mainly served to protect the low part of the island, in 1975 it was decided to protect the oceanward facing ridge. Early Japanese maps when compared with the air photographs of the 1970s, indicate that between 192:

Fig. 14. Low oblique photo of Kojima, Japan (photo courtesy of Public Works Division, Akkeshi Branch, Hokkaido).

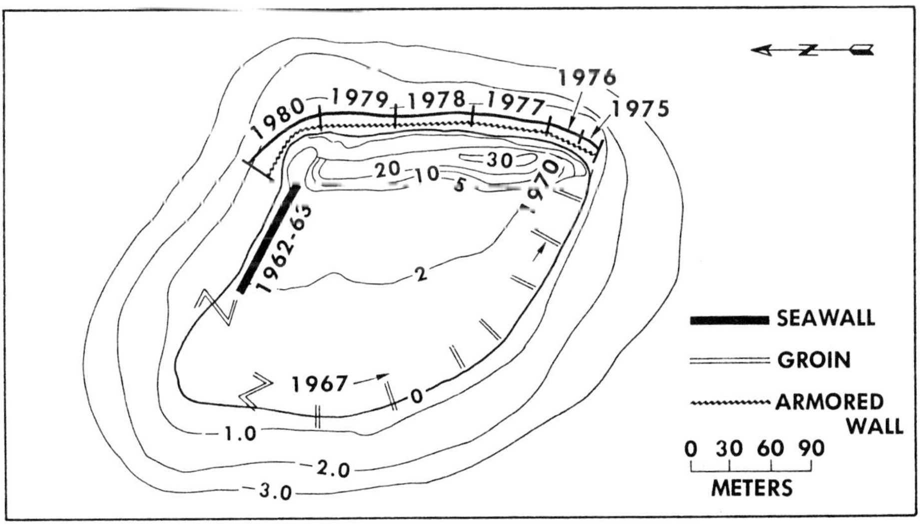

Fig. 15. Defense structures on Kojima (adapted from a map prepared by the Public Works Division, Akkeshi Branch, Hokkaido).

and 1958 the ridge had been reduced in length from about 400 m to 220 m. Armoring of the ridge was begun in 1975 and completed in 1980 (Fig. 15). This protective structure is composed of hollow triangular blocks for about one-third of its distance and tetrapods (Fig. 16) for the rest. Together their total length is 276 m.

4. Aquaculture: An Aboriginal Hawaiian Example

The conversion of low-lying coastal areas into fish farms has a long history. Nonetheless, within the past few decades its expansion has been rapid. In many parts of the world, especially in the tropics and subtropics, vast areas of mangrove swamps, marsh lands and near-shore shallow ocean areas have been converted into ponds for raising both fin fish, crustaceans and seaweed. Some of the areas converted to ponds are extensive as for example in South China and Hong Kong (Fig. 17, 18).

Although modern technology and equipment is now being used in the building of ponds, in earlier times construction was relatively simple but none-the-less effective. One such example was that used in Hawaii long before its discovery by Captain James Cook in 1778.

The Polynesians living on the Islands at the time of discovery depended to a large extent on the sea as a major source of food, although they did grow some crops and had pigs, dogs and chickens. Not surprisingly, most Hawaiians, lived near the shore and were good swimmers and seamen. The importance of the sea was emphasized even in land ownership. Property boundaries extended across the shoreline into the sea and even out to coral reefs offshore.

Because nearshore and offshore fishing could be variable, the Hawaiians preserved fish by drying and salting. In addition, they also took steps to insure a ready supply of fresh fish, no matter the weather or season and that was to impound them. They built both fresh water and seawater ponds. Although there are various estimates as to the number of fish ponds present in the islands at the time of Cook, Farber[11, 12] writes that there may have been more than 480. Thirty-three years earlier Summers[13] wrote that the number of ". . . ponds for all the islands, . . . not including the many small inland ponds, was about 210." Thus, it appears that the islands had at least 200 shoreline aquaculture ponds in 1778.

The appropriate Hawaiian terms are *loko* (pond) and *loko i'a* (fish pond), although today the term *loko* by itself is most often used. Of the several types of shore ponds, the two most important were *loko kuapa* and *loko 'umeiki* referring respectively to a fishpond with a seawall (Fig. 19) and a pond with a wall that has numerous openings (Fig. 20).

The size of these ponds ranged from about 0.5 ha to more than 200 ha. They were built along sheltered shorelines such as those found on the south coast of Moloka'i (Fig. 21), which retains the best preserved of fish ponds today, and along the shoreline of Pearl Harbor in Oahu where there were more than 40 of them. Often they were located where inland streams flowed into them bringing in nutrients from the land. *Loko kuapa* were built with their wall enclosing a small bay or out from the shoreline forming a semi-circular enclosure (Fig. 19). This latter type is especially common along the relatively straight coastline of Moloka'i.

Fig. 16. Tetrapods protecting the base of the ridge on Kojima.

Fig. 17. Farm pond in Shijiang delta, China. This area was reclaimed ca 1500 years ago.

Fig. 18. Details of recently established aquaculture in Hong Kong.

The walls of the Hawaiian *loco* ranged in length to more than 1.5 km, as exampled by He'eia pond in Oahu, a pond that is still in use (Fig. 22). The walls range from 1 to 6 m in width with an average of about 2 m although the widths do vary. The height of *loko kuapa* walls, which are not submerged at high tide, average between 1 and 2 m depending on pond depth which is usually less than 1 m. The walls were constructed out of the material that was most available, most often coral or basalt. The stones varied in size with some weighing as much as 500 kg (Summers 1964). Because of the geologic history of the various islands, the materials used varied among the islands. As Apple and Kikuchi write, "One kind of living lime-secreting coralline algae (*Porites* spp.) may have been used intentionally by the Hawaiians in some ponds on Moloka'i Island as cementing agents in the walls"[14].

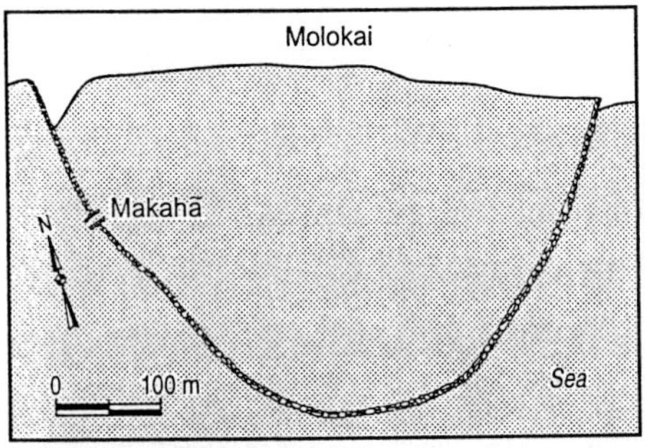

Fig. 19. *Loko kuapa* on Moloka'i[9].

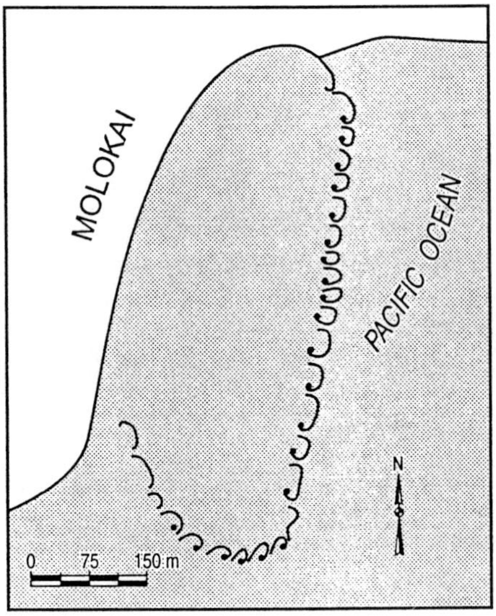

Fig. 20. *Loko 'umeiki* on Moloka'i. Note the platform locations for netting fish as the tide changes[13].

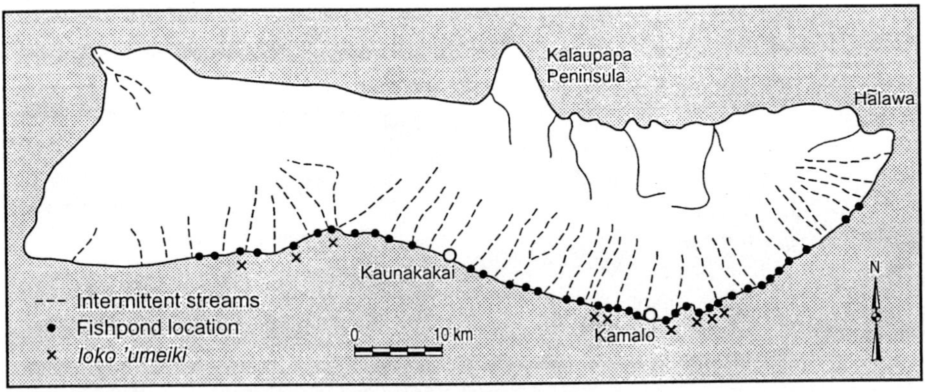

Fig. 21. Map of Moloka'i showing the location of both *loko kuapa* and *loko 'umeiki*[13].

The walls that separated the seas from the pond were well built (Fig. 23), although occasional tsunami or hurricane surges would damage them. For example, the tsunami of April 1, 1946 damaged or destroyed many *loko* walls on Moloka'i and also partially filled the ponds with silt and sand[15].

The walls are built so that the base is wider than the top. Often the seaward face has more slope to it than the pond face, presumably because the greater amount of wave energy that impinges on the outer-facing wall (Fig. 24c). One of the most important parts of a *loko kuapa* are the sluice grates known as *makaha* (Fig. 24a). Made of wood, they allowed a natural exchange of water between the pond and sea. The number of *makaha* per *loko kuapa* ranged from one to seven—most often one or two. Other frequent additions in the sluice grate area was a guard house and a netting location (Fig. 24b)[14].

The *loko 'umeiki* or fish trap type of pond depends on fish entering the pond on rising tide and exiting on ebbing tide. The walls are constructed so that the entrance and exit lanes narrow in the direction of the flowing water to allow easy netting. One such 13 ha pond on Moloka'i had 16 entrance lanes and 9 exit lanes. Often each lane had its own name[13].

One of the curious aspects associated with the Hawaiian *loko* is the "engineers" involved in their construction. Apple and Kikuchi report that only 23 fishponds in Hawaii were considered (by the Hawaiians) as having been made by humans[14]. The rest, according to folklore, were constructed by the *menehune* who were (are) believed to be supernaturally endowed dwarfs. Further, they were supposed to have completed their construction in the course of one night.

Although most of the Hawaiian *loko* are not being used at present there is some effort being made to rejuvenate many of them especially on Moloka'i[11]. Unfortunately, disuse and misuse has seen a rapid deterioration of the walls and also the ponds. Increased siltation along with the introduction of such plants as mangroves has hastened their deterioration. On the Big Island (Hawaii), some have been covered by lava.

5. Tsunami Protective Structures

Many locations along the shorelines of the world are occasionally impacted by large waves generated by typhoons (hurricanes, tropical cyclones) and tsunami Although there are some locations where seawalls have been constructed primarily to defend against typhoon waves (Fig. 25), this discussion deals mainly with tsunami protection.

Tsunami are waves that have periods of from five to 60 or more minutes that are ". . . primarily created by disturbances in the crust of the Earth underlying bodies of water, . . ." [16]. The word is a combination of the two Japanese words <u>tsu</u> meaning harbor and <u>nami</u> meaning wave. It is within bays (where many harbors are located) that such waves are especially strong. Tsunami are unpredictable and, although they can develop along any coast, they are most common along the Pacific Rim where earthquakes are frequent.

The time between generation and coastal impact varies from almost instantaneous in the case of local generation to several hours in the case of those that develop thousands of kilometers away. For example, a tsunami generated in Chile (as happened

Fig. 22. He'eia pond on Oahu.

Fig. 23. *Makaha* on a *loco kuapa* on Moloka'i.

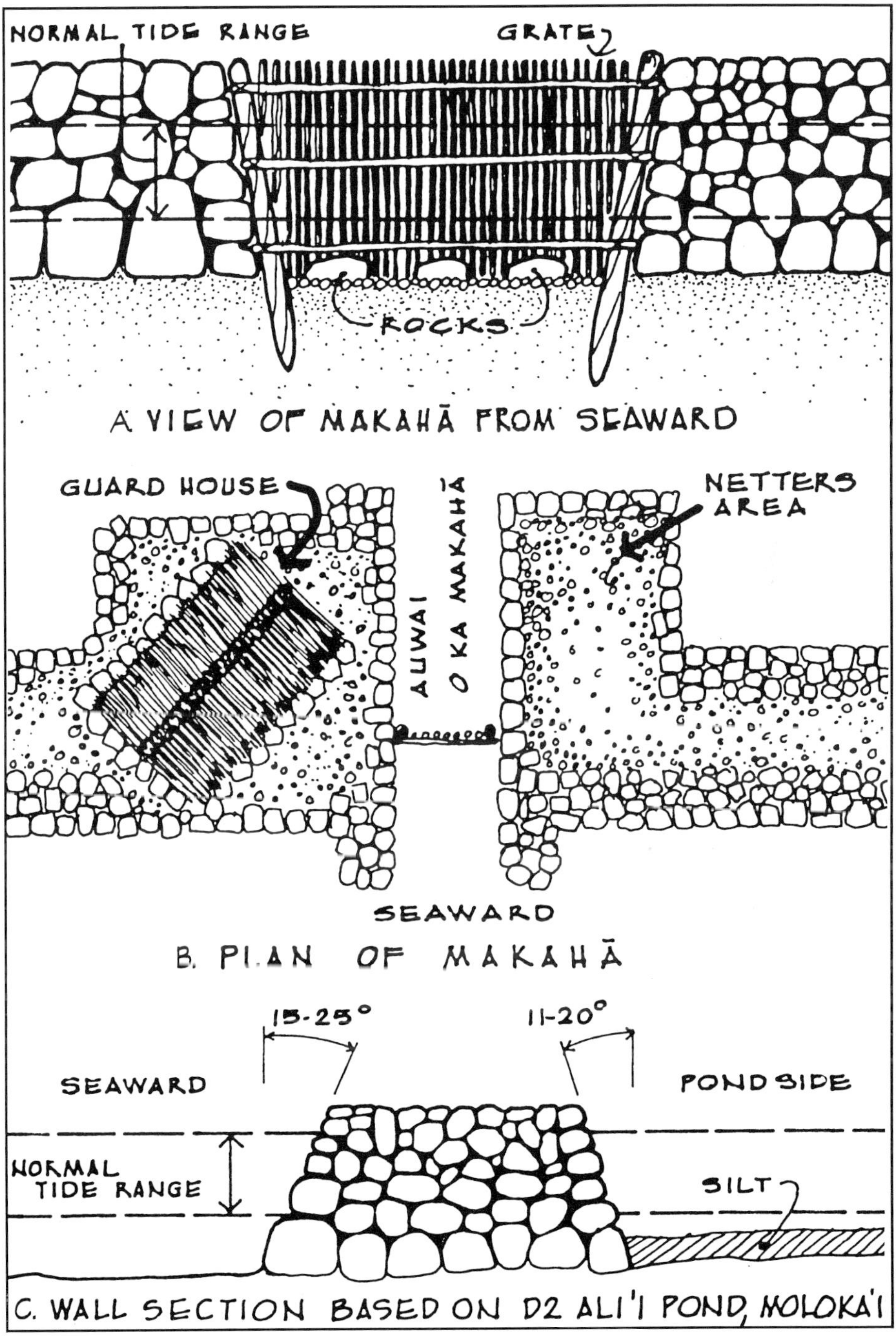

Fig. 24. A schematic of pond features[14].

May 24, 1960) took 15 hours to reach Hawaii and 22 hours to reach Japan. With modern warning systems evacuation in such cases can result in the saving of lives. In the past no such warnings were available and in the case of generation nearby, warnings often are too late to completely avoid loss of life.

Several factors of a coastline affect tsunami run-up and area inundation, including: orientation of the coast, submarine topography, presence or absence of coral reefs, and impediments such as buildings and vegetation[17]. The shape of impacted bays is critical because of the funneling effect they may have on the advancing waves.

The Sanriku coast in northeastern Honshu, Japan is especially vulnerable to damaging tsunami because of the irregularity of its coast and the frequent seismic actions that occur in the vicinity. Thus the citizens of the area have responded in a variety of ways to tsunami. Two of the most common responses, in earlier years, were the planting of trees along the shoreline and moving to higher ground. For example, after the severe 1933 tsunami, 35 villages (more than 3000 houses) in two prefectures were moved to higher ground. As a result of the same tsunami, one village, because no space was available for relocation of the town, raised the ground level on which it was built by 3 m, just as did Galveston, Texas, USA after a 1900 hurricane virtually destroyed the city.

After the 1933 tsunami, large seawalls, some more than 10 m high, were built at vulnerable locations (Fig. 26). In addition breakwaters were constructed across a number of bays (Fig. 27). Ofunato Bay, which is 740 m wide, is one example (Fig. 28). Breakwaters, 12 to 14 m wide at the top, which was 5 m above water level, were constructed from each side of the bay in water depths of as much as 37 m. The opening between, which allowed ships to enter, is 200 m wide. These bay entrance breakwaters are designed to reduce the wave heights of tsunami by more than 2 m[17, 18].

At some locations seawalls are built in the town itself usually to separate waterfront activities from residential and business concerns (Fig. 29). Such seawalls may be massive and always have large gates through which vehicular traffic can move. The gates are constructed so that they can be closed with a minimum of effort. Further some of the seawalls are sufficiently wide on the top to be used as roadways.

Another location where tsunami have attracted much attention is in the Hawaiian Islands and especially the Big Island. The Hawaiian tsunami may be the result of local (i.e. Hawaiian) earthquakes as well as distant ones the same as is true of the Sanriku coast. Although the Hawaiians knew of tsunami and have legends about them, there is no record of their occurrence before early in the 19th century in Hawaii.

Subsequent to 1837 there have only been some 11 tsunami that have caused severe damage in Hawaii although the water gauge in Honolulu records about one per year. Those of local origin have been caused by both volcanic eruptions and earthquakes and are especially common on the Big Island[19].

One of the strongest tsunami to hit the Hawaiian Islands was that of April 1, 1946 which originated in Alaska. The one that originated in Chile in 1960 also did much damage. In the 1946 tsunami waves reached a maximum height of 17 m and caused much damage in Hilo on the northeast coast of the Big Island. The breakwater that extends nearly three-fifths of the way across Hilo Bay was severely damaged (Fig. 30). Although the breakwaters of other harbors in the islands suffered some damage it was

Fig. 25. Twelve meter high seawall on typhoon prone shore on the east coast of Honshu, Japan.

Fig. 26. Tsunami wall illustrating the easily closed gate.

Fig. 27. Tsunami wave protection breakwater on the Sanriku coast, Japan.

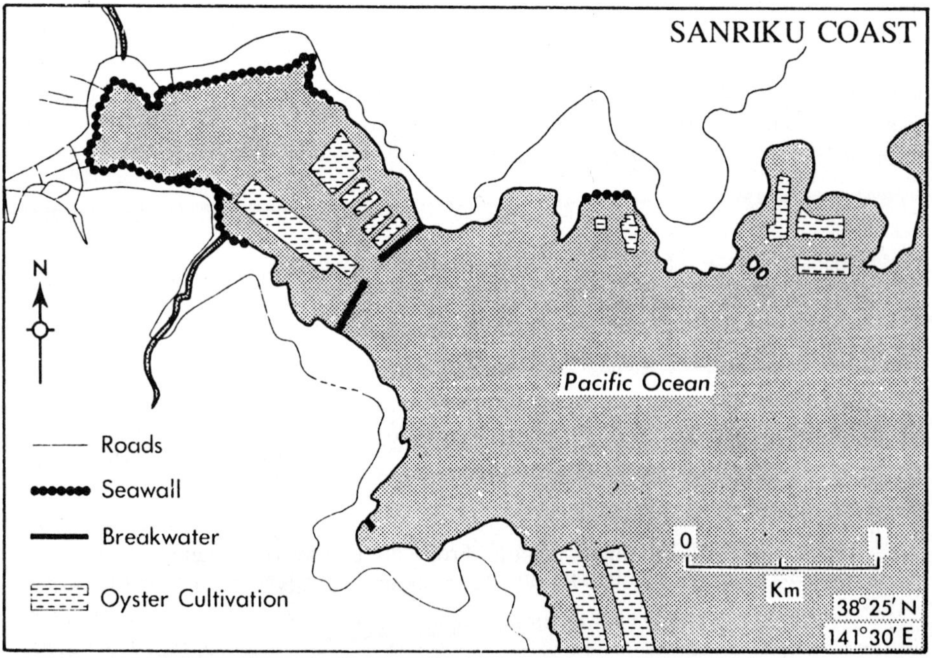

Fig. 28. Map of a tsunami prone bay on the Sanriku coast of Japan showing the location of the breakwater and seawalls.

minor in comparison to that at Hilo. About 1800 m (61%) of the 3000 m long breakwater was damaged. Shepard et al wrote:

> The cap and outside face were composed of rocks weighing 8 tons or more, and the inside face of rocks weighing 3 tons or more. The rocks were thrown both shoreward and seaward by the waves. The average depth of scour in the gaps of the breakwater was 3 feet[15].

They also noted that, despite the damage to the breakwater, it ". . . probably reduced greatly the severity of the attack of the waves . . ." inside the harbor[15]. Well constructed seawalls were little damaged, although some erosion occurred behind them because of overtopping. Model studies conducted by the U. S. Army Corps of Engineers determined that a continuous protective barrier would be effective at controlling tsunami in the Hilo area[20].

However, unlike the Japanese, who, as noted above, often constructed sea walls and breakwaters to protect against tsunami, the Americans (State of Hawaii and city of Hilo) opted to leave the formerly developed section of the shore zone as a green belt.

6. Solar Salt Pans

One of the earliest industries to capitalize on the chemical characteristics of sea water was the production of salt (Fig. 31). With an average salinity of $35^0/oo$, the sea is a veritable storehouse of salt, a condition that has been capitalized on for thousands of years. Among the several procedures used for extracting salt from sea water, the most widespread is by solar evaporation. This process involves the creation of ponds within which sea water is trapped allowing it to evaporate. If the trapped water evaporates completely, all of the salts are precipitated. Of the $35^0/oo$ of total salt in the average sea about $25^0/oo$ (or more than 70%) is sodium chloride, the most in demand for human use. It also is the first of the many salts to precipitate, so that the other salts known as bitterns, can be removed while still in a brine state.

Because the effectiveness of solar salt production depends largely on the sun's energy, many coastal locations do not lend themselves to development. In areas with adverse weather conditions, the boiling of sea water and the leaching of the ash derived from the burning of sea plants and peat among others, are frequently used to obtain salt. Those areas where cloudiness and rainfall are minimal are best suited for solar salt production. However, because transportation costs may be high, locations adjacent to consumers are desirable. The transportation of salt is credited with being responsible for the development of some of the earliest roads[21].

Along some coastlines, lagoons dry up seasonally leaving behind a layer of salt that is harvested prior to again filling. From an engineering standpoint it is the construction of ponds into which sea water is diverted that is relevant. Just when such structures were first made is unknown. However, it was early around the Mediterranean Sea and along the coast of China and elsewhere (Fig. 32). Although in many nations, e.g. Japan, a number of solar salt operations have been discontinued, it is still of major importance elsewhere.

Fig. 29. A Sanriku coastal town with tsunami protective walls running through it.

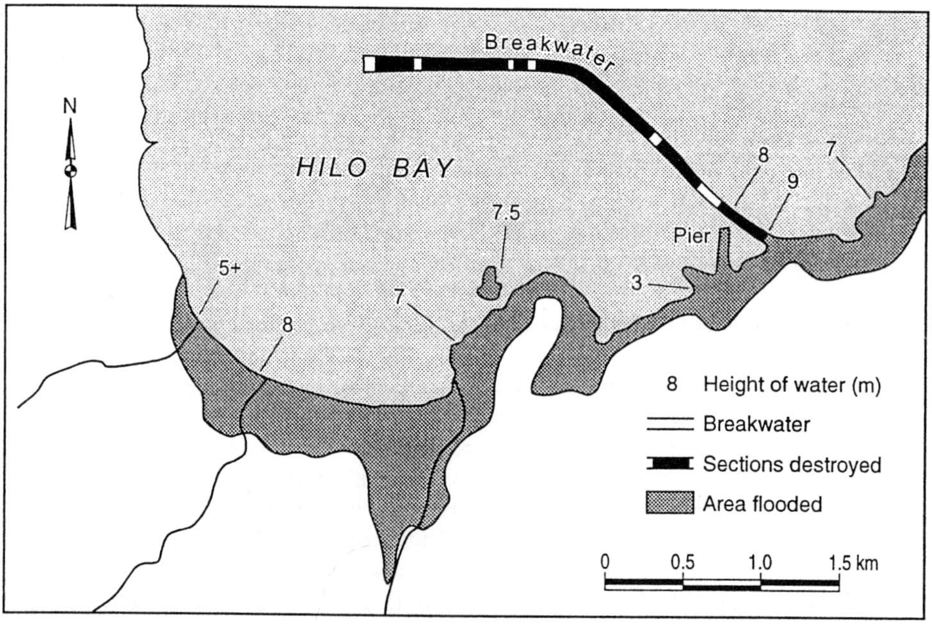

Fig. 30. Hilo harbor showing heights of the 1946 tsunami wave, extent of flooding and destruction of the breakwater at Hilo, Hawaii (modified from Shepard et al[15]).

Given favorable marketing conditions and good evaporation rates, there are several other factors that enhance production. The presence of a low-lying and relatively flat surface underlain by impervious soil onto which sea water can be "funnelled" lead to economical production. In some locations the pond base is covered with an impervious material to reduce loss through percolation. A major benefit accrues when the tide can be used to furnish sea water to the initial pond (Fig. 33) and when gravity can be used to provide ". . . an orderly brine flow through the system"[22] (Fig. 34). Thus, not only is there needed a seawall or dike with gates in order to protect the pond area from the sea and permit the introduction of sea water into the system but also there is needed within the area a series of canals to transfer the concentrating brines and roads to allow for maintenance and haulage of salt[22].

Although many solar salt industries around the world are small operations, some can be very large. For example, in 1958, Ver Planck[23] wrote that "By far the most important source of salt in California is the solar salt industry, which is centered in the south end of San Francisco Bay." During the middle of the 19th century, numerous operations began the processing of solar salt but by the time of WWII consolidation had occurred. Because of the nature of San Francisco Bay, minimal dike protection is needed from wave action. However, as with other locations the ". . . concentrating ponds are located between the high and low tide marks so the intake can be by means of tidal gates to minimize pumping" [23].

7. Small Ports, Fisherina and Marinas

Once humans began to use the sea for more than collecting shellfish or other edible or usable materials that washed up on shore, they began to modify it. With the development of boats, shelter from surf and storm waves became important. Such protection originally must have been little more than finding a protective bay or rearranging rocks along the shoreline (Fig. 35). As boats became larger and trips to sea more commonplace, protective structures became more elaborate.

During most of seafaring history, fishing boats dominated and harbors tended to remain small. However, more than five millennia ago maritime commerce began to take shape and by 2000 years ago ports were commonplace around the Mediterranean Sea (Fig. 2).

With this increase in dependence on the sea, the technology associated with port construction advanced apace. Tsinker writes, for example, that for sheet piling:

> . . . the Phoenicians used planks made from Lebanon cedar. Various types of timber sheet-piling techniques (e.g., tongue and groove, laminated, and others) were used. . . . The Phoenicians also used heavy blocks locked together with copper dowels for construction of the open-sea port at Tyre. This type of construction was also used by the Romans. In the 1800s, both materials still played a major role in port construction[24].

Ports have been classified in a variety of ways including by location (for example, seaports and inland ports) and by function which includes those dominated by and for commerce, the military, fishing (Fig. 36), small craft and refuge. Many ports serve

Fig. 31. A statue dedicated to solar salt production, illustrating the layout of the salt pans in France.

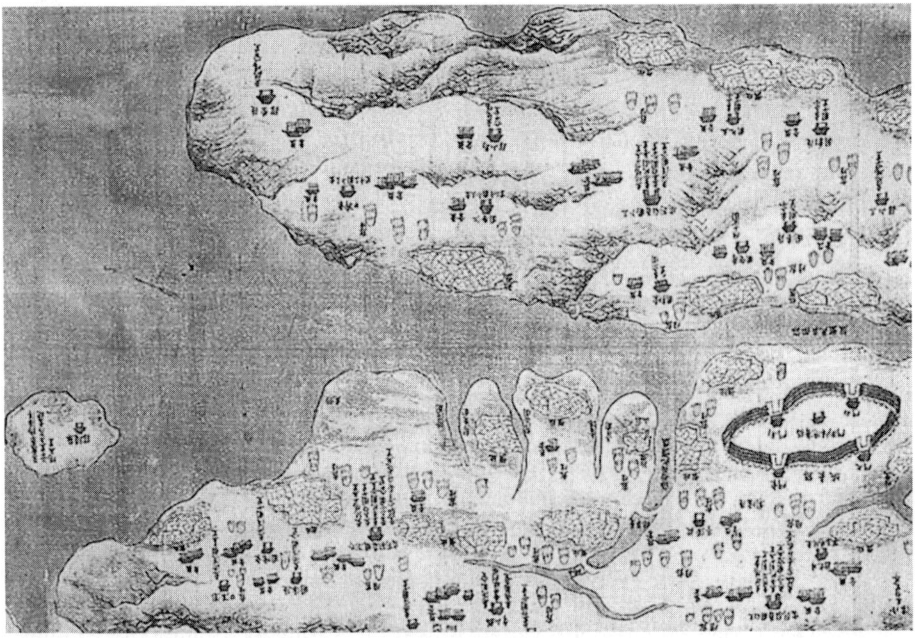

Fig. 32. Solar salt pans as depicted on an early Chinese map (original in the Geography and Map Division, Library of Congress, USA).

Fig. 33. Solar salt pans in the high tidal range region of west Korea at low tide.

Fig. 34. Method of transferring water from one evaporating pan to another in Korea.

Fig. 35. Modification of shoreline to form boat ramp in Hokkaido, Japan.

more than one function. Refuge, probably the earliest reason to seek a safe harbor, is still a main priority in seamanship. Hershman, for example, writes that "Along the coasts and Great Lakes of the United States, there are 1435 navigable waterways, harbors, and river stretches—places where boats or ships can take refuge from open water" [25].

Although small ports for fishing and light commerce have been in existence for millennia, their use for recreational purposes is relatively new having developed almost wholly within the last century. This section deals mainly with small ports especially fishing ports and marinas.

Around the coast of Japan, which is about 32,000 km long, are located more than 4000 harbors, an average of about one for every 8 km of coastline.

> They range from those developed on the open coast to those in naturally protected bays, and they range in size from those capable of sheltering only a few fishing boats to those dredged into sand dunes capable of handling supertankers[26].

7.1 FISHERINA

Whereas, throughout most of history, small Japanese ports were used entirely by the fishing industry, recent rising interest in marine recreation has begun to place pressure on them. This pressure has led to the development of fisherina which are designed to allow fishing and recreational activities to co-exist.* It revolves around a plan that provides separate areas for commercial fishing vessels and for sports fishing and pleasure craft. Fisherina developments are taking the local environment around the harbor and waterfront into account as well as making improvements in the ports themselves. They are tailor made to the needs of each unique environment. Those near urban areas are linked to the city's infra-structures. The main objective is to ensure that fishermen and recreational users both benefit from the arrangement. In all, Japan has five fisherinas completed or in development. One of the largest is that established as the Marine Pier Kobe project under the Kobe City Fishing Renaissance Plan. It includes plans for a marine farm, canal walks, a resort hotel, and a fisherina with moorings for both fishing and recreational boats.

An expanded version of this idea is exampled by Tokyo. The Tokyo metropolitan government which, as part of a plan to encourage the co-existence of commercial vessels and general-use vessels as well as promote marine sports, has designed the largest marina in Japan. Because of the recent increase in pleasure craft and a big decrease in the lumber industry which formerly demanded storage space in water (Fig. 37), Tokyo converted an 18 ha lumber storage facility on the calmest section of Tokyo port into a 5.7 ha marina. It has moorings for 640 boats with related facilities including a promenade for the casual visitor. Its adjacency to central Tokyo makes it ideal for a yacht owner to spend time with his boat with a minimum of travel time.

7.2 MARINAS

Although pleasure boating must have been engaged in by the elite in the earliest of civilizations, it is probably valid to suggest that it became a major endeavor only after

WWII as a ". . . result of social and economic changes in our society and technological advances in the boat manufacturing industry"[27]. He further stated:

> If this trend continues, as experts believe it will, there will be mounting pressure on public and private enterprises to provide the harbor facilities—channels, breakwaters, docks, floats, and such—to accommodate the swelling recreational small craft fleet and provide access to the nation's coastal and inland waterways[27].

Although Goodwin was writing specifically about the United States, his comments are valid for many countries around the world as suggested in the discussion about fisherinas above. Unlike the fisherina concept, however, most marinas are specifically designed for recreational craft. The extent of this development is indicated by N. W. Ross (reported in Goodwin[27]) that in the United States about 11,000 enterprises provided moorage and storage for recreational craft in 1985. Of these, 78% were managed by private firms.

Two main aspects of marina development that are critical for this discussion of engineered coastlines are: 1) the modification of the shoreline (coastal zone) by marina construction and 2) the structures used to allow access to marinas and to protect them from the sea.

Because of their function (i.e. recreation), marinas tend to be located where demographic, economic and environmental factors favor them. The extent of shore modification varies greatly depending on whether the marina is made on the open coastline (Fig. 38), within a natural bay or built (dredged) out of the subaerial part of a shore zone as for example in a coastal marsh or a mangrove swamp (Fig. 39).

One of the key elements of most marinas is the breakwater that is needed to protect the marina itself and/or the entrance channel to the marina from the ". . . adverse affects of waves, currents, migrated sediments, and drifting ice and to create a calm water area within the harbor"[24]. They may be shore-connected or located offshore (Fig. 40) in which case their main function is wave dissipation. The position of the breakwater in relation to the marina will affect its form, effectiveness and cost. As Adie notes wrong positioning may encourage pollution, limit protection from waves, result in damage to the breakwater, and promote siltation and erosion[28].

The type of breakwaters used in making marinas vary with many factors. Usually cost, because they are privately financed, is a critical consideration. Materials used include timber piles, riprap, gabions, hydraulic fill, poured concrete, armor units (tetrapods, hexalegs, akmons, doloesse etc.) (Fig.41) and floating attenuators among others[30].

An example that illustrates many of these aspects of marina location and construction is the case of the Club Mykonos Langebaan Pleasure Craft Harbor in Saldanha Bay on the west coast of South Africa[31]. First, the name stems from the similarity in climate of the South African coast to that of the Aegean Sea in the Mediterranean. It was the first privately owned marina to be developed in South Africa and was completed in 1989 (Fig. 42). It has space for 130 vessels with all of the necessary facilities. The basin is 4 ha in area and was deepened by dredging; the dredged material was used to reclaim 1.4 ha adjacent to the basin. The shoreline of the bay, with spring tides of 1.5 m and a wind direction mainly from the SSW, receives

Fig. 36. Small fishing port in Hokkaido, Japan with concrete breakwaters and armor blocks.

Fig. 37. The port of Tokyo illustrating subsidence of the city to the left and water level with floating logs to the right.

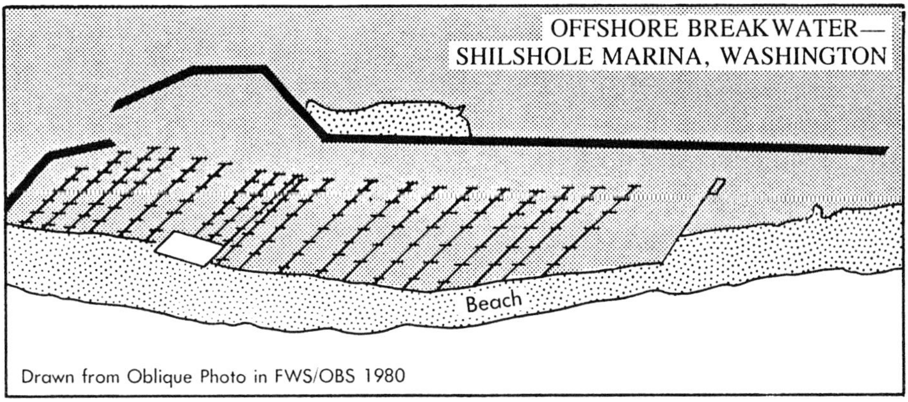

Fig. 38. Example of a small marina in Washington State, USA.

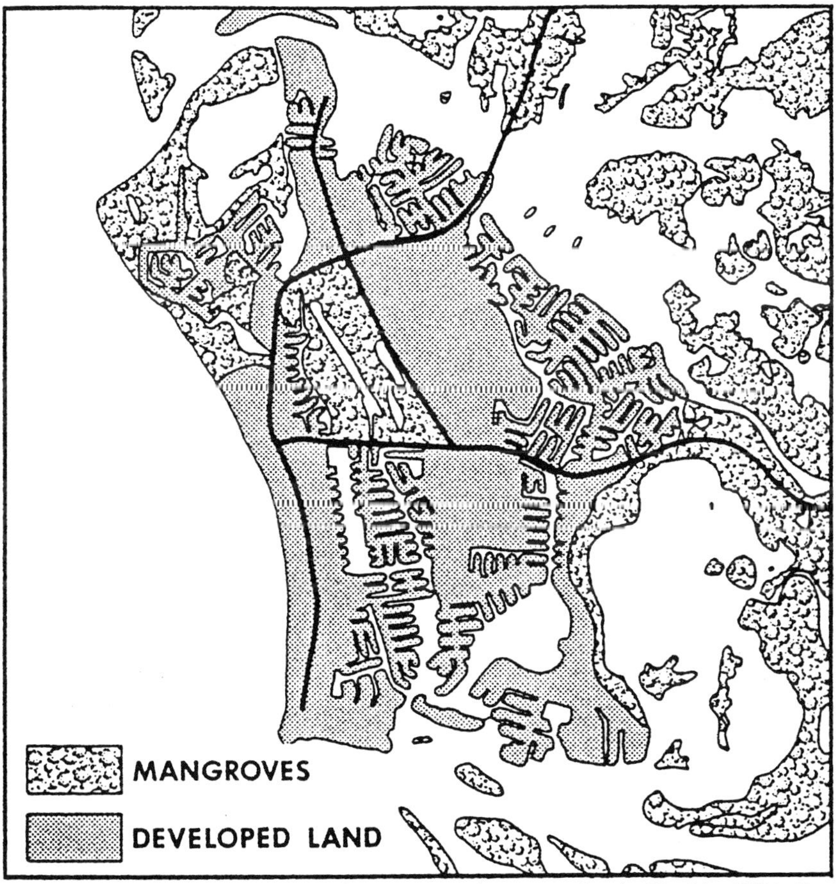

Fig. 39. Conversion of an area of mangroves into a marina in Florida, USA. In the process the shoreline has become even more contorted than before reclamation[29].

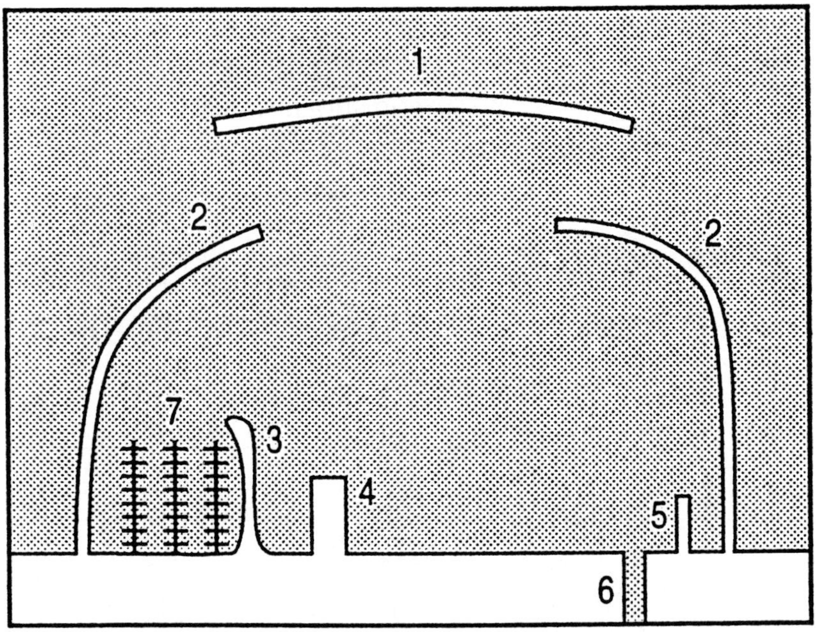

Fig. 40. The elements of a small port: 1) detached breakwater, 2) land-tied breakwater, 3) in-harbor breakwater, 4) pier, 5) wharf, 6) pier, 7) dry dock, 8) marina (modified from Tsinker[24])

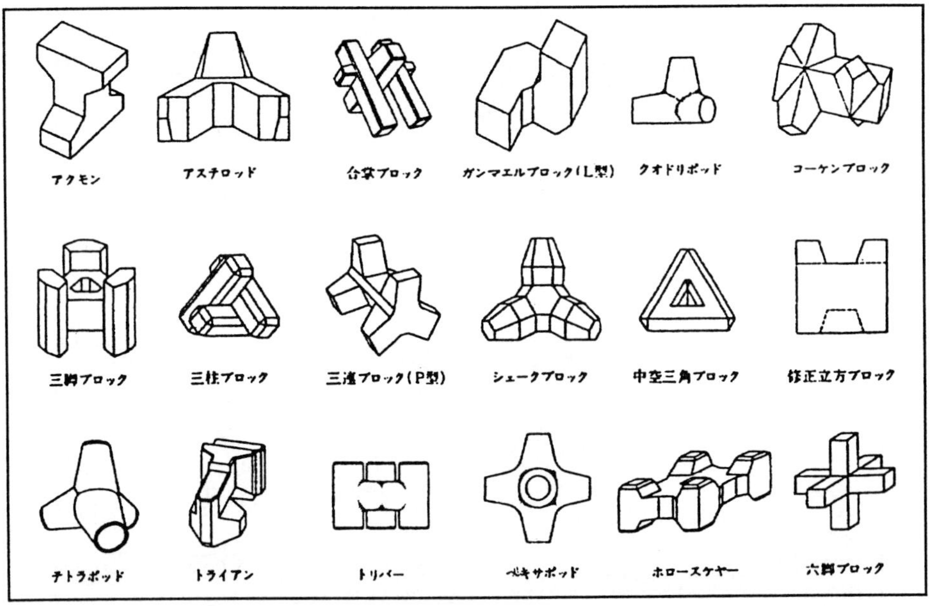

Fig. 41. Sketches of 18 different armor blocks as depicted in a Japanese engineering manual.

oceanic waves that, because of refraction, are normally reduced in height by 40%. After considering three separate designs, it was decided to use a conventional rubble mound protected with Accropode units[31]. The reason for selecting the Accropode design, even though more expensive than a berm type breakwater, was because of the fear that storm events would be less likely to cause crest damage. The main breakwater is made of 45,000 m^3 of rock, 1400 Accropodes and 3000 m^3 of concrete. The secondary breakwater is made of rubble with armor units on the seaward side.

As often happens during engineering projects along coastlines, unanticipated problems arise. In the case of the South African marina during construction of the secondary breakwater sedimentation occurred so rapidly that construction had to be delayed until a spur breakwater (Fig. 42) could be built in order to deviate the rip current that was causing sedimentation.

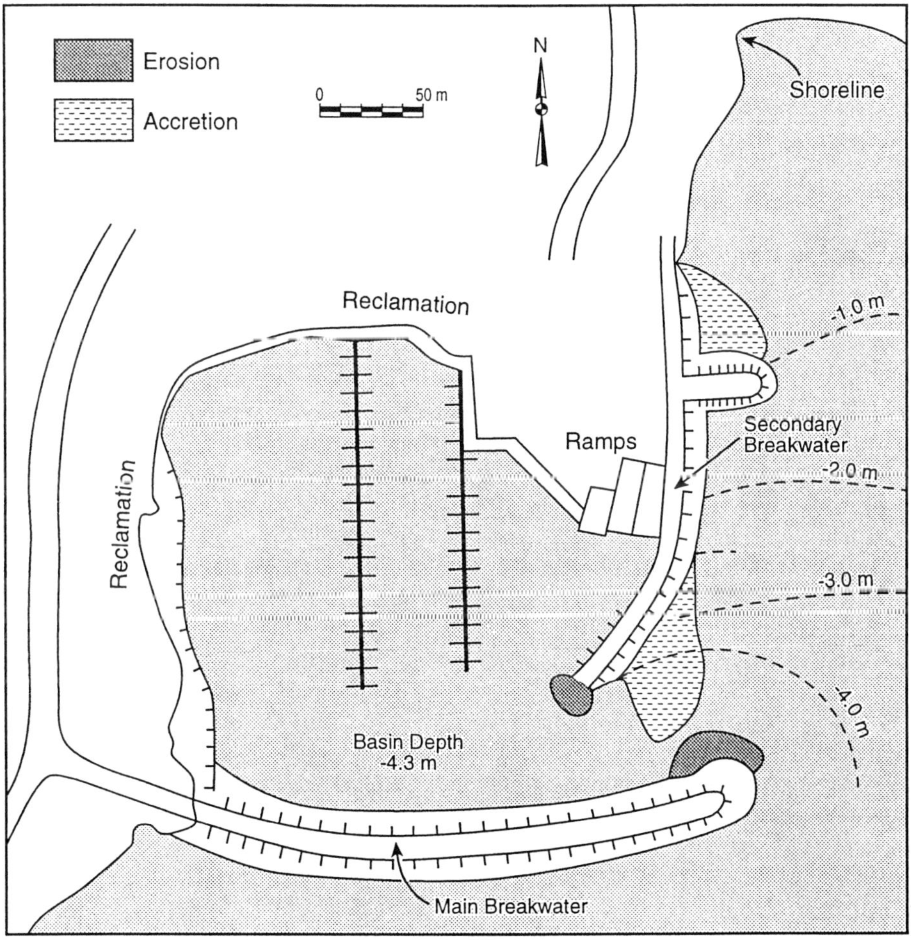

Fig. 42. Harbor layout of the Club Mykonos Langebaan Marina (modified from Bosman et al[31])

During the first year after construction some erosion occurred at the ends of both breakwaters and deposition occurred to the lee of the spur breakwater and at the entrance to the marina. This deposition is controlled by dredging[31].

8. Conclusion

Although there are innumerable examples of how humans have modified the shorelines of the world, we hope that the six examples in this chapter along with the 14 longer entries have given the reader a "feel" for the extent and intensity to which humans have "engineered" the coast. Despite the fact that restrictions have been placed on shoreline engineering in recent years in many countries, there is little doubt that humans will continue to have a major impact on the coastal zone.

This chapter, as well as the book as a whole, has emphasized the direct modification of the shoreline. Nonetheless, there are many ways it has been and is being modified indirectly and usually unintentionally by human activity. For example, the construction of dams on many of the rivers of the world altered the type and amount of sediment delivered to the sea and the placement of groins along one section of the coast have affected adjacent coastal areas and usually adversely.

9. References

1. Volman, T. P. 1978. Early archeological evidence for shellfish collecting. *Science*, 201, 911-913.
2. Kerr, R. A. 2000. A victim of the Black Sea flood found. *Science*. 289, 2021.
3. Walker, H. J. 1981. In defense of an island: Japanese style. *Scientific Bulletin*, 6(1), 1-10.
4. Russell, R. J. 1967. Aspects of coastal morphology. *Geografiska Annaler*, 49 (Ser. A), 299-309.
5. Inman, D. L. 1974. Ancient and modern harbors: a repeating phylogeny. *Coastal Engineering*, 15, 2049-2067.
6. Walker, H. J. 1992. Sea level change – environmental and socio-economic impacts. *Geo Journal,* 6(4), 511-520.
7. Singer, C. et al 1956. *A history of technology*. Vol. II. Oxford University Press, Oxford.
8. Takahashi, K. 1987. *Coastal modification in Hawaii: A case study of the reef runway*. Term Paper, The Coastal Environment, University of Hawaii, Manoa.
9. Walker, H. J. 1980. Reservoirs from the sea: Hong Kong's answer to its water supply demands. *Scientific Bulletin*, 5(1), 19-25.
10. Hong Kong Government. 1979. *Hong Kong Yearbook 1979*. Government Publications Center, GPO Building, Connaught Place, Hong Kong.
11. Farber, J. M. 1997. *Ancient Hawaiian fishponds: can restoration succeed on Moloka'i?* Neptune House, Hawaii.
12. Farber, J. M. 1998. *The historic fishponds of Moloka'i, Hawaii.* Internet: http://www.epa.gov/owow/estuaries/coastlines/spring98/fishpond.html.

13. Summers, C. C. 1964. *Hawaiian fishponds*. Bishop Museum Press, Special publication 52.
14. Apple, R. A. and Kikuchi, W. K. 1975. *Ancient Hawaii shore zone fishponds: an evaluation of survivors for historical preservation*. Manuscript. National Park Service. U. S. Department of the Interior.
15. Shepard, F. P., MacDonald, G. A. and Cox, D. C. 1950. *The Tsunami of April 1, 1946*. University of California Press, Berkeley, 5(6), 391-528.
16. Camfield, F. F. 1980. *Tsunami engineering*. Coastal Engineering Research Center, U. S. Army Corps of Engineers, Special Report No. 6.
17. Morgan, J. R. 1978. *The tsunami hazard in the Pacific: Institutional responses in Japan and the United States*. PhD dissertation, University of Hawaii.
18. Fukuuchi, H. and Ito, Y. 1967. On the effect of breakwaters against tsunami. *Coastal Engineering, Proceedings*. Vol. II, 821-839.
19. Cox, D. C. and Morgan, J. R. 1977. *Local tsunamis and possible local tsunamis in Hawaii*. Hawaii Institute of Geophysics, University of Hawaii, Manoa, HI.
20. Palmer, R. Q. and Funasaki, G. T. 1967. The Hilo harbor tsunami model. *Coastal Engineering, Proceedings*. Vol. II, 1227-1248.
21. Multhauf, R. P. 1978. *Neptune's gift: A history of common salt*. Johns Hopkins Press, Baltimore.
22. Garrett, D. E. 1966. Factors in the design of solar salt plants. Part 1. Pond lagoons and construction. In: Rau, J. L. Ed. *Second symposium on salt*. The Northern Ohio Geological Society, Inc. Cleveland, Ohio. 168-175.
23. Ver Planck, W. E. 1958. *Salt in California*. Division of Mines, State of California, San Francisco. Bull. 175.
24. Tsinker, G. P. 1997. *Handbook of port and harbor engineering* . Chapman & Hall, New York.
25. Hershman, M. J. 1988. Harbor Management. A new role for the public port. In: Hershman, M. J. Ed. *Urban ports and harbor management*. Taylor & Francis, New York. 3-25.
26. Walker, H. J. and Mossa, J. 1986. Human modification of the shoreline of Japan. *Physical Geography*, 7(2), 116-139.
27. Goodwin, R. F. 1988. Small-boat marinas: the new professionalism. In: Hershman, M. J. Ed. *Urban ports and harbor management*. Taylor & Francis, New York. 195-216.
28. Adie, D. W. 1977. *Marinas: a working guide to their development and design*. The Architectural Press, Ltd. London.
29. Walker, H. J. 1991. Anthropogenic landforms in the coastal zone. *Proceedings of the International First Regional Conference of Geomorphology*. Ankara, Turkey.
30. Tobiasson, B. O. and Kollmeyer, R. C. 1991. *Marinas and small craft harbors*. Van Nostrand Reinhold, New York.
31. Bosman, D. E., Retief, G. de F., Kapp, J. F., Kloos, M. and Ridge, A. B. 1990. Design and construction of pleasure craft harbour-Club Mykonos Langebaan. *Twenty-Second Coastal Engineering Conference. Proceedings*. Vol. 3, 3239-3253.

REGULATION OF THE CHANGJIANG ESTUARY: PAST, PRESENT AND FUTURE

CHEN JIYU
LI DAOJI
State Key Laboratory of Estuarine & Coastal Research
Institute of Estuarine and Coastal Research
East China Normal University
Shanghai, China 200062

1. General Introduction to the Changjiang Estuary

The Changjiang is the largest river in China. Its mouth is advancing seaward gradually. For example, in the 7th century B.C. the mouth was in the reach from Zhenjiang to Yangzhou. Then in the 17th century A.D., it moved eastward to the reach of Jiangying, and in the middle of the 20th century to the area of Xuliujing (Fig. 1). Now downstream from Xuliujing, the Changjiang mouth is firstly divided into the North and the South Branches by Chongming Island. The South Branch is likewise sub-divided into the North and the South Channels by Changxing Island. Furthermore, the South Channel is again divided into the North and the South Passages by Jiuduansha Island. Thus, today, there exists a three-order bifurcated estuary with four outlets to the sea (Fig. 2).

The Changjiang estuary has a meso-tidal range. The highest tidal level is 5.20 m and the lowest is -0.58 m, with an average tidal range of 2.70 m. Its tidal prism varies between 1.3×10^9 to 5.3×10^9 m^3. The discharge at 2.93×10^4 m^3/s totals 9.24×10^9 m^3/a. The discharge varies seasonally, the maximum is 9.26×10^4 m^3/s and the minimum is 4.62×10^3 m^3/s. The annually averaged sediment load totals to 4.86×10^9 tons, 87% of which is discharged during the flood period (from May to October). Additionally, 1 or 2 typhoons per year affect the Changjiang estuary. The typhoon of August 17, 1997 increased the water level by 5.99 m at Wusong.

Because of the interaction of the great upstream discharge and the huge tidal runoff, the evolution of the riverbed is very complicated. The difference between the directions of the flood and the ebb currents, and the swinging of the main channel between the flood and dry periods are the intricate dynamic factors in the evolution of the estuary

The position of the main channel in the estuary often changes because of erosion and deposition. During the past several centuries, a number of changes occurred in the Changjiang estuary. For example, in the 18th century the main stream of the Changjiang moved from the North Branch to the South Branch. The floods in 1860 and 1870 resulted in the formation of the North Channel and major floods from 1949 to 1954 resulted in the formation of the North Passage. The swinging of the main stream makes the shoals in the river move quickly.

An estuary is a mixing area of fresh and saline waters. Every year the turbidity maximum zone near the apex of the salt wedge moves between 121°55'E and 122°30'E. In this area, because the flood wave disperses, and suspended sediment coagulates and is deposited, a sand bar system 90 km wide forms. It results in the formation of a submerged delta with an area of more than 10,000 km^2.

A large delta plain formed near the Changjiang estuary. Its area is up to 4×10^4 km^2, and includes Taihu Lake. It is one of the most active regions in China with a flourishing economy, prosperous trade and aggregate culture. Shanghai, located near the Changjiang estuary, is

J. Chen et al. (eds.), Engineered Coasts, 185–197.
© 2002 *Kluwer Academic Publishers. Printed in the Netherlands.*

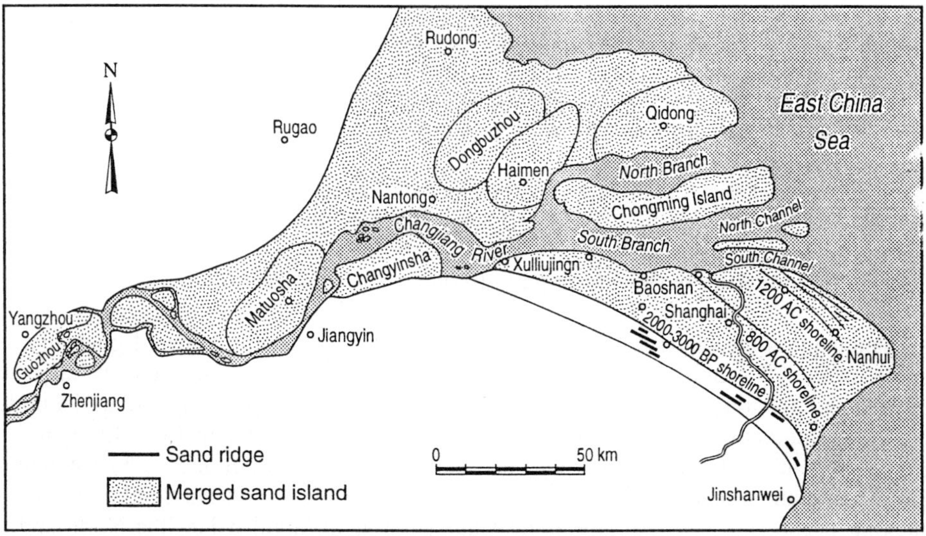

Fig. 1. The Changjiang estuary.

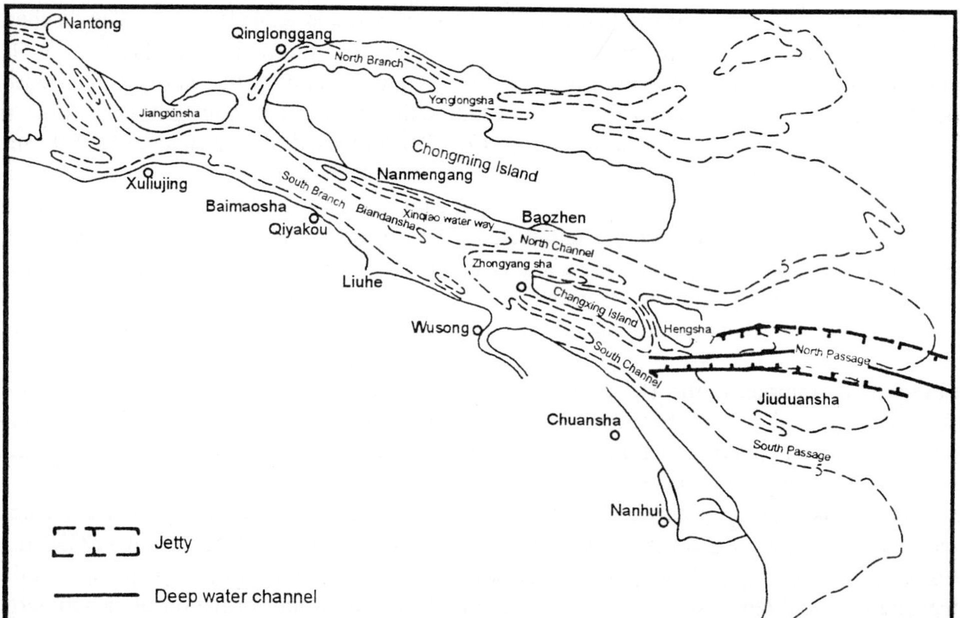

Fig. 2. Location map of the Changjiang, its estuary and its deepwater navigation channel.

China's largest city and has its largest port. Thus the Changjiang is a major economic lifeline in China. Any changes in the Changjiang estuary will directly affect the country's economy and oceanic and river transport. Important factors are river bank erosion, river channel switching and navigation channel deposition. In addition, storm surges threaten lives and property. How to prevent floods and stabilize and deepen the navigation channel are among the most important problems facing China.

China has a long history of and experience in combating sea hazards. But a long course and much work lies ahead in estuary regulation between defense and regulation. In this paper we discuss the regulation of the Changjiang estuary: its past (before 1950), its present (1950 to 2000) and its future.

2. A Brief Review of the Regulation of Changjiang Estuary

Throughout history, the regulation of Changjiang estuary was devoted simply against floods and storm surges. A few regulation projects were attempted during the last part of the 19th century and the first half of the 20th century.

In the past, many disasters occurred because of storm surges, with great life loss and property damage. The earliest such disaster recorded was that in Nanjing City in 251 A. D., during which several hundred people either died or were lost(1). However, the most severe disaster was in Chongming and Jiading in 1696, during which about 100,000 people were killed by seawater.

In the early years, before systematic embankments were constructed, 'cheniers' marked the coastline at the front of Changjiang delta (Fig. 3). During these periods 'mounds' were built in order to save lives during storm surges(2), a practice used along other coasts in the world. On the islands in the Changjiang estuary, mounds were still in use in the 18th century. For example, around the year of 1730, several mounds with heights of 6-7 m and areas of 140 m^2 were built on the sand islands in the lower reaches downstream of Jiangyin. And in 1733, 42 mounds were built on Chongming Island. Nothing could exist on those islands without mounds when storm surges came as occurred on Hensha Island in 1905.

One soil dike, with a length of more than 50 km, was built by Wu between 1052-1054 from Wusong to Jinshan in the Changjiang estuary(3). Soon after, a dike was built in Tongzhou to prevent the collapse of the riverbank. Since the extension of the coastline, new sea dikes were built. But they were often damaged by storm surges. The dikes in Changjiang estuary were repaired many times. By the first half of century, three layers of dikes had been built along the south bank of the estuary. Namely, (1) the Qin Gong dike, completely repaired by Qin Lian in 1733 in Qing Dynasty; (2) the Wang Gong dike, built in 1883; and (3) the Li Gong dike, built in 1906. It was used as a defensive barrier until the middle of the 20th century (Fig. 3).

The dikes in Changjiang estuary are different from those in Qiantangjiang estuary, which were built with large stones. Framed with stone pickets, the dikes in Changjiang estuary were built with soil and small rock as the main ingredients. This construction method was used by Yi Jishan first in 1730 (Fig. 4).

'To protect the shoals in order to protect the bank', was a principle known to the dike engineers in ancient China. The earliest use of the T-shaped groin to defend against the tidal wave was at the Shishun shoal of Nanhui, recorded in the history of Shanghai City in the 15th century. Some 330 m long T-shaped groins were built in the 18th century. In addition to groins, there also were some longitudinal dikes, which were constructed farther from tidal flats (Fig. 5).

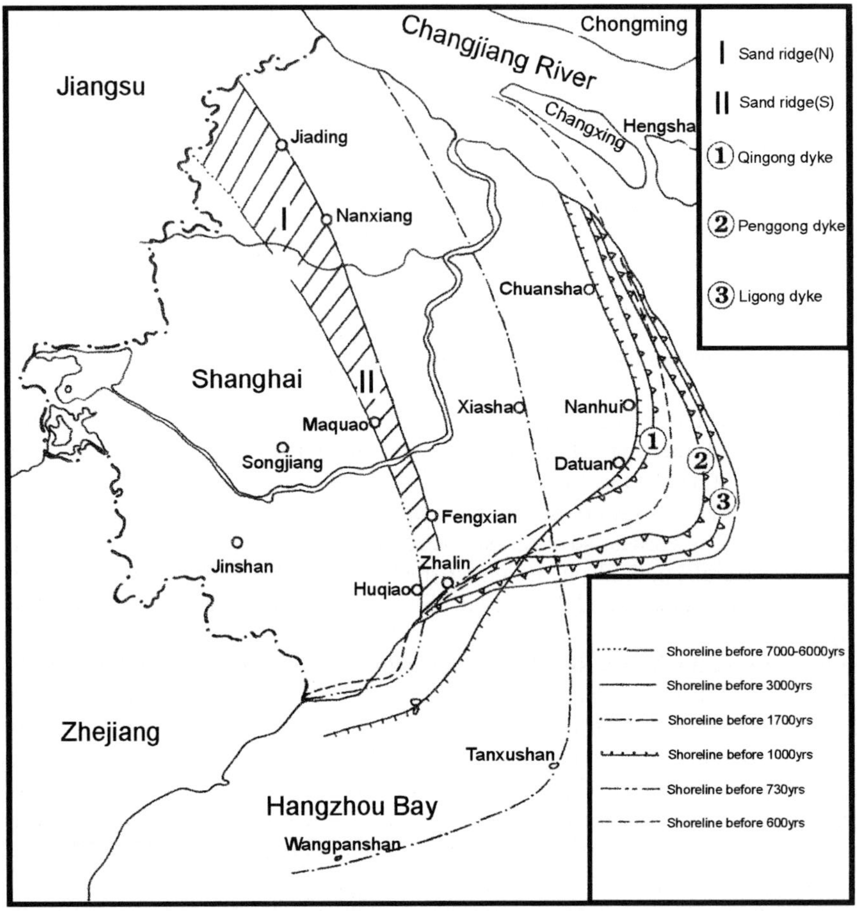

Fig. 3. Land-forming processes and main historical sea walls in Shanghai area.

In the middle of the 19th century, a sub-water topography chart of the Changjiang estuary was plotted; the earliest was drawn in 1842. These charts were used to analyze erosion and deposition in the river channel. Then some essential measures were conducted in the dangerous reaches. For example, 18 wooden structures were built in the river bend near Nantong during 1916 to 1926, which allayed the erosion of the riverbank. At the end of 19th century the Changjiang divided into three branches at Duansha section, the North Branch was the main discharge channel and the South Branch was the low discharge channel. In 1917 the Middle Branch was blocked first, then in 1922 it was possible to cultivate some of the sand fields because of the channel blocking projects.

In the middle of the 19th century, Shanghai harbor was formed. There were more and more ships navigating the Changjiang estuary to the Huangpu River. In 1906 the navigation channel of Huangpu River was regulated. A 1395 m long dam was built along the western side of the Wusong mouth. Another dam was built along the eastside, which changed the entrance of Wusong to the Changjiang into an arc. This resulted in an increase of water depth from 2.4 m to 5.8 m. It was a successful work in the regulations of estuaries[(4)] (Fig.6).

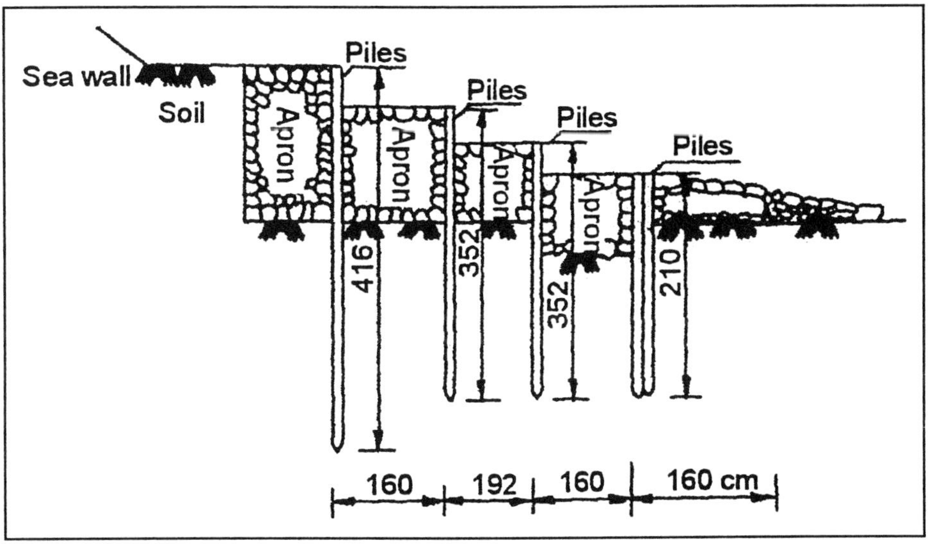

Fig. 4. The structural diagram of the protective dike in the Qing Dynasty.

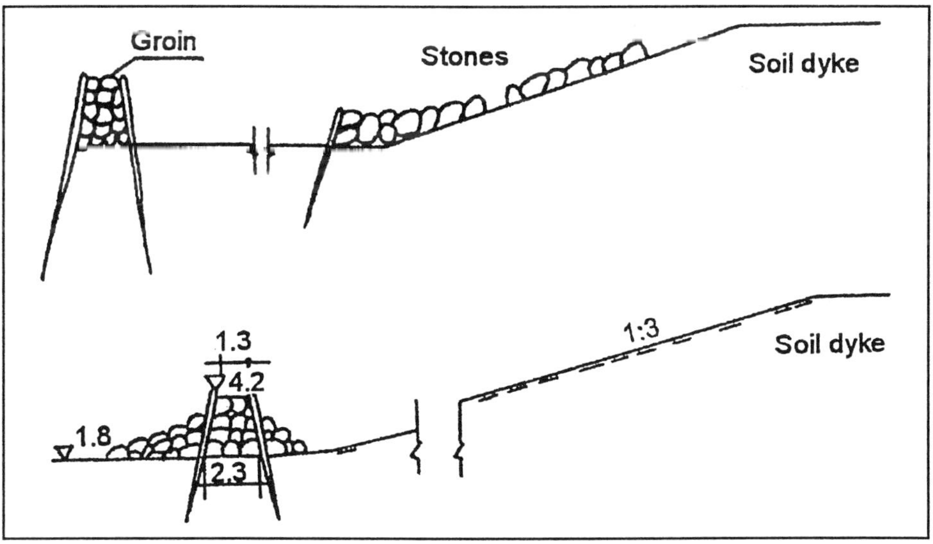

Fig. 5. Barrages: A. with double piles, B. protective revetments.

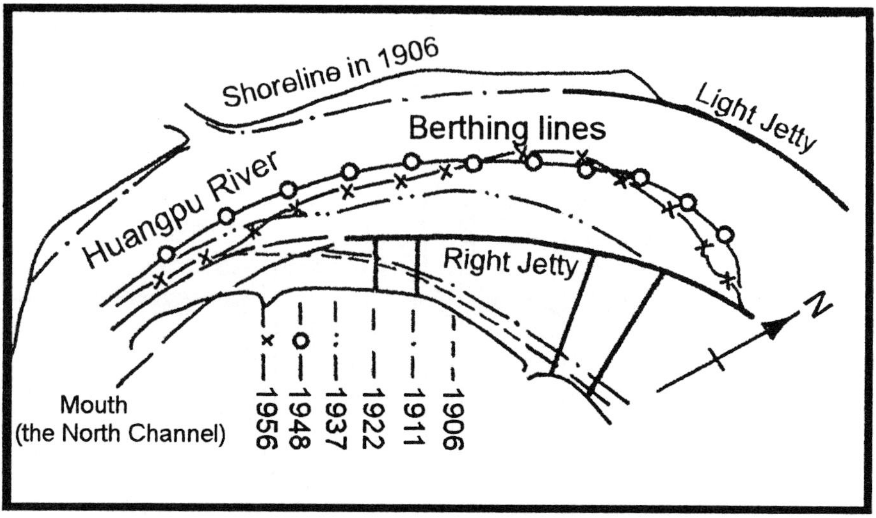

Fig. 6. The regulation engineering diagram of Huangpu River.

As scientific progress and the financial benefits increased, toward the end of the 19th century and early in the 20th century, estuary regulation was remarkably developed. A concrete dike was first built at Baoshan in 1900. Raft sinking technology was applied at Nantong in 1931. In 1882, dredgers were used at the mouth of Huangpu River. In 1935, the dredging tests for the sandbar were begun, but had to be stopped later because of the war. Hence, the Changjiang estuary was still in a state of natural evolution in 1945.

3. The Regulation of the Changjiang Estuary Since 1950

After World War II, the navigational channel in Changjiang estuary was not suitable for an ever increasing size of ships, because of the existence of a sand bar with a depth of 5.5-6.0 m. On the other hand, after the great flood in 1954, sediment eroded from shoals and was deposited in the channel. In 1959, the Shitou sand and the head of the Ruifeng sand were washed out by the new Chongming waterway. Meanwhile a large amount of sediment was deposited on the riverbed. Then the Zhongyan sand waterway became shallower and navigation difficult. Hence the government became concerned about the regulation of the Changjiang estuary.

In 1959, the Department of Communication organized a research group for the regulation of the Changjiang estuary. Systematic research work was done to understand water and sediment movement, evolution of the river bed, changes of erosion and deposition, and coagulation during the mixing of river water and seawater. This research was done with numerical models, physical models and field tests.

After being studied for 40 years, a development model for the Changjiang estuary has been completed. The south shoal continuously developed, and the sand islands moved toward the north bank. Then the channel formed. The riverbed became deeper and its mouth narrower. Finally the delta extended seaward[5]. The Changjiang estuary has grown from its original funnel-shape to its present multi-grade bifurcating-shape. Meanwhile its width narrowed from 180 km (Qidong mouth) to 90 km (Nanhui mouth).

The modern processes of river channels, based on the relation of the sediment transport capacity and sediment concentration, determine the plain part of the Changjiang which belongs to the sandbar type and extends from Yichang to Baimao shoal in the Changjiang estuary except for the Jingjiang meander. The Xuliujing reach is an artificial node formed between the 1950s and 1970s, which is a fine control to the modern processes of the Changjiang estuary. Now the North Channel of the Changjiang estuary is declining, because its volume was reduced by 8.3 $x10^6$ m^3 from 1958 to 1982. In the South Channel, there are several underwater shoals, including those of the Biandan, Liuhe and Zhongyang. If one of them moves, all of the others also shift. All these sandbars are unstable factors in the incessant changes of the mouths of the North and the South Channels. Their changes influence the stability of the North and South Channels directly.

At the beginning of 1960s, the Zhongyang shoal moved down rapidly and resulted in hundreds of millions of tons of sand being moved into the South Channel. This made the South Channel vary from a single river channel in the 1950s to duplicate channels. From the end of the 1970s to the early 1980s, because of the change in the bifurcation at the mouth between the North and the South Channel, the Xinkaihe spit moved and caused the water-depth at the Baozheng reach to deteriorate. Although the water-depth of the sandbar is comparatively stable, the navigational channel to the sea varied with the water and sand supply. In the 1970s, a great amount of sand in the North Channel moved along the east flat of Hengsha Island and silted the lower reach of the North Passage. From the end of the 1970s to the beginning of the 1980s, the Jiuduansha sandbar moved southward and menaced the navigational channel of the South Passage. The navigational channel moved to the North Passage of the South Channel at that time. Among all the waterways of the Changjiang estuary, the North Passage has the shortest history, only 40 years ago its –5 m bathymetric line ran all the way through. Presently, the situation of the North Passage is good. Its water-depth is maintained at –7 m by artificial dredging. Therefore, adding the 3 m tidal range, ten thousand ships can pass in and out smoothly.

The Changjiang estuary is mostly in a natural state and is self-fitting and self-controlling, but the influence of artificial activities has shown that the artificial control has exerted its effect to some extent under certain conditions, especially regulation plays an increasing role in the Changjiang. From 1958 to 1972, the regulation of Tonghaisha and Jiangxinsha sandbar narrowed the Xuliujing reach from 13 km to 4.5 km and 5.7 km. This caused the main flow to pass through the South Channel, controlling the bifurcation mouth of the North-South Branch. Because river discharge was reduced in the North Branch, it became silted and narrowed. Another noticeable artificial control concerns the estuarine sandbars, which always swing. They became stable after a number of jetties and other regulation controls were carried out after the 1950s to prevent collapse of the riverbank(6). In the last 40 years, there were 341 groins constructed on the three islands, i.e. Chongming, Changxing and Hengsha. It is very important to control sandbar growth in the stabilization of the river channels(7).

Based on the aforementioned studies on the self-fitting and self-controlling rules of the Changjiang estuary and the artificial control experiences of riverbanks, we have developed the following regulation principles in the case of the Changjiang estuary:

1) At present, the Changjiang estuary is still in a narrowing mode. The narrowing is a natural rule that applies to the development of an estuary. This is a basic point in the regulation of an estuary.
2) Controlling the swing of sandbars is an important measure in controlling a river channel. For the sandbars in the flat of an estuary, it shall be taken as the principle in the regulation of the emerged sand, the stabilization of shoals and the reduction of deposition, so as to make the river channel stable.

3) Controlling bifurcation points is an important measure in stabilizing the river. One of the Chinese proverbs is: "to pull a bull you must pull it by the nose." The 'bull nose' idea is the key to a reach in the river. The Changjiang estuary has three 'bull noses'. These 'bull noses' are the three bifurcation mouths: the bifurcation mouth of the North-South Branch, the bifurcation mouth of North-South Channel, and the bifurcation mouth of the North-South Passage.
4) "Integrative regulation at the navigational channel head " is an important principle in the regulation of Changjiang estuary. The resource exploitation of Changjiang estuary is many-sided and includes land, traffic, water, and plants and animals. These resources must be integratively exploited to bring their potential into play. However, according to the economic development in the drainage area, the construction of a port city, and the three centers of economy, finance and trade, the chief task is to develop waterway traffic. Therefore, the navigational channel must precede others in the regulation of the estuary.
5) The management of the estuary needs the combination of regulating, dredging and dike construction. Keeping the navigational channel in balance with sandbars, dredging will be needed to increase the depth of the channel and break the natural balance. However, sediment will again be deposited and eventually resume the channel's balance. Therefore, it is difficult to maintain the navigational water-depth only by dredging in the Changjiang estuary. The only way to maintain an effective channel is to combine regulation and dredging. In such a large estuarine regulation project, it will be necessary to construct and maintain large earthworks. If these earthworks are used to produce land, the city area will increase on one hand, and the sources of sand that shift back to the navigational channel will be reduced on the other hand. Therefore, one of the important principles in the regulation of the Changjiang estuary is combining regulation, dredging and dike construction.

Owing to the rapid development of the container transportation in the Shanghai Port, the request to increase the depth of navigational channel becomes more and more urgent. In 1992, based on the research works of the past 35 years, a study on "the evolution of the navigational channel and the regulation plan of the deep-water channel" was conducted. In 1993, the plan was put forward to deepen the South Channel–North Passage of the Changjiang estuary to -12.5 m[8] (Fig.1). On the aforementioned base, several studies were carried out, including the feasibility of the project, the forecast of the river's freight capacity, the forecast of ship sizes, the estimate of investment, and the comparison of plans. In December 1997, the government audited the scale of the construction, the regulation of buildings, the engineering of the navigational channel, the protection of the environment, the period of the construction and the general budget, among others. On the January 27, 1998, the Navigational Channel Construction Ltd. Co. was established and began work.

The engineering plan contains the following five parts:

1) Constructing a North Jetty 49.2 km long.
2) Constructing a South Jetty 48.0 km long.
3) Constructing a bifurcation jetty 1.6 km long, and a submerged dam 3.2 km long.
4) Constructing 22 T-shape groins in order to narrow the water between the South and North Jetties.
5) Dredging 75 km of the navigational channel.

This project was divided into three stages:

1) Dredge to a water-depth of 8.5 m, build the North Jetty 16.5 km, construct three T-shape dams with a total length of 5.41 km, build the South Jetty 20 km, build six T-shape groins with a total length 5.44 km, build a bifurcation mouth of 4.8 km, and

dredge 42.5 km. The time limited for this stage is 3 years.

2) Increase the water-depth to 10 m, extend the North Jetty by 32.66 km, build seven T-shape dams with a total length of 2.45 km, extend the South Jetty by 28.08 km, build six T-shape dams with total length of 9.58 km, and dredge for 57 km. The time designated is 3 years.

3) Dredge for 75 km, supplement and adjust the jetties and T-shape dams. The time planned is 4 years.

In sum, the project is to take 10 years.

The first stage utilizes large engineering fabric interlocked with concrete blocks and the structure of sand-supported soft rows to protect the bottom; the North Jetty uses the optimized structure of the slope on the interior of which are fabric bags filled with sand and held down by many stones. The bank of the South Jetty utilizes the new type of the semicircular combined forms. The process utilizes many innovations in construction. The first stage proceeded smoothly and was finished by August 2000.

During the last 50 years the application of modern science, scientific programming, using advanced designs and effective construction techniques allowed the regulation project of the Changjiang estuary to proceed successfully.

4. Prospects for Changjiang Estuarine Regulation

As mentioned above, the first stage work on the deepwater channel was completed in the last year of the 20th century. It already has improved the condition of the navigation channel for Shanghai harbor. The works during the second and the third stages will be completed in the first ten years of the 21st century. By that time the water depth of the navigation channel will reach -10 meters, which will allow third and fourth generation container ships to go through the Changjiang estuary every day and will play an important role in Shanghai's harbor becoming a hub for container ships. This engineering is a very important step in the regulation of the Changjiang estuary in reaching the central aim of navigation channel construction. But, some problems need to be resolved in order to make navigation water depth meet the demand of fifth and sixth generation container ships. In the new century, besides making navigation channel construction a central aim, the integrative regulation of the Changjiang estuary is needed for comprehensive development. We must implement the philosophical thought of 'combining nature and humans' in the integrative regulation between artificial control and natural processes of the Changjiang estuary[9].

Artificial estuaries made from medium and small estuaries in the world have reached this regulation aim. But in China, making the Qiantang estuary into an example of artificial estuary has only been pursued during the last 30 or 40 years. Under the guide of 'combining reclamation and estuarine regulation', there were 870 km^2 of land reclaimed from the Qiantang estuary during this period so that the width of Yanguan river section was narrowed from 12 km to 2.3 km. Both banks of the river were controlled. The extreme unstable events of riverbanks and tidal flats, caused by the tidal bore, were eliminated. The same attempt for artificial regulation also was carried on to form a node point of Xuliujing in the Changjiang estuary. Therefore, it is possible to convert the Changjiang estuary into an artificial one.

Certainly, the three 'bull noses' are the keys to the regulation of the Changjiang estuary. The two bull noses of both North-South Branches and the North-South Passages were completed basically in the last part of the 20th century. The last bifurcation is of the North-South Channels and should be regulated in the 21st century.

Reclamation is an important measurement in making an artificial estuary. The

Changjiang estuary had 750 km^2 reclaimed during the last half 20th century (Fig. 7). Some of the reclaimed land became the node point of the river, which narrowed the river course by 4.15-4.5 km and sped up the flow of water. Chongming Island has had its area expanded since 1951 from 602 km^2 to 1400 km^2. Changxing Island was formed from the combination of six shallow sand islands. Reclamation also expanded the tidal flat area at Nanhui by 4 km eastward. From the reclamation, the land shortage in Shanghai is relaxed, but at the same time it also has an impact on the wetland ecosystem.

Shanghai is the dragonhead in the economic cycle of Changjiang delta and the cities in Changjiang river basin. Shanghai will become one of the international centers of economy, finance and trade, one of the international centers of navigation and one of the international great metropolises. So when planning the regulation of Changjiang estuary, we must consider the large amount of land and water demands for Shanghai's development in the 21st century, on one hand. On the other hand, we must also consider the severe challenge to population, resources and the environment from the standpoint of sustainable development in Shanghai, as well as the whole of China. Therefore, the implementation of this regulation planning must depend on reclamation and artificial estuarine construction. In the meantime, by implementing the regulation and dredging, the artificial navigation channel will meet the demand of economic development. Further, a high quality environment can be achieved through development and protection. The population in Shanghai will be 15 million and it is estimated that Shanghai will need an additional 70,000 to 100,000 ha of land by 2025. Therefore, the solution of the large amount of land requirement should be combined with estuarine regulation. We think that it is possible in the new century to reclaim 70,000 to 100,000 ha of land from the Changjiang estuary. The reasons are as follows: 1) the tidal flat area above 0 m is 90,000 to 100,000 ha, 2) sediment from the river (the sediment load is about 500 million tons) adds 150 to 300 ha/a year, 3) the sandbar is at the same location as the turbidity maximum zone in the Changjiang estuary and catches high amounts of sediment, and 4) in addition to the sediment in the submerged delta that comes from the river basin, the submerged delta itself is also a sediment source for the Changjiang estuary, because sediment is resuspended under the action of waves and currents and transported to the estuary by tidal currents. Thus, there is a sediment exchange between the inner part and the outer part of the Changjiang estuary, and it provides sediment for tidal flat reclamation.

The very problems pointed out here are in connection with the lower tidal flat reclamation. During the past 50 years reclamation mainly focused on the higher tidal flats where 750 km^2 were reclaimed. From now on, the reclamation in the lower tidal flats becomes a new tendency. But, engineering will be necessary and will include both biological siltation and engineering siltation. For example, recently we reclaimed 700 ha of tidal flats for Pudong International Airport construction. After the cofferdams were made, it only took three years for the tidal flats to rise up 3 m. The siltation engineering for the artificial peninsula construction of east tidal flat of Nanhui also took three years to produce 1300 ha of land. As for the biological siltation, we have carried out ecological engineering on Jiuduansha Island located between North and South Passages, by planting *Spartina alterniflora* and *Phragmites communis.* After one year, the grass wetland raised by 30 cm. Due to the obvious progress made by the application of new materials and techniques for reclamation in the last 20 years, we have been able to make cofferdams in the lower tidal flats.

The Shanghai Survey Design Research Institute has worked out an integral plan for the regulation of the Changjiang estuary(10), which is as follows: 1) build north and south jetties at Xuliujing to enhance the artificial node point to control bifurcation between North and South Branches; 2) build a dike to control the oscillation of Baimaosha sand bar to keep reasonable bifurcation between north and south waterways of Baimaosha sand bar; 3) build dikes at

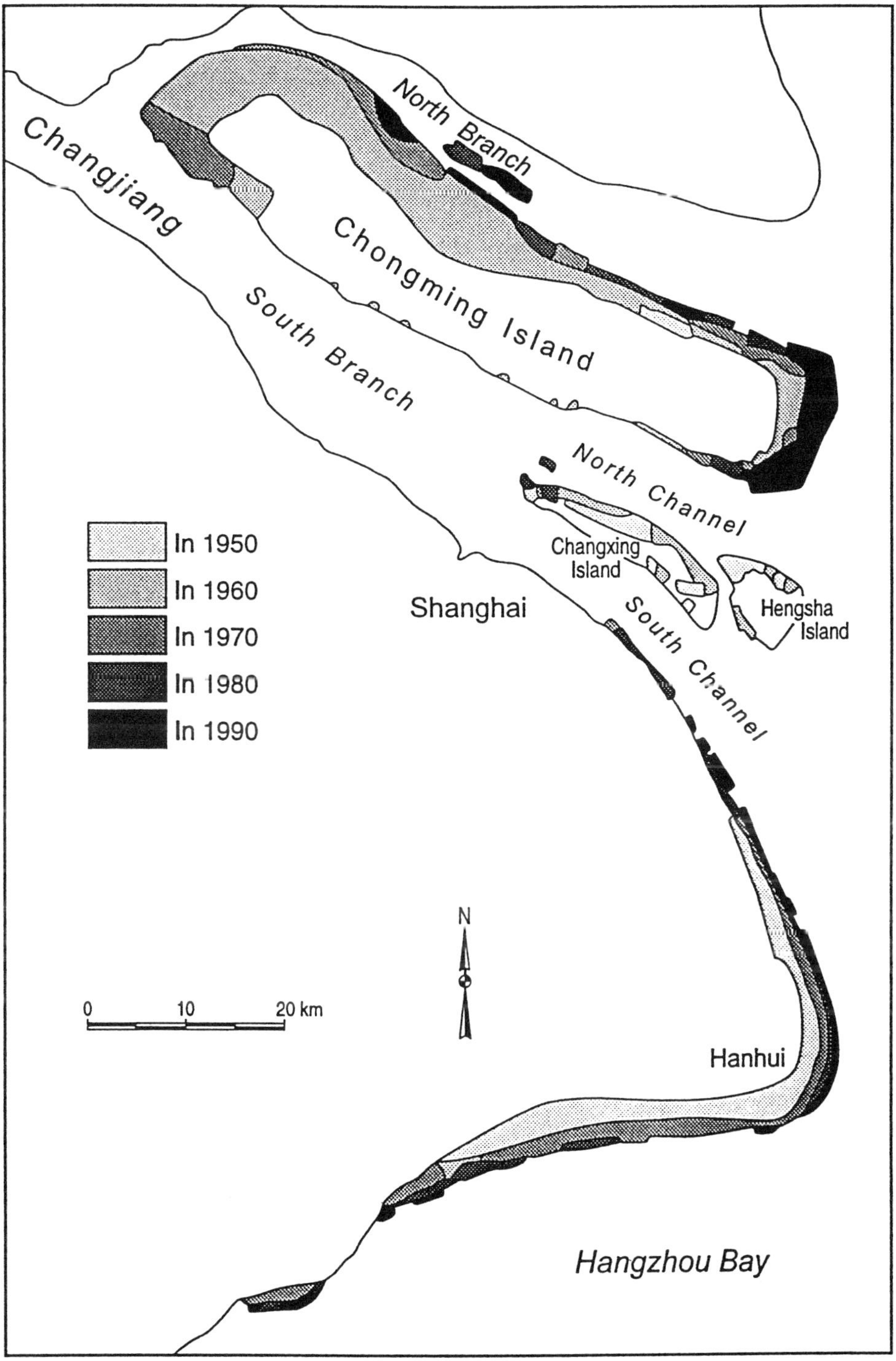

Fig. 7. The diagram of reclamation in the Changjiang estuary.

Yujiaosha sand bar and Huangguasha sand bar (Xinglongsha) of the North Branch to control tidal flood volume of the North Branch and to reclaim 3,000 ha of land and make an 85×10^6 m^3 reservoir; and 4) reclaim east and west Dongfengsha sand bars in the upper part of Biandansha sand bar, keep the waterway of Gelonggang open to connect the South Branch, and keep the Xinqiao waterway open to maintain the Xingqiao waterway.

As mentioned above, the bifurcation engineering between the North and the South Passages, which will distribute the ebb tidal flux reasonably by a bifurcation jetty (in Chinese "fish mouth") and submerged dike, is being carried out by the Ltd. Company of the Changjiang Estuarine Deepwater Navigation Channel Construction, the Ministry of Transportation. But, the bifurcation of the North and South Channels, the second 'bull nose', needs to be planned further. In 1991, the first author suggested that the Qingchaosha sand bar reservoir be built to provide high quality water to Shanghai. Recently, following this suggestion, concerned departments have completed a preliminary feasibility study. But, the reservoir planned is on the higher tidal flat, which will restrict its storage capacity. In addition, the impact of saltwater intrusion because of the deepwater navigation channel and the water adjustment project from the south to the north needs to be demonstrated. For this reason, studies are being done as to where the reservoir in mid river should be placed. It seems to be around the bifurcation of the South and North Channels. Therefore, the combination between the bifurcation and the reservoir engineering will save a great amount of investment, and will easily resolve the problem of the third 'bull nose'. In addition, the Qingchaosha reservoir with 30 km^2 of land planned originally will be available for reclamation.

These main-engineering constructions will form a basic framework for making the Changjiang estuary into an artificial estuary. This plan will be realized during the first 50 years of the 21st century.

One of the problems in this plan is how to deal with the wetland ecosystem? But the concerned departments in Shanghai are also thinking about that and have taken a number of dynamic measures to protect the wetland ecosystem effectively. On the reclamation and the wetland ecosystem in Shanghai, on the one hand, the green tidal flats expand from lower tidal flat to higher tidal flats by cofferdams enhancing siltation, which can maintain the ecosystem for a long time. On the other hand, there is an abundance of sediment in the Changjiang estuary and the reclamation of tidal flats play a role of some degree in the tidal flat siltation. Cofferdams will change tidal flat profiles and the sediment deposition will adjust to new profiles and form a new green tidal flat. Therefore, exploitation when protecting and protection when exploiting reflect concretely the coordination between reclamation development and ecosystem protection in the Changjiang estuary.

5. References

1. Wu, T.J. 1949. *Water Reservation Books of Jiangsu Province*. Nanjing Water Reservation Engineering Department, p.200. (in Chinese)
2. Yang, J. The Southern Song Dynasty. "Wujun Tujing" from "Yunjian Zhi" ed. By Zhu B.Y. p. 78. (in Chinese)
3. Zhang, X.G. 1998. *Several key points on land forming processes in Shanghai*. Historical Geography Serials 14. (in Chinese)
4. Yan, K. 1992. *Coastal Engineering of China*. Hehai University Press. (in Chinese)
5. Chen, J.Y. 1979. Development model of the Changjiang estuary. *Acta Oceanologica Sinica* No.1. (in Chinese)

6. Jin, Z.X., Su, D.Y. and Zhou, J.D. 1997. *The investigation of coastal resources utilization of Shanhai. Shanghai Water Reservation* 1997, No.2 (in Chinese)
7. *Shanghai Water Reservation Annals*. Shanghai Social Scientific Press, 1998, Shanghai (in Chinese)
8. Luo, J.Z. 1997. *Introduction to the regulation of deep water navigation channel in the Changjiang estuary*. Shanghai Water Reservation 1997, No.2 (in Chinese)
9. Chen, J.Y. 1995. The natural adaptation and artificial control of the Changjiang estuary. Selections of the studies of thc turbidity maximum and estuarine fronts in the Changjiang estuary. *Journal of East China Normal University* 1995. (in Chinese)
10. Shanghai Survey and Design Institute, 1997. *Report of comprehensive regulation planning of the Changjiang estuary* (in Chinese)

RECLAMATION AND REGULATION IN THE PEARL RIVER DELTA, CHINA

LEI YAPING
WU CHAO-YU
Institute of Coastal and Estuarine Studies
Zhongshan University, Guangzhou
PR China 510275

CHEN JIYU
State Key Laboratory of Estuarine & Coastal Research
Institute of Estuarine & Coastal Research
East China Normal University
Shanghai, China 200062

1. Development and Evolution of the Pearl River Delta

1.1 DELTAIC DEVELOPMENT PROCESSES

Quaternary sea level changes have played a key role in the development and evolution of the Pearl River delta. About 10,000 years ago, sea level was 100 m lower than today. There were deeply incised river valleys in the modern Pearl River delta area at that time. They formed the delta base topography and large amounts of coarse riverine sediment were deposited on top of that base. Approximately 6000 years ago the Holocene transgression reached its maximum and sea level approached today's level. Upstream deposition in the backwater reaches of the river valleys ended at that time; these deposits are widely distributed beneath modern network channels. The topography was more or less ria in form, consisted of shallow bays with a weak hydrodynamic sedimentary environment. Many rocky islands were scattered in the ancient drowned estuary[1,2]. During that period, the most northward deltaic coastline was located along a line linking Daojiao, Huangpu, Shiqiao, Chencun, Shunde, Jiangmen and Shafu[3].

The position of the mouths or the estuarine boundaries of the main input rivers—the West, North, East, Liuxi and Tan Rivers—gradually changed and the deltaic coastline expanded seaward from the head of the ancient bay. The area of the delta was small and there were large areas of wetland and salt marsh on the deltaic plain. It was the aggradation phase of the Pearl River delta and marked a long transition period during which deposition changed from retro-gradation to progradation and the predominant dynamics changed from oceanic to fluvial[4,5]. Riverine silt and clay carried by the flood waters formed the alluvial deltaic plain. The altitude of the plain was higher than the water level so that it could be directly enclosed by dikes for cultivation and to prevent flooding.

J. Chen et al. (eds.), Engineered Coasts, 199–228.
© 2002 *Kluwer Academic Publishers. Printed in the Netherlands.*

Thus dike-enclosed fields in the upper deltaic plain were formed. Around 2500 B.P. the ecological balance of the drainage basin began to be destroyed by human activities and the sediment discharge of the Pearl River increased, causing large amounts of sediment to be transported out of the river mouths and deposited in the estuaries and along the coast. The ancient drowned Pearl River estuarine bay was reduced in size, while sand bodies were formed and merged with each other, becoming broader as well as shallower.

The distributary river mouths and their estuarine boundaries moved quickly seaward. The newly formed sand flats and shallow water areas in the lower deltaic plain could not be used for dike-enclosed fields but had to be reclaimed from the sea (and thus are designated 'reclaimed fields')[6]. Therefore, there are two distinct phases in the development of the Pearl River delta: one is the earlier slow transition period with aggradation at the bay head, the other is the quick progradation phase after 2500 B.P. The progression of the deltaic coastline is shown in Figure 1.

1.2 THE GENERAL CHARACTERISTICS OF THE DELTA

The sediment of the Pearl River delta is mainly derived from five input rivers : the West, North, East, Liuxe and Tan Rivers. The seaward moving rate of the distributary river mouths and the estuarine boundary differs with the spatial imbalance of Land-Ocean interaction. Scattered islands act as deposition centers for the supplied sediment. This causes floodwater diversion and the generation of branches along the major channels. As a result, as the delta progressed seaward, the ancient Pearl River bay was reduced in size because of the presence of many deposition centers, each with its own expansion rate.

This complex delta system and its spatial imbalance are a distinctive feature of the Pearl River delta, which makes it a special case in the world (Fig.1). The present complex Pearl River delta includes the West and North River delta, the East River delta and the lower reaches of the Liuxi and Tan River valleys. The West and North River delta is the triangular area between the Sixianjiao waterway and the Human and Yamen tidal gates, which form the central part, or main body of the Pearl River delta. The two tide-dominated estuarine bays, Lingdingyang and Huangmaohai, are located at the east and west sides respectively of the main body. The East River delta is the triangular area between the Shilong, Huangpu and Humen tidal gates, and is located to the east of the West and North River delta separated by the Shiziyang tidal channel.

The general geomorphological features of the Pearl River delta are :

1) A deltaic sedimentary base which is undulating, the Quaternary deposits being less than 40 m thick in general;

2) A plain which covers 80% of the total river network area—its elevation is about 0.9-1.7 m (Pearl River datum), rocky hills cover 13.3%;

3) More than one thousand islands distributed over the deltaic plain and the offshore zone, some of which are up to several hundred meters high (such as Huangyang and Wuguishan islands);

4) There are four or five grades of distributary channels along the major channels; they constitute a fine river network system;

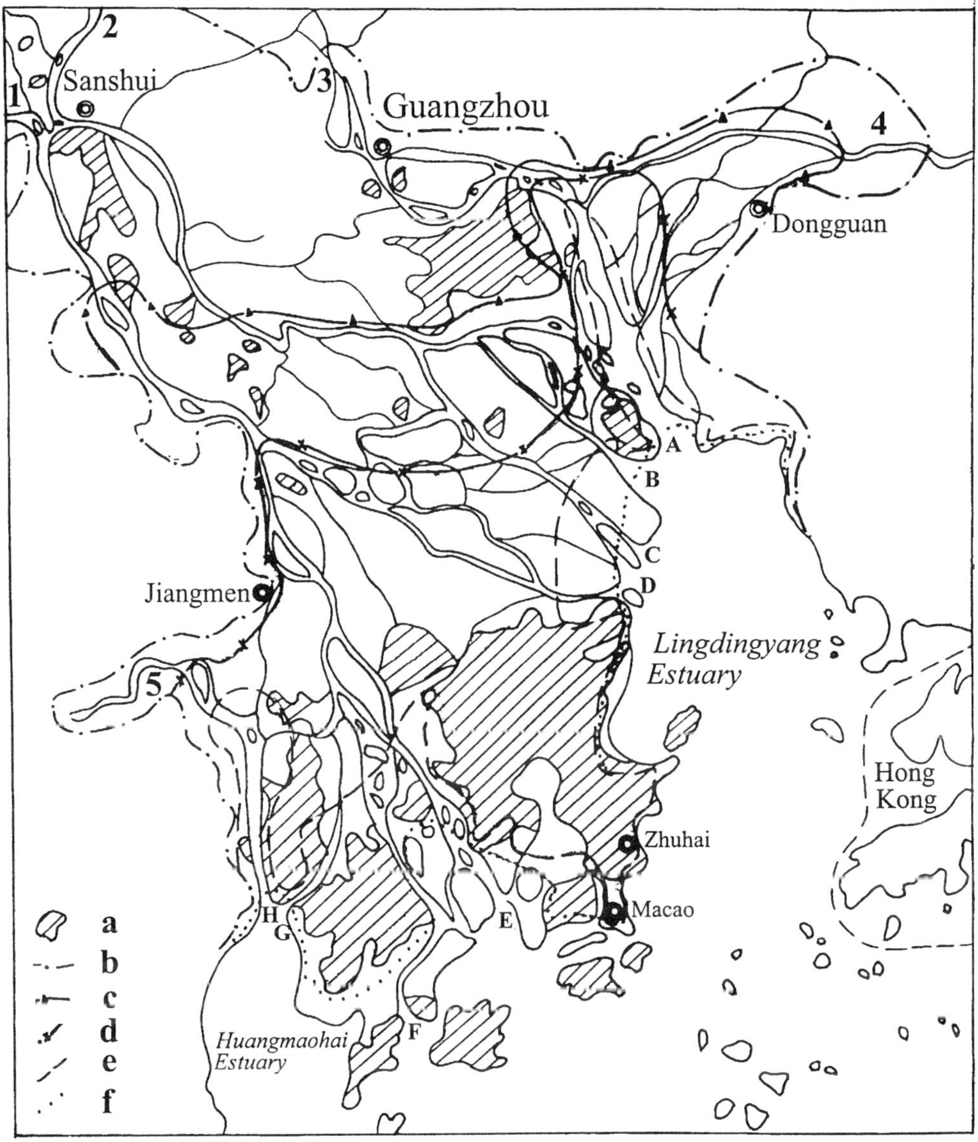

Fig. 1. Map of Pearl River development. The main rivers are: 1. West River; 2. North River; 3. Liuxi River; 4. East River; 5. Tan River. The eight outlets in the Pearl River delta are from east to west: A. Humen tidal gate; B. Jiaomen mouth; C. Hongqili mouth; D. Hengmen mouth; E. Modaomen mouth; F. Jitimen mouth; G. Hutiaomen mouth; H. Yamen tidal gate. Map symbols: a. hilly land; b-f represent the location of the delta front at different periods: b. ca 6000 BP; c. 6th century; d. 13th century; e. 17th century; f. 19th century.

5) The river discharges to the sea through eight outlets: Humen, Jiaomen, Hongqili, Hengmen, Modaomen, Jitimen, Jutiaomen and Yamen from east to west along the deltaic coast.

Because hills or islands are located along both sides of a distributary river (except at the Hongqili River mouth), the suffix 'men' is added, meaning 'gate' in English. Humen and Yamen are two tidal gates that are located at the heads of the Lingdingyang and Huangmaohai estuarine bays. The others are river-dominated distributary mouths (Table 1). The Pearl River delta is located in the south subtropic maritime monsoon climatic zone. Mean annual precipitation is 1600-2300 mm. Precipitation during the flood season (April to September) comprises 82-85% of the yearly total. The mean annual discharge into the network channels of the delta is 3.124×10^{11} m^3. The mean sediment concentration is 0.284 kg/m^3 and the mean annual sediment load is 8.872×10^7 tons.

Table 1. Discharge, sediment transport and tidal range of the eight outlets of the Pearl River delta[7].

Outlets (Measure station)	Mean annual discharge unit:$\times 10^8 m^3$ (Percentage)	Mean annual sediment transportation unit: $\times 10^4 t$ (Percentage)	Annual tidal range (Maximum tidal range)	Ratio between discharge and flood tidal prism
Humen tidal gate (Dahu)	603 (18.5)	658 (9.3)	1.63 (3.39)	0.51
Jiaomen River mouth (Nansha)	565 (17.3)	1289 (18.1)	1.36 (2.81)	3.32
Hongqili River mouth (WanQingshaxi)	209 (6.4)	517 (7.3)	1.21 (2.79)	4.12
Hengmen River mouth (Hengmen)	365 (11.2)	925 (13.0)	1.11 (2.48)	5.26
Modaomen River mouth (Denglongshan)	923 (28.3)	2341 (33.0)	0.86 (2.29)	11.5
Jitimen River mouth (Huangjin)	197 (6.1)	496 (7.0)	1.01 (2.71)	5.64
Hutiaomen River mouth (Xipaotai)	202 (6.2)	509 (7.2)	1.20 (2.66)	6.81
Yamen tidal gate (Huangchong)	196 (6.0)	363 (5.1)	1.24 (2.95)	0.59

Compared to other large rivers in China the Pearl River has a high discharge but not a high sediment concentration. It has five times the discharge of the Yellow River, while the sediment load is about 50% of the Yangtze load. The mean tidal range along the Pearl River coast is less than 2 m. The maximum is 1.69 m at Zhouzaiwei station inward of the Humen tidal gate. The minimum is 0.86 m at Denglongsha station near the Madaomen river mouth (Table 1). There is a southwestward littoral drift which veers to the southwest and is slightly associated with a coriolis effect.

1.3 EVOLUTIONARY TRENDS OF THE MODERN PEARL RIVER DELTA.

Modaomen is the largest outlet of water and sediment among the eight outlets of the modern delta. Because of a combined effect of reclamation and regulation the inner sea behind the Hengqin and Sanzao islands has nearly disappeared due to the large sediment input. The mean annual rate of the Modaomen mouth moving seaward is 110 m. Now there is only one major channel (the Henzhou channel) discharging water and sediment, while the Modaomen mouth has declined[8]. Consequently the fluvial component has been strengthened while the tidal component has decreased to some degree. Comparing the period before 1970, when the reclamation decreased, with the present, the mean annual tidal range at the Denglongshan station on the inner Modaomen mouth decreased from 0.860 m to 0.853 m., while the mean annual low tide level shifted from -0.45 m to -0.40 m. The mean tidal range at the Sanzao island station outside the river mouth decreased from 1.116 m to 1.097 m[6].

The Jiaomen, Hongqili and Hengmen distributaries flow into the Lingdingyang estuary. The total water and sediment discharge of these three river mouths amounts to 54 and 50% of the total input from the West and North rivers respectively. This considerably affects the Lingdingyang estuary: on the charts of 1861 to 1883, the area of shoals below the 0 m isobath in the Lingdingyang estuary was only 132 km^2. In 1936 it was up to 229 km^2 ; in 1953 to 1955 the area had increased to 317 km^2 and the shoals became shallower. In 1974-1975 the area was 379 km^2, which is nearly three times the area during 1861-1883[6]. The mean eastward displacement of the -5 m isobath was 200 m in this period. In recent years, because of the reclamation and regulation programme of the Lingdingyang and the outlets, the reclamation of the west shoal is proceeding quickly. The expansion southward and eastward of shallow water and subaqueous shoals is a continuous process.

The Liuxi and Tan Rivers are two small rivers that discharge into the Lingdingyang and Huangmaohai estuaries through the Humen and Yamen tidal gates respectively. The tidal currents are strong and the bed load is retained at the upper reaches of the tidal current limit. The somewhat coarser sediment is discharged and moves Humen and Yamen in a seaward direction. The Shiziyang and Yinzhouhu channels, the two tidal channels inward of the Humen and Yamen tidal gates, are the remnants of a former drowned bay[9]. The rate of seaward displacement of the Modaomen mouth is slowing down, because it is moving offshore into deeper water which is devoid of islands. In contrast, the seaward-moving rates of the distributary mouths that lie to the west of the Lingdingyang estuary and east of the Huangmaohai estuary are increasing.

Lingdingyang and Huangmaohai are very important estuarine harbours in southern China. The trend is that at both harbours only two long and narrow funnel-shaped tide-dominated channels will remain deep enough for sea going vessels. It is therefore necessary to regulate and reclaim the estuaries and distributary mouths of the Pearl River delta.

2. The Channel Network System of the Pearl River Delta

During the development of the delta, branches are regularly produced from the major river channels and a complex network system has resulted; four or five grades of distributaries can be distinguished (Fig.2). In total there are 324 channels in the Pearl River delta with a total length of about 1600 km. The most important waterways have a total length of 1200 km. The density of the waterway system is up to 0.81 km/km^2[8]. At the lower reaches of the Liuxihe and Tanjiang Rivers (including the Shiziyang and Yinzhouhu tidal channels) the tidal current is strong. There are many sinuous branch tidal channels and creeks on the plain and the tidal flats along both sides of the wide and straight main tidal channels, where the tides coming in through the Humen and Yamen tidal gates can be propagated. The West, North and East Rivers have developed many branches that together constitute a network channel system. The West and North River's network system of interconnected channels begins at the Sixianjiao waterway. The width of the longitudinal channels is about 400-1800 m and their curvature is between 1.03 and 1.26, while the width and curvature of the transverse waterways is about 10-300 m and between 1.26 and 1.46 respectively. The East River network system begins at the town of Shilong. There five major river channels flow southwestward into the Shiziyang tidal channel with a length of about 138 km and a width of about 300-500 m. The width of the transverse branches is about 10 m. The mean hydraulic gradient of the West and North River's network systems is about 0.023°/oo and 0.037°/oo respectively, that of the East River network system up to 0.26°/oo (Fig. 2).

The formation mechanisms of the river network systems are:

1) The main river channels and their orientation are controlled by the faults that were also followed by the rivers valleys of the former West, North, East and Tan Rivers before the delta was formed. The main river valleys form the skeleton of the Pearl River delta, not only in the past but also at present and are the main passages of river flow and sediment discharge. The secondary channels are newly formed branch channels that originated during the progressive development of the delta.

2) The rocky hills and islands act as deposit centers that are beneficial to the generation of river channel branches.

3) Essentially branch channels are generated by the hydrodynamic and sedimentary processes at the river mouth. The volume and scale of the river mouth bar is associated with the main floodwater discharge of the river. The bar is retarding the flood discharge flowing out into the sea. Consequently, in order to reach the sea along the largest hydraulic gradient, a short cut is made: the flood breaks through the river bank generating a new branch channel or speeding up the development of an existing branch channel.

4) Because of the large discharge, the West and North Rivers may be regarded as the 'high-pressure regions' of the river level[10]. Because the water level is near sea level, the Shiziyang and Yinzhouhu tidal channels are the 'low-pressure regions', situated along high-pressure regions to the east and the west. The flood hydraulic power of the high-pressure regions is 2 m higher than that of the low-pressure regions during the flood season. Also inside the high-pressure region the discharge of the West River is larger than the discharge of the North River and the flood hydraulic power is higher than that of the North River in general. For example the hydraulic pressure at the west entrance of the Ganzhuxi waterway, that connects the two rivers, is 1.5 m higher than at the east side (the exit). Therefore the major West River channel may be regarded as the 'hydraulic ridge' of the Pearl River delta as a whole.

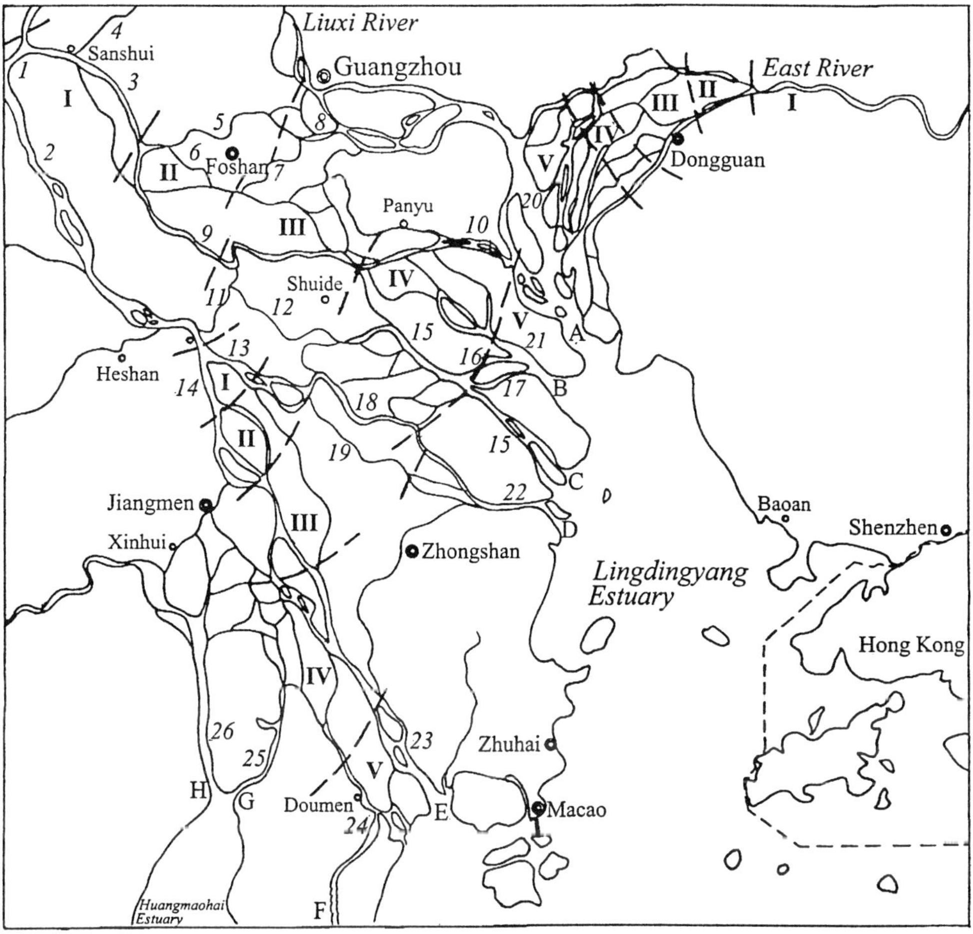

Fig. 2. The Pearl River delta channel system. A-H indicate the outlets as listed in Fig. 1. I-V indicate the grade number of the distributaries. The 26 channels and distributaries are: 1. Sixianjiao waterway; 2. main West River channel; 3. main North River channel; 4. Xinanchong branch; 5. Foshan waterway; 6. Tanzhou waterway; 7. Pingzhou waterway; 8. Chencun waterway; 9. Shuide waterway; 10. Shawan waterway; 11. Ganzhuxi waterway; 12. Shuide branch; 13. Donghai waterway; 14. Xihai waterway; 15. Hongqili channel; 16. Shanghengli waterway; 17. Xiahengli waterway; 18. Jiya waterway; 19. Xiaolan waterway; 20. Shiziyang tidal channel; 21. Jiaomen channel; 22. Hengmen channel; 23. Modaomen channel; 24. Jitimen channel; 25. Hutiaomen channel; 26. Yinzhouhu tidal channel.

5) The branches extend from this 'hydraulic ridge'[10] towards both sides with a large acute angle between its main channels. Because of the shortcut, once it exists, the hydrodynamics of the branches increase several times and they may become even stronger than the channel they are derived from.

3. Reclamation in the Delta

3.1 HISTORY OF RECLAMATION

The history of reclamation is closely related to the development of deltaic deposition and human culture in the Pearl River delta. By the end of the Holocene transgression approximately 6000 B.P., most parts of the deltaic plain were wetlands or flourishing forests. The aborigines lived by fishing and hunting. After a long period of aggradation at the head of the bay the fluvial influence became predominant after about 2500 B.P. By then the ancient city of Guangzhou was established and iron was made into work tools such as ploughs and spades. With the increase in sediment discharge of the West, North and East Rivers the deltaic plain was quickly expanding seawards which resulted in new land. Thus was the beginning of farming in the Pearl River delta.

The ancient drowned Pearl River estuary was very wide and large during the Sui dynasty (581-618 A.D.) and the Tang dynasty (618-907 A.D.) The advanced culture that had originated along the Yellow River and Yangtze River valleys had not yet been introduced in the Pearl River delta. During this period the farmers in this area used fields that were well above water level and high enough to cultivate safely without dikes to prevent flood disaster.

During the Song dynasty (960-1279 A.D.) and the Yuan dynasty (1279-1368 A.D.) large numbers of immigrants introduced advanced techniques of farming and water control into the Pearl River delta. In 996 A.D. the first dikes were constructed to defend homes and fields against flooding. At the beginning a dike was built along both banks of a river and opened downstream to let the flood waters out. The farmland behind the dikes was an open area, mainly in the upper delta plain. Because the demand of farmland increased in relation to the population, it became necessary to reclaim land from the subaqueous shoals and shallow water areas and dikes were built before the shoal had accumulated high enough to allow farming. In this way the lower and subaqueous deltaic plain, formed with riverine sediment since 2500 B.C., could be used. There were a number of gaps or sluices in the dikes for irrigation, restraining salt water and the passage of small boats. The total length of dikes that were built in these periods was 388 km, mainly at the west or north of the modern delta.

The expansion of the shoal area was fastest during the Ming dynasty (1368-1644 A.D.) and Qing dynasty (1644-1911 A.D.). This was the period with the maximum reclamation from river banks and the sea. The target for reclamation changed from the subaerial shoals to the subaqueous shoals and then to reclamation of shallow water areas. The length of the dikes constructed during this period adds up to 1121 km[11]. These dikes and enclosed farmland are mainly present at both sides of outlets (e.g. the Modaomen, Jitimen, Hengmen, Jiaomen River mouths and the Humen tidal gate) and along the coastal zone of the West and North delta, also including the south part of the East River delta and the lower reaches of the Tan River.

After the independence of the People's Republic of China in 1949, the government organized a large campaign of building sluices and merging dikes into large-scale units. According to the demands of standards of design, small areas were merged into large enclosed areas and earth dikes were made into stone dikes. The height and tightness of all dikes was improved and closed-circle dike systems were formed. As a result, more than 20,000 short and low dikes were merged into about 400 strengthened defense dikes along the river banks and the delta coast.

3.2 THE STEPS FOR BUILDING A DIKE FOR CULTIVATION.

The steps for building dikes in shoals or shallow water along the Pearl River delta coast are associated with the accumulation process of the subaqueous delta. The bottom layer of shoals in the Holocene delta is coarse-grained land facies sediment. The central layer is a marine facies deposit overlaid by mud or fine silt in the topmost layer. Because the weight-load capacity of the sediment is very low, reclamation of the subaqueous shoals and shallow water areas can not neglect this. At the beginning it is beneficial to accelerate the accumulation process by dumping loose stones around the backwater region of the site. The dikes should not be built before the subaqueous shoal or shallow water area has been turned into a subaerial shoal by deposition of fine-grained sediment. There are five steps to be taken based on observation of the local biological community evolution and the sedimentary processes on the shoal or the shallow water areas[12]:

1) Fish swimming stage. The water depth is about 1.5-2.0 m. Shoals of fish can be seen during ebb tide. If dikes are built during this stage, expenses are very high.

2) Stage of paddles touching the bed. The water depth is 1.0-1.5 m. It is the time to bring loose stones with small light boats and to build the base of the dike. During ebb tide boat paddles often touch the sea bed.

3) Crane standing stage. The shoal has become gradually shallower and the water depth is about 0.2-0.4 m. Part of the shoal is above the water during slack low tide. Waterfowl, e.g. cranes, often come looking for food. It is the appropriate time to plant herbage, e.g. reeds, in order to accelerate sedimentation and to increase the soil quality.

4) Grassy marsh stage. All of the shoal is above water with a large area of grassy marsh visible at slack low tide. Reclamation begins to be possible (Fig. 3).

These first four natural stages in delta growth are summed up in Chinese by “yii you, lu po, huo li, cao pu” which translates to “fish swim, rudders touch bottom, cranes stand, grass spreads”[13].

5) Stage of enclosed dikes. Two or three years after the grassy marsh stage, when the height of the shoal has increased, the groundwater level becomes lower. Now is the most appropriate time to close the dike to obtain farmland with a number of ditches and small gates for irrigation and outflow correlated with the rhythm of the tides.

3.3 DISTINCTIVE RIDGE-POND ECOLOGICAL AGRICULTURAL SYSTEM

The elevation of the enclosed farmland is near mean tide level. At the initial flood stage of the tides, with water level rising, the small sluices are opened leading the upper layer of fresh water into the ditches or creeks to irrigate the inner farmland. Then the sluices are closed to prevent excess irrigation and the influx of more saline water. During the ebb, the sluices are opened to drain off inner water when the water level of the channel outside is lower than the water level in the ditches and creeks inside the dike. The cycle of freshwater being let in for irrigation during the flood tide and drainage during the ebb tide is distinctive in China for the Pearl River deltaic plain. In the large-scale dike-enclosed areas there are many separate small areas bordered by small ditches and creeks, where light boats can pass for the transport of farm products, and which also can be developed for aquaculture.

Because of the abundant precipitation and the low elevation of the deltaic plain, there are many large low-lying farms within the enclosed areas. Farmers dig ponds for fish breeding in the moist low-lying land; the ponds are rectangular with water depths of 2-3 m (Fig. 4). The mud that is dug out is used to build ridges around the pond. The central part of the cross-section of the ridge is steeply inclined towards both sides. Mulberry, sugarcane and orchard trees are planted on the ridges and are called mulberry-ridge pond, sugarcane-ridge pond and orchard-ridge pond, respectively. The area ratio between ridge and pond is commonly 4:6, although some may be 3:7 or 6:4[12]. This kind of ecological agricultural system is suited to local conditions and not only gives economic and ecological benefits but also reduces waterlogging in the delta plain.

3.4 THE EFFECT OF RECLAMATION ON CHANNEL NETWORK SYSTEMS

Reclamation was developed in the Pearl River delta for two reasons. One is the solid discharge increase caused by human activity in the Pearl River basin so that the delta expanded quickly seaward and a large area of land became available. The other is economic growth and increase of population. This gave a large impulse to the demand for land and thus to reclamation as well as flood prevention. The long period of reclamation of the subaerial and subaqueous deltaic plain had the following positive effects:

1) It enlarged the area of the deltaic plain, accelerated the accumulation rate of the subaqueous shoals, and increased the rate at which distributary mouths and the deltaic coast are moving seaward.

2) The subaqueous shoals were merged through reclamation of the shallow water areas between them. Also the waterway network was simplified by dikes and sluices that blocked the distributaries and branch waterways.

3) The dikes concentrated the flow in the channels and restricted the flood overflow effectively, while the riverbanks became more stabilized. Such changes not only defend the land from flooding but also are beneficial to controlling salt water intrusion.

However, because reclamation in the past converted water to land with minimal planning and management—and at a too fast pace in localized areas—disadvantages of this type of land reclamation were accentuated:

1) The process of sediment-loaded flood water overflow was interrupted, so that no sediment could accumulate on the inner farmlands and the low-lying wetlands could not become higher. At present 2300 km^2 of the deltaic plain is lower than mean high water level and often suffers waterlogging[14].

Fig. 3. Acceleration of deposition in the grassy-marsh stage by planting marsh grasses such as Cyperus malaccensis.*

Fig. 4. Well established aquaculture/agriculture complex of the mulberry-ridge pond variety at Le Liu near Shuide in the upper delta.

*All photos by J. Walker

2) The high density and intensity of local human activities, e.g. building dikes, constructing sluices, blocking branch waterways and concentrating the main channel flow, reduces the flood retention area and the lateral areas of the main and branch channels. Therefore the natural capacity for holding flood water is decreased. Confined in the channels with dikes along both sides, the floodwater level increases continuously, resulting in a large potential for disaster. Furthermore the capability of the dikes to control the flood water is reduced, especially when floods occur during spring tide. In about 1960, the flood water (tidal) level in the lower reaches of the North and West Rivers was raised 0.3-0.6 m and 0.1-0.3 m respectively[6,15,16].

3) Reclamation of subaqueous shoals and shallow water areas speeds up the rates of seaward movement of the river mouths and increases the length of river flow. Therefore the outflow into the sea through the mouth is less effective because the mouth bar becomes shallower. Consequently, once a new distributary has been formed, it is scoured deeper as the flow is stronger than in the main channel[10].

4. Regulation Engineering in the Channel Network Region.

More than 75% of the total annual discharge of the rivers into the Pearl River estuary is during the flood season. The percentages of discharge during flood season of the West (Makou), North (Sanshui), East (Boluo) and Liuxihe (Niuxinling) Rivers are 77.7, 88.8, 74.3 and 84 respectively. Thus, the discharge during the low water season is only 11.2 to 25.7% of the total discharge[17]. Therefore, during the flood seasons, flood disasters are liable to happen because of a high peak discharge, a high flood water level, or a long period of flood. Parts of the deltaic plain outside the enclosing dikes are below flood water level. When heavy rains occur, areas within the dikes can suffer waterlogging because the rainwater can not drain out quickly by gravity, or may even be completely blocked. In addition there are yearly two or three tropical storms that have a direct impact on the Pearl River deltaic plain, and one violent typhoon hits there every two or three years. When a violent typhoon occurs in combination with a spring flood, the seawalls, dikes and other embankment installations risk breakage[8].

Because of the large effects of different regional natural conditions and long-term artificial changes, the channel evolution in the river network region is very complex. Fluvial forces and tidal energy determine the hydrodynamic features in the river network system. Because every waterway is connected with another, there are tidal currents at both ends downstream of the tidal limit. At one end there is ebb when at the other there is flood, or there may be flood at both ends. Thus a flood current convergence point may be formed in the waterway, which complicates the flow. At the distributing or converging site between the main channel and the branch channel, the river bank at one side is scoured and a deep, curved channel is formed by strong flow, whereas at the opposite bank a shoal is formed by lateral flow and is attached to the bank. Because the erosion resistance of the deltaic riverbanks is very weak, once a curved channel is formed, the curvature increases and the riverbank is in danger of collapsing, which threatens flood control and shipping.

The embankment projects and the regulation projects in the Pearl River network region have three purposes: flood control, dealing with waterlogging, and meeting the demands of shipping in the waterways. The measures to prevent flooding are not only to build dams and reservoirs in the upper drainage basin, but also to construct dikes in the waterways region. According to local conditions diverse measures are taken, which for waterlogging include pumping with electrical pumps, dredging of the irrigation ditches in the fields, digging new ditches and lowering the groundwater level. The measures for channel regulation include restricting small distributaries to enhance the flow in the main channels, simplifying the waterway network system, dredging waterways, renovating dangerous riverbanks, and cutting off shoals to straighten curved channels.

4.1 EMBANKMENT PROJECTS

Building dikes to prevent flooding and saline tides is a primary measure for protecting homes and fields against disaster. Generally the dikes along the riverbanks in the Pearl River delta can withstand 20 year floods. Large-scale combined embankments can resist even those occurring every 50 or 100 years. The design standards for seawalls are made to withstand typhoons of rank 10.

Beijiang embankment extends along the left side of the lower reaches of the North River. It is the most important embankment in Guangdong province and guards 666.7 km^2 of farmland as well as a population of three million in the cities of Guangzhou, Feshan, Qingyuan and Sanshui. It is also regarded as one of the seven most important embankments in China where safety must be ensured. The North embankment begins at Qihuling hill located on the left bank of the Dayanhe branch, extends through the towns of Shijiao, Lubao and Xinan, and finally ends at Nanhai city. The total length is 63.34 km. There are two large flood-diversion sluices, one each at Lubao and Xinan. In addition there are 35 sluices for irrigation, drainage and water transport. The width of the Beijiang embankment is 7 m at the top, at 1.5 m above flood water level. On the outer side the slope is 1:3. On the inner slope there is a platform 3 m below the top, and above and below the platform the slope is 1:2 and 1:3 respectively[8]. The main water-permeable bases are properly strengthened and stabilized. Beijiang embankment can sustain a 100 year flood. It mainly resists floodwater from the Beijiang River. If a large flood occurs in the North and West Rivers at the same time, the safety of the Beijiang embankment is severely endangered. It is now planned to carry out a key water control project at the Sixianjiao waterway in order to change the dangerous pattern of flood diversion and to adjust especially the diversion of the flood from the West River to the North River during the flood season.

The five largest enclosure dikes are at Zhongshui, Feshan, Qiaosang, Jiangxi and Jingfeng. The combined dikes safeguard 1114.27 km^2 of farmland and a population of 2,150,000. The region is not only a base of marketable grain production but also of advanced industry and commerce. If these five combined dikes break, serious flooding and waterlogging will occur in the dike-enclosed area, and the Beijiang embankment will be endangered as well as Quangzhou and other areas in the Pearl River delta. According to their design standard these five dikes should resist the largest flood in 50 years. The total length of the dikes that have now been reinforced, both in height and in width, is 423.67 km. In addition, 52.85 km of endangered channel reaches have been regulated, 64 endangered sluices have been rebuilt, 192 sluices reinforced and 15 newly constructed. If, in the future, dams and reservoirs are built in the upstream drainage basin, the five dikes together after the improvements will be able to resist a 100 year flood.

4.2 IMPORTANT CONTROL SLUICE PROJECTS

There are more than 30 large sluices in the network region of the Pearl River delta. They mainly regulate and adjust flood discharge, control the distributary channels and deal with waterlogging. Among these sluices, the most important for adjusting flood discharge are the Lubao and Xinan sluices and the planned Sixianjiao Key Water Control Project.

Lubaochong and Xinanchong are two branches through which water from the North River flows into the Guangzhou waterways. The Lubao and Xinan sluices are part of the Beijiang embankment and are essential to control and adjust the flood discharge of the lower North River entering the Guangzhou waterways as well as the flow into other distributary channels. Lubao and Xinan are eight- and nine-holed sluices respectively, the width of every hole being 10 m. The basis and support of these sluices is impermeable: together with the Beijiang embankment they form the most important flood prevention shelter of Guangzhou.

The Sixianjiao waterway is the beginning of the network system and the first junction between the West and North Rivers. It is 1200 m long and 500 m wide. The water levels of the Xijiang and Beijiang Rivers at the western and eastern ends of the Sixianjiao waterway are very near to each other. Especially during the medium and low water season, the difference is less than 0.30 m, with the water level at the eastern end being generally higher than at the western end[18, 17]. Therefore, during most of the year, a small part of the North River discharge flows through the Sixianjiao waterway into the West River, but because of the tides during several hours the flow is in the reverse direction and a small part of the West River discharge flows into the North River. During the high water season more discharge from the West River flows into the North River. Because of the small difference in water level the flow is slow and there is no bedload exchange through the Sixianjiao waterway. If the diversion of flood water from the West River into the North River is large, it has a negative influence on the flood water flow and the channel evolution of the Dongping waterways, but during the low water season the discharge through the Dongping waterways is reduced and water depth is low, which is not beneficial to navigation.

The Dongping waterways comprise the whole of the Sixianjiao waterway, part of the lower reaches of the main North River, the Tanzhou waterway and the Pingzhou waterway, which form the shortcut to Guangzhou with a total length of 76 km. Five hundred ton boats can navigate there the entire year, so that it is an important passage in the Pearl River network system. With the plans for the Sixianjiao key water control project it is hoped to actualize a rational adjustment and control of the West and North River discharges during both the high and low water season. In order to ensure the safety of Guangzhou and the dikes along the distributaries of the lower reaches of the North River, part of the floodwater can then be diverted from the West River to scour the North River channels. Also it can ensure a navigable water depth in the branch waterways of the North River during the low water season.

In addition to the large waterworks there are several important reservoirs and hydropower stations such as the Feilaixian key water control project on North River, and the Xinfengjiang, Fengshuba and Baipeizhu reservoirs in the East River basin. Combined with the dikes and sluices in the Pearl River delta they form a reservoir-dike-flood control system. Within the Pearl River delta the Ganzhuxi power station is on the Ganzhuxi waterway, a distributary going from the main West River channel to (the water-level-ridge) to the North River with a difference of l.5 m in water level between both ends. The (low water head) power station not only controls the discharge of the Shuide distributary flowing from the West River, but also improves the conditions of navigation by a shipping gate to the right of the dam.

4.3 WATERWAY REGULATION IN THE DELTAIC NETWORK AREA

The purpose of waterway regulation is to solve problems in the complicated network: accumulation in channels, dikes too long to prevent flooding and the safety of navigation. After rock has been removed, bends cut off, shoals dredged away, the channels dredged and diversion jetties built in the main waterways, boats of 100-1000 tons can pass these channels. There are nearly 100 ports in the entire Pearl River area (Fig. 5). It constitutes a thriving water transport network (Fig. 6) with water-land transshipment, river-ocean transshipment and oceangoing shipping. The main measures that were or will be taken are:

1) Simplification of the network system. In order to control flood diversion and reinforce the main channel flow momentum, sluices on branch waterways have been built and dikes have been linked by engineering measures. The network has been reduced from several hundred main flood diversion channels to only 20 or 30 and the total length of the dikes is reduced as well, which is beneficial in preventing flooding. Large amounts of water and solid discharge are concentrated in the main channels and go seaward through the eight distributary mouths or tidal gates, which lead, as seen above, to a quick development of subaqueous shoals and a larger area of potential land. Water quality, however, and sediment deposition are deteriorating conditions in the sluice-controlled branches. The sluices need to be opened or electric pumping carried out to remove the waste water and sediment.

2) Dredging and the removal of rock with explosives. These engineering measures are mainly concentrated in the Chencun, Tanzhou, Pingzhou, Lijiao, Shawan and Ganzhuxi waterways in order to meet the demands of navigation and to ensure flood flow safety. The dredged mud and sand are mainly used for filling in low fields, for reinforcing dangerous dikes and for other engineering works.

3) Cutting off bends and removal of shoals. Most of the waterways in the deltaic network are sinuous and winding due to the fluvial and tidal interaction. Together with numerous shoals this is not beneficial for navigation. Improvement was made by straightening the channels and removing shoals that obstruct shipping. Thus the curved lower reach of the Xizhijiang branch, which was 8 km long, was cut off by a straight channel with a length of 856 m. A strongly curved bend of the Dadaowan reach (part of the Dongping waterway) with a curvature radius of 180 made the passage for ships difficult. After removal of the shoal and moving the dike 263 m backwards, the curvature radius is now 289 m.

4) Reinforcing dangerous dikes. Most dangerous dikes are at sharp bends or at channel confluents. Because the channel is close to the riverbank and scouring the bed, the

Fig. 5. An example of one of the distributary gates in the Pearl River delta.

Fig. 6. The channels in the Pearl River delta are major routes for river boats of all sizes.

dikes risk collapsing or bursting. This is prevented by dumping stones, building spur jetties and constructing dikes that follow the flow.

4.4 THE RIVERBANK REGULATION PROGRAMME FROM GIUANG-ZHOU TO THE HUMEN TIDAL GATE

Guangzhou is the largest open harbour in South China. The waterway from Guangzhou to Humen is not only threatened by upriver floods, but also by the tidal currents. At present the following difficulties occur:

1) The annual highest flood water level, which has an impact on the city embankment, is rising;

2) The water depth and the harbour installations can not keep up with the economic development;

3) Pollution is increasing and the water quality is deteriorating;

4) The local utilization of the riverbank is not reasonable.

The aim of the regulation programme is to coordinate the problems of flood control, harbour and channel development as well as urban construction.

The waterlogged areas in Guangzhou have been calculated to be 3.3, 8.84, 11.45, and 13.0 km^2 at water levels of 2.0, 2.5, 2.8 and 3.0 m respectively[16]. The highest flood tide water level at the Fubiaochang tidal gauge within 100 years is only 2.79 m. The flood-preventing dikes along the riverbanks near Guangzhou have been constructed on this basis. Underground wastewater drainpipes and drain stations have been renovated or rebuilt and part of the urban wastewater is led through a pipeline into the Shiziyang waterway far downstream from Guangzhou. These measures improve the water quality in the Guangzhou waterways to some extent. When the Feilaixia, Datengxia and Longtan water control projects in the North River basin have all been realized, this will greatly reduce the threat from floodwater. In addition to this the flood control capability of the dikes along Guangzhou can be greatly improved.

Figure 7 is the planning map for the waterways regulation project from Guangzhou to the Humen tidal gate. The East waterway riverbanks are regulated through broadening the channel by 1.5% starting at the Haizhu bridge, where the channel is 180 m wide and ships of 1000 tons can pass. For filling in the area between the present riverbanks and the planned 1.5% broadening of the banks more than 2.0×10^7 m^3 of sediment are needed for backfill and 9.0×10^5 m^3 for dike construction.

Besides the advantage of using the sediment produced by dredging and regulating the channels, 2.59-3.93 km^2 of additional land is created between Guangzhou and Huangpu for urban development[19]. The wide and deep channel from Huangpu to Humen has better potential for harbours and shipping development. Regulation in this reach is mainly focussed on constructing deepwater docks and shipping channels. At present a dock has been built at Xisha with ten berths for ships of 35,000 tons. Construction of a dock for ships of 50,000 tons is planned at the east bank of the Humen tidal gate (Fig.7).

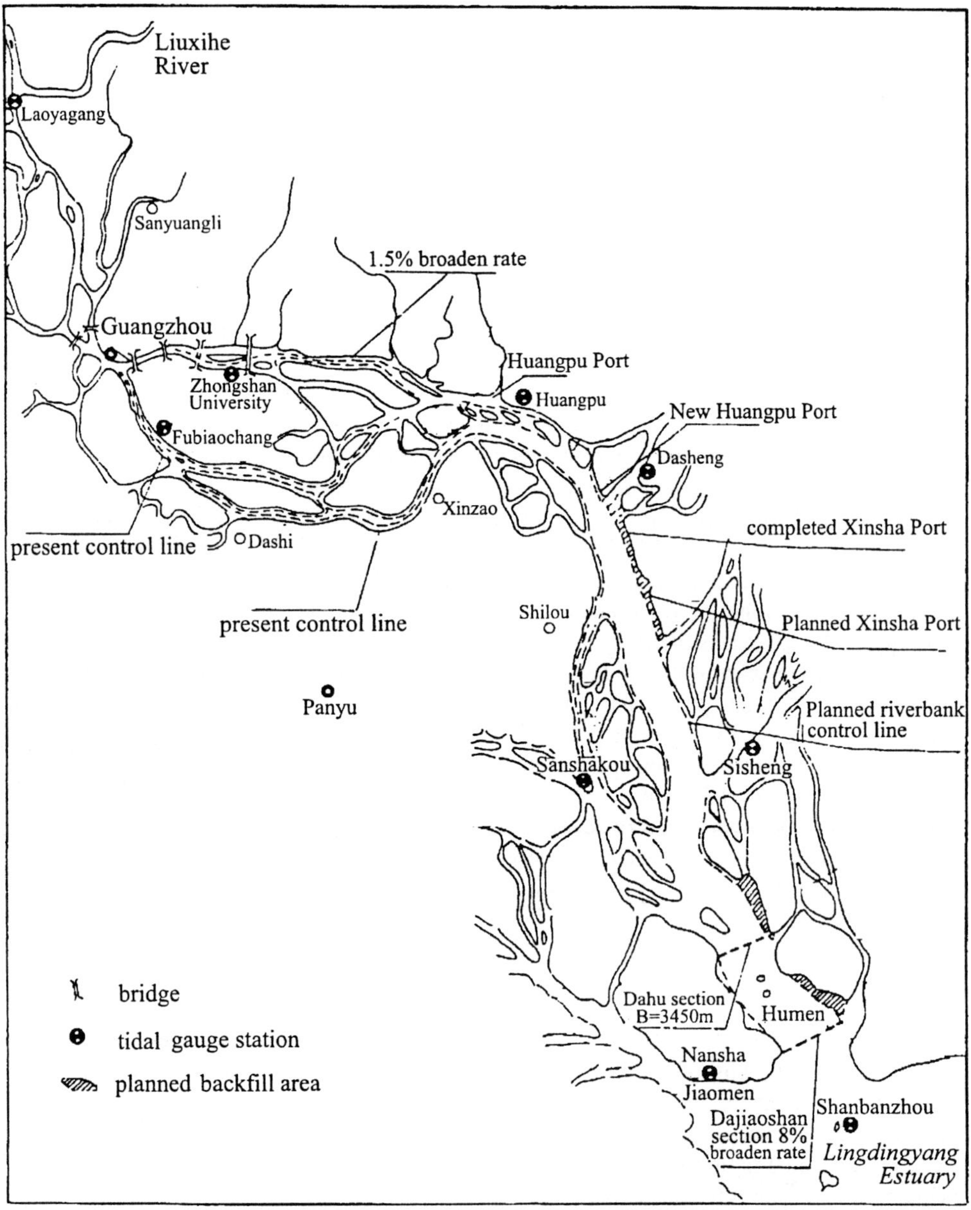

Fig. 7. Sketch map of the regulation plan for the shipping channel from Guangzhou to Humen.

5. Planning of Regulation and Reclamation in the Modern Pearl River Estuary

The modern Pearl River estuary is 150 km wide from the Jiulong Peninsula in the east to the Chixi Peninsula in the west. The coastline is 457 km long and 197 islands are scattered in water depths of less than 30 m. The total length of the island coastline is 1260 km. These islands are natural barriers that restrain the impact of waves and storm tides on the deltaic coast. The delta is expanding seaward at a mean rate of 100 m per year and subaqueous shoals develop quickly because of the large amounts of sediment deposited in the estuary and the coastal sea. At the outer edge of the delta the area of subaerial and subaqueous shoals shallower than -3 m is in total 762.2 km^2, the area shallower than -2 m being 547.2 km^2 (Table 2). Large areas of shoals are located in the Longdingyang estuary adjacent to Hongkong and Macao. These shoals can be used for agri- and aquaculture, as fertile land with a low salinity can be made out of them and freshwater is abundantly available.

Table 2. Area and distribution of shoals in Pearl River estuary (unit: km^2)[8]

Location area	West of Lingdingyang estuary	East of Lingdingyang estuary	Modaomen mouth	Niwanmen mouth	Huangmaohai estuary	Sum
Shallower than –2.0m	90.4	190.2	139.3	15.3	112.0	547.2
Shallower than –3.0m	108.4	291.1	166.0	30.0	166.7	762.2

There are two important problems in the Pearl River estuary today. First the distributary mouths and shoals are moving seaward so fast that it is hindering floodwater flow, drainage and irrigation. Second, because water flow and sediment transport in the West and North Rivers are different, large amounts of water and sediment are diverted into the Lingdingyang estuary and the shipping channel of Guangzhou harbour during the flood season. As this affects flood control and navigation, the aim of present regulation planning is to reduce the floods that occur because of this, and to maintain the deep water shipping channel, while at the same time reducing waterlogging, salt intrusion and storm tides, and improving the environmental quality and ecological balance.

5.1 REGULATION AND PLANNING IN LINGDINGYANG ESTUARY

5.1.1 *General characteristics and modern evolution.* At the Humen tidal gate, the head of the Lingdingyang estuary is only 4 km wide, whereas the outer bay mouth, between the Jiulong Peninsula of Hong Kong in the east to Macao in the west, is 65 km wide. The length of the estuary is 60 km, its area 2100 km^2. There are three extensive shoals separated from each other by two longitudinal channels with a mean water depth of -11.5 m in the east channel and -8.6 m in the west channel. At present the west channel is the outward bound passage for ships coming from Huangpu harbour, the outer port of Guangzhou. The shipping lane is -11.5 m deep, and is being dredged to -12.5 m. The water level in the Lingdingyang estuary during mean mid-tide is declining towards the southwest during flood, and to the southeast during ebb. The tidal range increases from the bay mouth to the head. During the river flood season the tidal range at Humen is larger, but it is lower during the period of low river discharge. The flood current velocity in the east channel is larger than in the west channel, but nearly equal during the ebb[20]. The region covered within the planning is the inner Lingdingyang estuary north of a line along Chiwan, Neilingding Island, Qijiao Island and Tangjiawan. The distribution of tidal discharge along the boundaries of this area is given in Table 3.

Table 3. Tidal discharge at input and output boundaries of the Lingdingyang estuary ($10^8 m^3/a$)[7]

Boundary	**Input Boundary**				**Output Boundary**			
Input mouths/ Output sections	**Humen**	**Jiaomen**	**Hongqili**	**Hengmen**	**Chiwan**	**Neilin-ding**	**Jinxing-men**	**Sum**
Mean annual flood tidal discharge	2288	325	97	133	3244.3	5362.6	935.1	9542
Mean annual ebb tidal discharge	2891	890	306	498	3994.5	6375.5	914.0	11284
Mean annual net output discharge	603	565	209	365	750.2	1012.9	-21.1	1742

The area of the inner Lingdingyang estuary covered by the planning is 1041 km^2. The mean annual input of river discharge is $1742x10^{11}m^3/a$, which is 53.4% of the entire Pearl River network system. The mean annual sediment input is $3389x10^7$ t. The mean annual sediment output is $151x10^7$ t. The sediment deposited in the inner Lingdingyang estuary thus amounts to 1878×10^7 t., which is 55.4% of the entire Pearl River network (Table 1). The mean accumulation rate is about 0.2 cm/a. At present the west shoals are relatively quickly extending southeastward and the west channel is slightly narrowing. The length of the central shoal is increasing and there is little narrowing of the east channel. The east shoal is stable.

5.1.2 *The regulation plan for the input boundary.* The Humen gate is a strategic passage and its channel mainly depends on the strong tidal currents. Jiaomen, Hongqili and Hengmen are fluvial dominated distributary mouths and their main channels are not very stable because of the flood diversion branches from the hydraulic ridge to the low-pressure region. Therefore, near the distributary mouths, the evolution of the eastward branches is more variable than the development of the main channels, and once they are there they become deeper and deeper. This also causes increasing problems of sediment

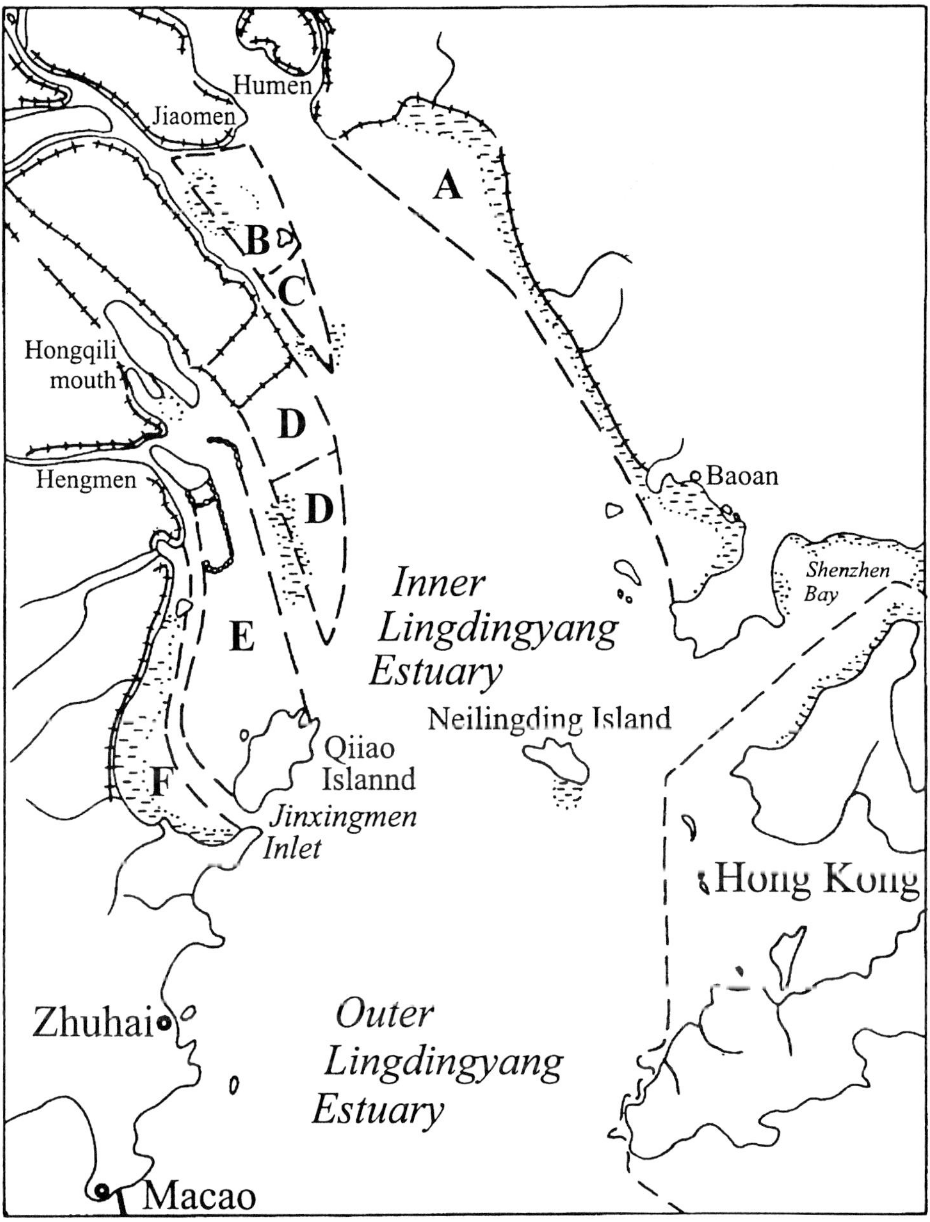

Fig. 8. Sketch map of the Lingdingyang estuary regulation plan[21].

accumulation, drainage and irrigation at the three distributary mouths. The regulation projects at the four input mouths (Fig. 8) are as follows:

1) The Humen gate is controlled by flood and ebb jet flow and can be regarded as a tidal inlet. The reclaimable shoals are mainly distributed in the shallow water regions directly inside and outside the gate/inlet. During 1976-1984 4.02 km^2 was reclaimed at the Xidatan shoal inside the gate near Dahu island; during 1980-1987 14.1 km^2 was reclaimed at the east shoal outside the gate. Based on the favourable geographical position, the economy and industry are developing quickly in the area near Humen gate. The Shajiao thermal power plant, the Chiwan deepwater port and Shenzhen airport have been constructed. According to the regulation program, also 4.0 km^2 of valuable land can be obtained.

2) The Jiaomen mouth is 16 km long from Yanshawei to Nansha. Here three branches of the Shawan waterway and two branches of the Hongqili waterway (the Shanghengli and Xiahengli) converge, which results in the largest sediment input along the input boundary. From the Jiaomen mouth outwards two channels run into the Lingdingyang estuary. One is a southward (main) channel. already marked on a map of 1883; the other is the eastward Fuzhou channel formed in 1907. During nearly a hundred years the width and depth of this channel have developed from the initial several 100 m and 1.0-2.0 m respectively, to the present 2 km and 3.0-6.0 m, with a corresponding increase in discharge. Now, because the Fuzhou channel has a larger cross-sectional area, a shorter length and a steeper hydraulic gradient, the discharge is 85.4% of the discharge at Nansha station. The present trend is that the left channel is scouring while the right channel is narrowing. As a result, large amounts of sediment are transported into the Humen gate, reducing the stability of the shipping channel from Guangzhou harbour. The narrowing of the right channel also influences the irrigation and drainage of nearby farmland.

Therefore the key problem of the regulation of the Jiaomen mouth is controlling the diversion through Fuzhou channel and increasing the flow in the main channel. Building of underwater jetties in combination with shoal reclamation will lead to adjustment of the longitudinal gradient of the left channel. The flow in the right channel will be increased by dredging and by building diversion dikes at the conversion point of Shanghengli and the Xiahengli branches.

3) The Hongqili mouth was initially formed in 1670. In recent years the discharge through the mouth is declining because of scour in the eastward branches, resulting in greater depth and stronger flow. Thus the diversion ratio of the Shanghengli and Xiahengli branches increased from 49 and 21.6% in 1952 to 58 and 22.3% in recent years respectively. As a result, the sediment deposition at the Hongqili mouth is serious and the flow velocity is reduced. Here the aim is to dredge the Hingqili channel and to control the diversion ratio of the branches.

4) The Hengmen mouth is divided into north and south branches, which were formed before 1883. The north branch diverts up to 60% of the total discharge. The south branch flow into a wide water area south of Damao island and forms a large shoal. The north branch makes an almost right angle with Hongqili, which reduces the flow velocity considerably, which is not favourable for flood drainage. For the north branch the aim is therefore to smooth the confluence with the Hongqili channels; for the south branch the aim is to improve the stability of the main stream and the shoal area (Fig. 8).

5.1.3 *The overall plan for the regulation of the Lingdingyang estuary.* Figure 8 is a map of the overall plan for the regulation of the Lingdingyang estuary whereas Figure 9 shows examples of earlier reclamation at the Hongqili and Hengmen gates. An eastern regulation line begins at Shajiao and extends southward to Shenzhen Bay along the -5 m isobath (Pearl River datum). A western regulation line begins at Dajiao hill and extends southward to the northeastern cape of Qiiao Island by way of Shadui and Jiaobeisha. The lines indicate the limit to where land is planned to be reclaimed, leaving in the center a large trumpet-shaped inner Lingdingyang estuary. The east shoal region has a high value and in the near future 33.7 km^2 is planned to be reclaimed. Reclamation of a further 36.0 km^2 is planned for a later time. The north, center and south parts of the east shoal will be used for traditional aquaculture, as well as for further development of Shenzhen airport, and for urban and harbour construction. The west shoal zone has the advantage that there is much freshwater, low salinity, high soil fertility, and that it is suitable for both land and water transport. Therefore it is designated for agriculture, aquaculture (Figs. 10 and 11) and economic development. In the near future 12.0 km^2 will be reclaimed and somewhat later a further 66.7 km^2. The major areas are the Jibaosha, Wangqingshawei and Hengmengkou shoals.

5.2 THE REGULATION PLAN OF MODAOMEN MOUTH

5.2.1 *General characteristics and modern evolution.* The Modaomen mouth was an inwards curved drowned bay in 1883. At that time the Denglongshan shoal did not exist, Baijiao and Baiteng were isolated islands separated by Dahengqin, Xiaohengqin and Sanzao Islands (Fig. 12). At that time there was an inner shallow sea and an outer shelf sea, with the Niwanmen, Hengzhou and Hongwan channels running into the inner sea[6,3]. After the large Baiteng reclamation project was carried out in the inner sea in 1958, the discharge of water and sediment of the Niwanmen channel were completely diverted into the Jitimen distributary mouth and from there into the Huangmaohai estuary. Behind the closed dikes a large lake was made, Baiteng Lake. A large amount of sediment, discharged through the Modaomen channel was deposited in the inner sea, where large shoals developed. In the 1960's, 16.7 km^2 of these shoals was reclaimed and another 20.0 km^2 in 1972, but by 1977 there was again 11.3 km^2 of shoals available for reclamation[6] (Fig. 13). The shoal area continues to increase and the Longshicu, a scour channel formed by flood tidal currents, is shrinking considerably. Nowadays the inner sea has been nearly filled in and the Modaomen mouth is moving outward and extends towards the outer sea. Between 1946-1964 and 1964-1971 deposition and expansion of the shoal at the Modaomen mouth increased rapidly. In the near future the islands will be joined to the mainland and the river mouths as well as the subaqueous delta moved out to sea.

Modaomen is the distributary mouth with the strongest fluvial dynamics and the largest extension rate of the Pearl River. From 1883 to 1962 the annual rate of elongation was 121.7 m/a. At present there is a large crescent subaqueous mouth bar at the outer end of the major Hengzhou channel which is moving seaward at a rate of 80-100 m per year[6]. The back slope of the mouth bar is eroding, whereas the outer slope is accumulating. The

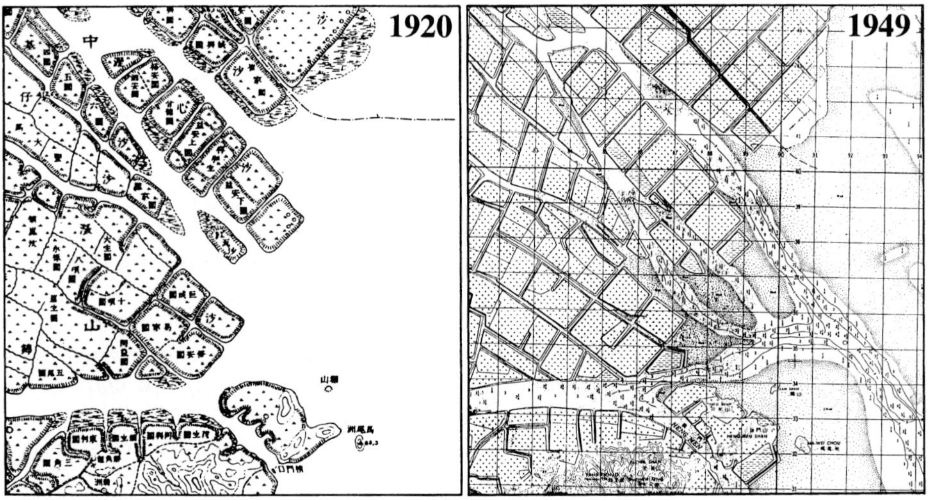

Fig. 9. Stages of reclamation at the Hongqili and Hengmen gates between 1920 and 1944.

water depth at the top of the crest is less than 3 m. Because of the mouth bar the -5 and –10 m isobaths are protruding seaward towards the southeast. In a lateral direction the volume of the west part is larger than that of the east part, which is partly due to the main flow at the west side being ebb flow and at the east side being flood flow. It also depends on reworking by the waves from the southeast and on the southwestward transport by the longshore currents[22].

Because Modaomen is the largest tributary mouth with the largest discharge of water and sediment, flood control is the key problem in this area. In recent years the flood water level has been slightly raised because of reclamation. Thus between 1950 and 1980 water level rose 0.5-1.0 m annually while the width was reduced by 100 m. But the water diversion ratio increased by 2% , and thus the fluvial dynamics was more centralised and strengthened. Considering these changes, more attention should be given to the development of the subaqueous shoals and the rise of flood water level when regulating the Modaomen distributary mouth.

5.2.2 *Regulation plan for the Modaomen mouth.* The area of the inner shallow sea near the Modaomen mouth is about 170 km^2. Dikes have been planned along the main channel and the branches (Fig. 13), in order to regulate the main flow through the Hengzhou channel, and to control the Hongwan branch for water supply and flood diversion. The distance between the planned east and west dikes is 2-2.3 km along the Modaomen channel, the dikes themselves being 9.6 and 18.6 km long respectively. The dike along the Hongwan waterway is planned to be about 500 m long. About 100 km^2 of farmland can be obtained behind the dikes in the inner sea area.

Fig. 10. Constructing oyster reef forms for use in the shallow waters offshore south of Zhuhai.

Fig. 11. Net fishing near the mouth of the Lingdingyang estuary with one of the many islands that will eventually be land tied by deltaic deposits.

Fig. 12. Sketch map of the Modaomen River regulation plan[23].

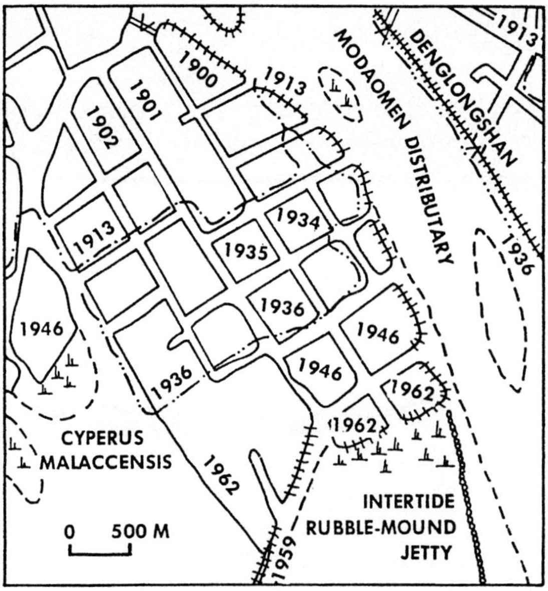

Fig. 13. Reclamation at the Modaomen River mouth illustrating the use of vegetation and jetties to aid in trapping river sediment.

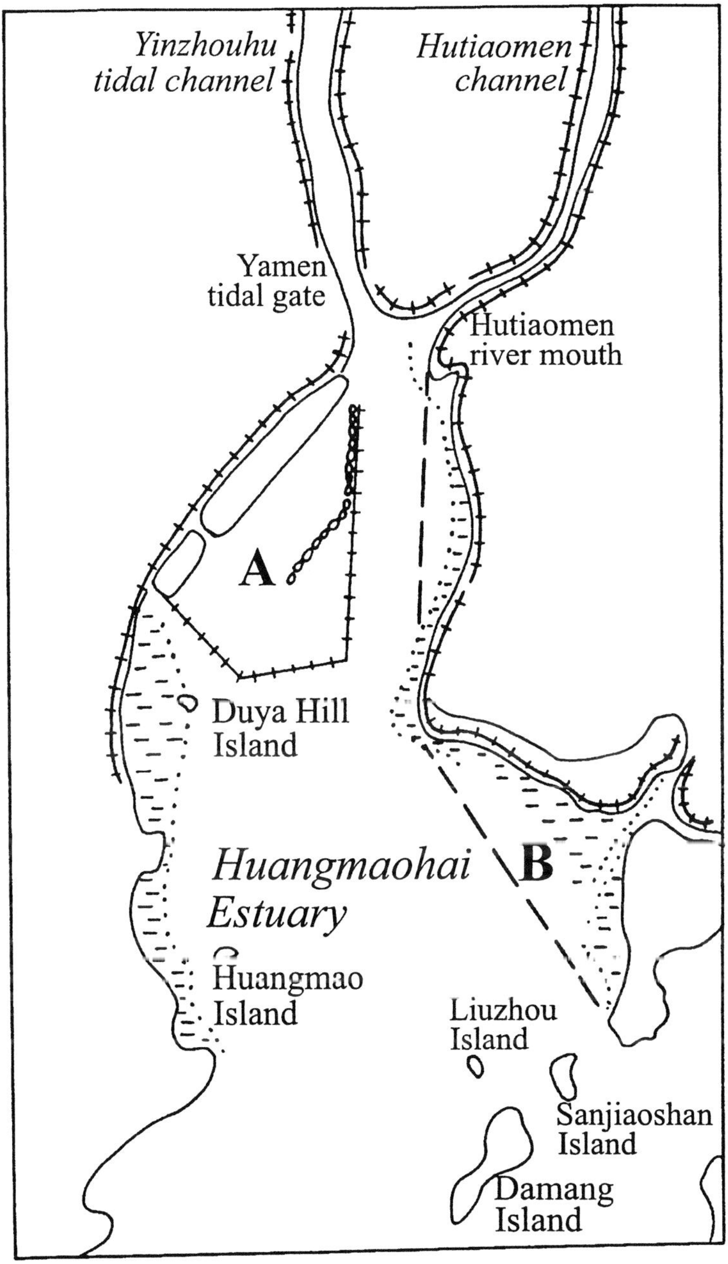

Fig. 14. Sketch map of the Huangmaohai estuary regulation plan[25].

5.3 THE REGULATION PLAN FOR THE HUANGMAOHAI ESTUARY.

The Yanmen tidal gate and the Hutiaomen and Jitmen mouths flow out into the Huangmaohai estuary (Fig. 14). The width of the bay head is 3.4 km, the width of the outer mouth 16 km, its length is 29 km and the surface area is 403 km^2. The Yizhouhu tidal channel is upstream of Yanmen gate, which is 1.5-2 km wide and 7-9 m deep. It is suited to the development of a port and shipping channel. The mean annual sediment discharge into the Huangmaohai estuary is about $9.0x10^6$ t[8]. Tidal currents dominate; the mean tidal range is 1.23 m. There is a large and shallow bar at the center of the estuary with across the west channel a depth at the crest of only 2.1 m and a depth of less than 4 m over a length of 10 km. At the east channel the depth over the crest is only 2.8 m and water depth is less than 4 m over a length of 7-8 km[24]. Therefore the mouth area blocks shipping going to the outer sea. Regulation is planned through dredging at the bar and reclaiming the estuarine shallow parts of the shore in order to strengthen the flow in the two main channels (Fig. 13).

At present there are 39 dikes around the estuary, 207 km of river dikes and 354 km of estuarine coastal dikes. The deepest channel is close to the eastern shoreline, which prevents extending the shoreline further outward by human activity. The planned estuarine area between the reclaimed areas will be somewhat funnel-shaped with a broadening rate of 5% in accordance with the dynamics of this tide-dominated estuary. About 167 km^2 of farmland will be obtained.

6. Acknowledgements

The authors are grateful to Prof. Doeke Eisma and Prof. H. Jesse Walker, who read an early version of the manuscript; their critical comments helped improve the paper. Other individuals who provided data and literature as background material for this paper include Zhaoying Dong, Zhangren Luo, and Ganran Yang.

7. References

1. Li Pingri and Qiao Pengnian, 1982. The model of evolution of the Pearl River Delta during lasts 6,000 years. *Journal of Sediment Research*. 3:33-42. (in Chinese)
2. Luo Kaifu, 1984. Analyses on the characteristics of Pearl River delta. *Pearl River*. 1:37-38. (in Chinese)
3. Zhao Huangting. 1990. *Evolution of Pearl River Estuary*. Beijing: marine publishing company.
4. Workroom of Estuarine and coastal studies in Zhongshan University. 1977, *Development and evolution of Pearl River Delta*. (in Chinese)
5. Long Yunzuo, Huo Chunlan , Ma Daoxiu, et al, 1985. On sedimentary characteristics and model of Pearl River Delta. *Marine Geology & Quaternary Geology*. 4: 49-57. (in Chinese)
6. Huang Zhenguo, Li Pingri and Qiao Pengnian, 1982. *Formation, development and*

evolution of Pearl River Delta. Guangzhou: Branch of Science Promulgation Publishing Company. (in Chinese)

7. Song Dingchang and Yuan Gusong, 1986. Study on tidal prism of the eight outlets in Pearl River Delta. In: *symposium on the investigation on resources of coastal and estuarine shore zone in Pearl River Bay (volume No.4)*. Guangzhou: science and technology publishing company of Guangdong province. 62~71. (in Chinese)
8. Pearl River Water Resource Commission, 1995. Chorography of Pearl River. Guangzhou: Science and Technology Publishing Company of Guangdong Province. (In Chinese)
9. Li Chunchu and Lei Yaping, 1998. Discussion on the characteristics and conservation of tidal-dominated channel from Guangzhou to Humen tidal gate. *Tropical Geography*. 18(1):24-28. (in Chinese)
10. Li Chunchu and Yang Ganran, 1988. Characteristics and evolution of channel network in Pearl River Delta. In: *The Evolution Process and Resources of Pearl River Delta*. Guangzhou: publishing company of Zhongshan University. Volume no.1, 48~65. (in Chinese)
11. SiTu Shangji, 1994. The *Historical Geography of South China*. Guangzhou: map publishing company of Guangdong province, 121~140. (in Chinese)
12. Pearl River Water Resources Commission, 1992. *Symposium on Embankment Engineering & Water Control History*. Guangzhou: Science and Technology publishing Company of Guangdong province. (in Chinese)
13. Walker, H. J. 1980. The Pearl River delta. *Scientific Bulletin*, 5(2), 1-6.
14. Li Zihao. 1985. The effects of united embankments on the flow shift and channel changes in the Pearl River Delta. *Tropical Geography*. 2: 99-107. (in Chinese)
15. He Zhuoxia, 1986. A study on the trend and reasons of floodwater level fluctuation in the Lower reach of Modaomen Channel. In: *symposium on the investigation on resources of coastal and estuarine shore zone in Pearl River Bay (volume No.4)*. Guangzhou: Science and Technology Publishing Company of Guangdong province.173-184. (in Chinese)
16. Zhang Shengcai, 1986. The reasons of flood and tidal water level rise in Guangzhou waterways. In: *symposium on the investigation on resources of coastal and estuarine shore zone in Pearl River Bay (volume No.4)*. Guangzhou: science and technology publishing company of Guangdong province. 136-144. (in Chinese)
17. Luo Zhangren, Luo Xianlin, Yang Ganran, et al, 1999. The influence of Human activities on the evolution of riverbed in the Pearl River Delta. *Tropical Geomorphology*, 20(2): 1-15. (in Chinese)
18. Ying Zhipu, Chen Zhiyong and Chen Shiguang, 1988. The evolution of Sixianjiao Waterway and it's effects on discharge and sediment transport. In: *The Evolution Process and Resources of Pearl River Delta*. Guangzhou: publishing company of Zhongshan University, volume no.1, 95~108 (in Chinese).
19. Dong Z Y, 1986. *The exploitation prospect of shallow resource in Pearl River Estuary*. In: *symposium on the investigation on resources of coastal and estuarine shore zone in Pearl River Bay (volume no.1)*. Guangzhou: Science and Technology Publishing Company of Guangdong Province. 75~81. (in Chinese)
20. Dong Z Y, 1986. Suspended sediment transport and erosion/deposition balance of

Lingdingyang Estuary. In: *Symposium on the investigation on resources of coastal and estuarine shore zone in Pearl River Bay (volume No.4)*. Guangzhou: Science and Technology Publishing Company of Guangdong Province. 210~230 (in Chinese).
21. Pearl River water Resources Commission, 1993. *Study report on Lingdingyang Estuary regulation plan.* (in Chinese)
22. Li Chunchu, 1993. The formation and evolution of Modaomen river mouth-bar and it's effects on regulation project. In: *The Collection of 7th National Coastal Engineering Symposium.* Beijing: Ocean publishing company, 1-9 (in Chinese).
23. Pearl River water Resources Commission, 1988. *Feasibility study reports on the regulation engineering of Modaomen river mouth.* (in Chinese)
24. Wu Chaoyu and Yuan Shuyao, 1995. Dynamic structures and their sedimentation effects in Huangmaohai Estuary, China. *Journal of Coastal Research*, 11(3): 808-820.
25. Pearl River Water Resources Commission, 1992. *Study report on Huangmaohia Estuary & Jitimen river mouth regulation plan.* (in Chinese)

COASTAL PROTECTION, STRUCTURES AND (SEA)DIKES
(as practized in the Netherlands)

JAN VAN DE GRAAFF
Senior scientific officer Faculty of Civil Engineering and Geosciences
Delft University of Technology (DUT)
Stevinweg 1, 2628 CN Delft, The Netherlands
Tel.: +31 15 2784846,-fax.: +31 15 2785124
e-mail: J.vandeGraaff@ct.tudelft.nl

1. Introduction

Coastal morphology changes continuously because of (changing) natural conditions. But also human interferences in the coastal system in past and at present (e.g. building man-made coastal structures like breakwaters to shelter port entrances) have affected the coastal morphology to a large extent. The coastal system is, however, a very vulnerable system, so one has to be aware of thoughtless interventions in the system.

Dunes, just like dikes, protect in many cases a low-lying mainland from flooding during severe storm surges, but unlike dikes unprotected dunes are expected to be reshaped (eroded) during a period of severe attack.

Large parts of sandy coasts all over the world suffer from structural erosion and/or dune and beach erosion during severe storm surges. The proper protection of these threatened coasts is an important aim in coastal engineering practice. Besides this type of protection new reclaimed areas sometimes have to be protected from the attacks by the sea. Coastal engineers have nowadays to select a proper protection tool out of the many available methods. Often, however, choices have already been made concerning methods of coastal protection in former times. The present coastal engineer has to deal with these choices and also with the effects of structures built along the coast in the past.

Protection with the help of 'hard' structures is often used. Series of groynes, series of offshore breakwaters, submerged breakwaters, dikes, and revetments or seawalls are examples of such 'hard' structures. With the help of these 'hard' methods one basically likes to interfere in the actual sediment transports in cross-shore and alongshore directions of the coast.

Furthermore 'soft' methods are possible, like artificial dunes, beach and shoreface nourishments. With the help of these 'soft' methods one principally does not want to interfere in the natural sediment transport processes. Only the losses of sediment as occurring in a limited stretch of coast are to be compensated on a regular basis.

J. Chen et al. (eds.), Engineered Coasts, 229–247.
© 2002 *Kluwer Academic Publishers. Printed in the Netherlands.*

Whether to select a ‘hard’ or a ‘soft’ method in a coastal protection scheme depends on the characteristics of the problem in each particular case and on economical considerations. In Coastal Zone Management practice the use of nourishments is increasingly popular. Many of the often occurring adverse side-effects of ‘hard’ structures can be avoided with artificial nourishments. However, the possible use of structures for coastal protection cannot and may not be disregarded. A coastal engineer should at least have a proper insight in the physical properties of ‘hard’ structures. The wanted effects (often: reduction or mitigation of the erosion potential of a given stretch of coast), but also the unwanted, often detrimental effects to adjacent coasts have to be considered with care. Only in that case an appropriate choice out of the many coastal protection methods can be made.

2. Coastal Protection Problems

Various problems in coastal engineering practice call for appropriate countermeasures. Some of them will be briefly discussed here; in Section 3 the use of structures as a tool in the solution of these problems will be outlined. The (restricted number of) problems as discussed are:

- structural erosion of coasts;
- beach and dune erosion during severe storm surges;
- protection of newly reclaimed areas;
- stabilization of dynamic tidal inlets.

Examples from the Dutch coast will be given.

2.1. STRUCTURAL EROSION OF COASTS

Coastal erosion is in fact a rather tricky notion. It is beyond doubt that the erosion of the sandy coast, which is often observed at the leeward side of port entrances sheltered by two breakwaters is a typical example of a so-called structural erosion problem (see Figure 1). Structural erosion will also occur as a result of long-term adaptation of the position of the coastline in an area where the sediment input from the river to the costal zone has reduced; e.g. because of up-stream damming of the river.

In the erosion area the volume of sand in an arbitrary cross-section between well-chosen boundaries in that cross-section (in m^3/m), apparently reduces gradually as a function of time. Such a reduction continues year after year; typical orders of magnitude of this type of erosion are 10 to 50 m^3/m *per year*. Structural erosion is a *permanent* erosion phenomenon.

The construction of a harbour at a sandy coast and the up-stream damming of a river are clearly man-made actions with adverse effects on coasts, but also entirely natural reasons can be associated with some structural erosion problems. E.g. the diversion of a river branch in an uninhabited and unbuilt region resulting in a change in position of the river mouth and consequently resulting in a different location of the source of river sediment in the coastal system.

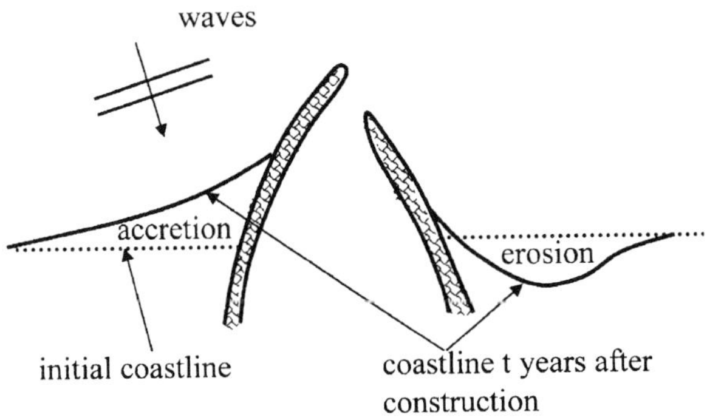

Fig. 1. Typical structural erosion problem at lee-side of harbour.

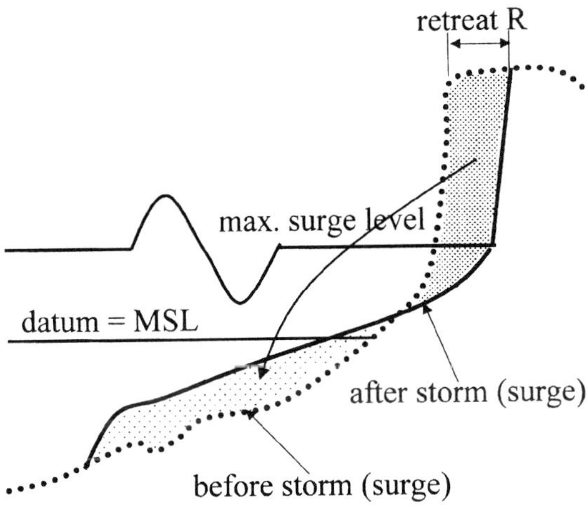

Fig. 2. Dune erosion during a severe storm surge.

The erosion of the beach and the dunes or the mainland during a severe storm surge (see next section) is often also considered as a typical erosion problem. Indeed, after the storm event the dunes and/or upper parts of the beaches may have lost sediment, and sediment has disappeared from its pre-storm position. Often, however, the lost volume is deposited at the bottom in the nearshore area (see Figure 2) so that the total volume of sand in a cross-shore profile has not basically changed during the storm surge. In fact only a redistribution of the sediment masses over the cross-sectional area has taken place. Depending of course on the severity of the storm surge, the volume of sand lost from the upper parts of the cross-section associated with this type of erosion is in the order of 10 to 100 m^3/m per event (say: *per day*). In the period after the storm surge event, often the ordinary natural conditions force a recovery of the beaches and dunes. By onshore sediment transport material is transported from deeper water to the beach and next transported back to the dunes or mainland by onshore winds. Dune erosion during a severe storm surge is often a *temporary* erosion phenomenon.

In a typical structural erosion problem ordinary conditions (as well as storm conditions) contribute to the eventual loss of sediment out of a cross-section. Often a *gradient* in the longshore sediment transport rate is the main reason. [Gradient: $dS/dx \neq 0$; S: net yearly longshore sediment transport; x: coordinate directed along the coast(line).] Structural erosion under ordinary conditions involves that the upper part of the profile (dry beach and slope of dune or mainland front) does not take part in the transport processes; the water and the waves do not reach this part of the cross-shore profile. The erosion of the foreshore, however, will continue under these conditions. Only if the waves can reach the dunes (under storm conditions; higher water level and higher waves) the upper part of the cross-section forms an integrated part of the total active profile. Erosion of dunes or mainland will occur.

While in a basically stable situation this erosion of dunes or mainland is only temporary, in a structural eroding case this erosion is (partly) permanent. Sediment originally from the upper part of the cross-section will during ordinary conditions not fully return into the dunes, but will be removed in a longshore direction. At the end of the day also gradual erosion will cause erosion of the dunes or mainland, but in this case with the 'help' of cross-shore transport processes. This distinction between dune erosion because of a severe storm surge and structural erosion is often not clear to coastal laymen. Ultimate permanent losses of dunes and mainland tend to be (wrongly) associated with storm surge events, while the basic problem is a structural erosion problem.

Structural eroding coasts are often very annoying to the various users of the coastal zone. Properties built close to the sea get ultimately lost; roads in that area disappear into the sea. Often society calls for taking action by the Coastal Zone Authorities in order to prevent the detrimental effects of structural erosion.

2.2. BEACH AND DUNE EROSION DURING SEVERE STORM SURGES

Also coasts which, seen over a number of years, are stable, may suffer from the effects of storm surge events (see e.g. Van de Graaff, 1986; Steetzel, 1993). It can be argued that during storm surge conditions the shape of the pre-storm profile is far out of the equilibrium shape which belongs to the severe storm surge conditions. Often a temporary increase of the still water level (surge) and far higher waves compared to ordinary conditions are associated with a storm surge. Erosion of the upper part of the cross-shore profile, while the nearshore zone is accreting, results in flatter slopes of the profile during the storm (see Figure 2). While the profile is flattening with time, the erosion process slows down.

As mentioned in the previous subsection the rate of erosion can be very large (of course depending on the actual conditions). Associated with the loss of volumes of sediment from the dunes or mainland, a retreat R of the dunes (see Figure 2) of tens of meters may occur during a single event. Nowadays methods are available for reliably quantifying the rate of dune erosion because of arbitrary boundary conditions during a storm surge (see e.g. Larson and Kraus, 1989; Larson et al., 1990; and Steetzel, 1993). The effects of permanent structural erosion or of temporary erosion because of a storm surge, on properties built (too?) close to the shoreline are eventually the same. In both cases the properties may be lost (see Figure 3). It is beyond doubt, however, that countermeasures meant to reduce or to obliterate any of the two types of erosion as mentioned, must be quite different.

Fig. 3. Damage because of dune erosion.

2.3. PROTECTION OF NEWLY RECLAIMED AREAS

The coastal zone is here simply defined as consisting out of a beach or a dike, an adjacent strip of sea and an adjacent strip of mainland parallel to the beach or dike. The width of the latter strip may be several kilometres. All over the world such coastal zones serve many different important functions.

Along some coasts (China, Holland) there is felt a shortage of this type of useful areas Reclamation of land areas from the sea may be considered an option to acquire new land; expansion of the land may be achieved by shifting the coastline in a seaward direction with the help of huge artificial sediment depositions. Also artificial islands are sometimes considered. In both cases structures can be used to mould the boundaries of the reclaimed area: either as a full dike or revetment protecting the core of the reclaimed area, or as some fixed stronghold points where in between protected sandy beaches may develop.

Artificial islands in open seas are often designed and built in relative deep water. (Chek Lap Kok airport Hong Kong; airport Osaka; plan under discussion in the Netherlands to build a new airport in open seas.) Such an island has to serve mostly one or a restricted number of special functions (e.g. hosting an offshore airport or heavy industry). Recreation is only seldom a primary function of an artificial island. Consequently the seaward borders of the island must only protect the core of the island, so that dikes and/or revetments are mostly chosen as limits for artificial islands rather than sandy beaches.

A large land reclamation project in front of, and connected with, the existing coast in open sea is another possibility to achieve new land in the coastal zone area, which can be used for many important functions (e.g. housing, transportation, recreation and industry).

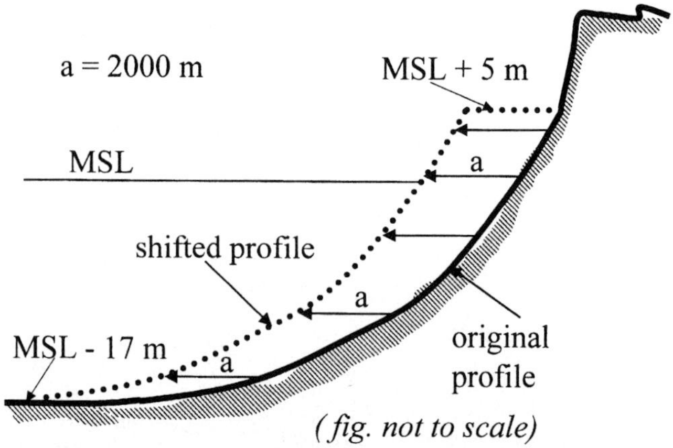

Fig. 4. Land reclamation with shifted depth contours.

The existing beaches are usually used for recreation purposes, so that one of the primary requirements of the 'new' coastline of the reclamation project will be the creation of 'new' beaches to serve recreation purposes again. That excludes the use of dikes or revetments as a seaward limit of the reclaimed area (e.g. the plan New Holland in the Netherlands).

Provided that the new situation requires indeed a beach as a sea front, a zero option for the execution of the project would be to shift all depth contours with the required distance (say 2 km) in an offshore direction. In that case the shape of the cross-shore profile is after the project the same as before (Figure 4). In a first approximation the coastal processes (longshore and cross-shore sediment transports) are expected to change only slightly by the reclamation project. Shifting all depth contours (say from MSL + 5 mtoMSL - 17 m; MSL: Mean Sea Level) means that for each m^2 of new land 22 m^3 of sand is required. Much of the eventual required volume is stored in deeper water in the 'toe' of the cross-shore profiles. If one would be able to avoid the fill of the toe, large savings in capital costs can be achieved.

2.4. STABILIZATION OF DYNAMIC TIDAL INLETS

The position of the main channel of tidal inlets is often not fixed. Natural conditions may cause a gradual shift in direction; sometimes the main channel is moving to and fro. A mobile behaviour of a tidal inlet often harms safe navigation and the integrity of property, built at both adjacent land areas, may be threatened. When the mobile behaviour of a tidal inlet system is felt to be unwanted, stabilization of the inlet system is then a genuine option. Structures may be erected to reach that aim.

2.5. DISCUSSION OF COASTAL PROTECTION PROBLEMS

A few examples of general coastal engineering problems have been briefly discussed in the preceding subsections. Different problems call for quite different solutions. Each problem is caused by specific conditions. In developing proper solutions to each specific problem, one has to meet problem-specific requirements. That calls for a proper insight into the physics of

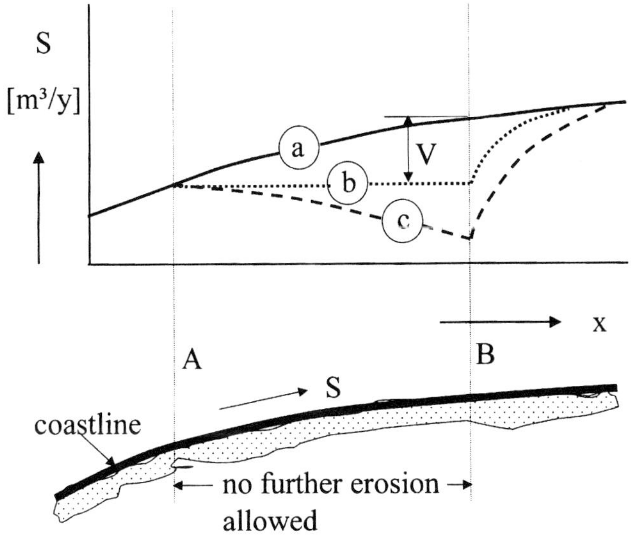

Fig. 5. Longshore sediment transport distribution along eroding coast.

the problem. In all problems as discussed so far, sediment transports play an important role. In solutions with structures ('hard' solutions), these structures must interfere in a subtle way with the sediment transports involved. In Section 3 this topic will be discussed further.

3. Use of Structures in Coastal Protection

3.1. INTRODUCTION

The use of structures as a tool in coastal protection relies in most cases on the ability of structures to interfere in the existing sediment transport processes. For stretches of coast suffering from structural erosion, often a *gradient* in the net longshore sediment transport along that stretch is the main reason of the erosion problem. Figure 5 shows in plan view such a stretch of coast (lower part). In the upper part of Figure 5 the net yearly sediment transports S along the coast (S: expressed in m^3/year) have been schematically indicated. Apparently the sediment transports increase with increasing distance x along the coast; curve *a* of Figure 5 represents this longshore sediment transport distribution. The increasing yearly transport from A to B (difference V m^3/year) is the very reason of the erosion problem in stretch A–B. (Along A–B: $dS/dx \neq 0$.)

Assume that one likes to protect only stretch A–B of the coast; for instance because of important investments have been made in section A–B, which are at stake, due to the structural erosion processes. Necessary and sufficient for that goal is that the existing sediment transport distribution curve *a* in Figure 5 is changed to curve *b*. (Along section A–B: $dS/dx = 0$ in that case.) At least in section A–B the erosion stops, if distribution *b* would be achieved. In the left-hand section from A the erosion will continue; the sediment transports did not change. In the right-hand section from B the existing erosion continues even at a higher rate, since the amounts of sediment passing through the cross-section in point B have been reduced, yielding steep gradients in the longshore sediment transport distribution in the section of the coast right from B. At the right-hand side of B consequently so-called lee-side erosion occurs.

To formulate the requirements for the sediment transport distribution in section A–B is rather simple; how to acquire curve *b*, however, is rather difficult. Structures which are in principle able to interfere in longshore sediment transport rates can be used. Series of groynes, series of (emerging) offshore breakwaters or submerged breakwaters will undoubtedly affect the existing longshore sediment transports. Properly tuning of these countermeasures is in any application a hard task. If the effectiveness of the countermeasures is less than expected, the erosion will be reduced but not fully stopped. If the design is too effective (the reduction of the sediment transports in stretch A–B is too large; see curve *c* in Figure 5), accretion in stretch A–B will be enforced. That may be beneficial for section A–B (accretion instead of the wanted stabilization), but the lee-side erosion in the section at the right-hand side of B will grow worse. Achieving curve c in Figure 5 represents in fact an 'over-kill' operation.

Even if the countermeasures in section A–B are well tuned, lee-side erosion beyond B will occur. Often this (increased) erosion beyond B is sooner or later also unacceptable, calling also for countermeasures in this section. The lee-side erosion is then shifted further downdrift, etc.

Lee-side erosion is in general unavoidable if an effective design is made for section A–B. In fact the solution of the erosion problem in section A–B with the help of structures which decrease the sediment transports along A–B, is at the expense of the stretch of coast beyond B; the problem has been *shifted*. Only if beyond B an accreting stretch of coast is present, or if position B represents the very end of a stretch of coast (beyond B e.g. a tidal inlet), lee-side erosion is less visible.

If in section A–B artificial nourishments had been selected as countermeasure, volume V (see Figure 5) had to be nourished on a regular time basis. In practice it is not useful to nourish V every year. Usually for a nourishment scheme a life-time of 5 till 10 years is chosen.

With artificial nourishment, the erosion problem is not solved; the nourishments do not reduce the sediment transports involved; the erosion continues. Consequently the nourishments have to be repeated. That seems a drawback. In many cases, however, artificial nourishments turn out to be more cost-effective than solutions with structures. A large advantage of artificial nourishments is that lee-side erosion does not take place.

Artificial nourishments are often placed at the (dry) beach in a cross-shore profile ('beach nourishments'). A rather new method is to artificially nourish the shoreface (see e.g. NOURTEC Final Report, 1997). The material is dumped in that case at a depth of, say, MSL – 5 m. It is expected that the sand as dumped, reaches eventually (partly) the beach by cross-shore transport processes. Shoreface nourishments are cheaper to execute than beach nourishments.

Many details related to the use of artificial nourishment are found in Coastal Engineering's Special Issue on Artificial Beach Nourishments (1991). In that Special Issue also many references are given to papers specially devoted to artificial nourishment.

Here the use of structures is the main point of interest. In this Introduction the structural erosion problem as indicated with the help of Figure 5 was chosen to clarify the possible use of structures in solving the erosion problem. Interference in existing sediment transports had become a necessary requirement for structures as countermeasure. From this requirement it can easily be understood that the following tools can in principle be used as countermeasures:

- series of groynes;
- series of detached offshore breakwaters;
- submerged breakwaters.

All these possibilities are able to affect the existing longshore sediment transports which take place throughout the year because of ordinary conditions, as well as under storm conditions.

The use of *(sea)dikes, shore parallel seawalls or revetments* (the latter two built along the front slope of the dunes or the mainland), has on purpose *not* been mentioned in the previous list as a possible countermeasure to the structural erosion problem in stretch A–B of Figure 5. It is a very bad solution to this type of erosion problems. Just after the construction of these structures the losses of dunes or mainland are effectively prevented indeed, but the cause of the underlying erosion problem is not cured. The erosion of the beach and nearshore will continue because of the existing gradient in the longshore sediment transports. Later on the attack to the structures will increase because of the disappearance of beaches. Damage to the structures will occur; the structures have to be repaired and strengthened. Finally all the beaches in front of the seawalls or revetments have disappeared and heavy structures remain to protect the mainland (see also Van de Graaff and Bijker, 1988).

From Figure 5 it had become clear that the existing transport distribution curve a must be changed into curve b in order to stop effectively the erosion along stretch A–B. When the beaches in front of the dike, seawall or revetment have finally disappeared and just in front of these shore parallel structures the bed level has become well below mean sea level, one could argue that the required reduction of the longshore sediment transport will occur indeed. So eventually this type of structures will 'work', however, only if the beaches are allowed to disappear. Many striking examples of this 'misuse' of coastal structures for the solution of structural erosion problems, can be found along coasts all over the world (see Figure 6).

> When one does not like beaches in front of his property, using seawalls or revetments is a perfect method to get rid of the beaches if the coast is suffering from structural erosion.

'Hard' tools which might 'work' if a structural erosion problem has to be resolved, are briefly discussed in the following.

Fig. 6. Damage of a revetment applied in a structural erosion problem.

Series of Groynes

Groynes (built more or less perpendicular to the coast) may reduce the wave-driven longshore currents directly. Possible shore parallel tidal currents are diverted to deeper water. Both effects contribute to a (wanted) decrease of the longshore sediment transports. If the groyne field fulfils the required behaviour in stretch A–B of Figure 5, always lee-side erosion has to be expected.

Series of Detached Offshore Breakwaters
Without going into details it can be argued that series of detached offshore breakwaters, built at some distance seaward from the shoreline, are able to reduce the wave heights in the zone landward of the structures. Reduced wave heights consequently yield a reduced longshore sediment transport. If series of offshore breakwaters are properly applied in de problem area A–B, lee-side erosion is unavoidable.

Submerged Breakwaters
Even submerged breakwaters (crest height below Mean Sea Level and built parallel to the shore) are able to reduce the wave heights in the zone landward of these structures. Also in this case a reduction of the longshore sediment transport rate in the area to be protected may be expected.

To finish this Introduction a few remarks can be made:

- From the erosion example of Figure 5 it has become clear that the solution of a coastal erosion problem always starts with a clear definition of the problem. This holds for the structural erosion problem, but similarly for all other coastal engineering problems.
- A clear definition of the requirements of a possible solution has to be given.

  • Should the erosion be stopped along the entire coast or only in a limited area?
  • Is halting the erosion in a limited area sufficient or is in fact a recovery (accretion) in that area desirable?

- In the next phase of a design, different alternatives have to be analyzed.

  • Which alternatives meet the requirements?
  • But also: what are possible unwanted detrimental effects side-effects?
  • And: what are the costs involved?

In the final selection phase the 'best' alternative has to be chosen.

- After implementation of the selected alternative in the field, it is strongly recommended to monitor the actual behaviour of the coast to be protected. Because 'art' and 'science' are still closely related in coastal engineering practice, it is not excluded that in some cases it has to be concluded afterwards that in fact an unlucky solution had been chosen. These experiences as obtained can be very helpful while designing new applications in future.
- Coastal protection as part of coastal engineering practice is a difficult task. Skilled and experienced professionals are required to do the job. In some countries special Institutes or Authorities have been appointed to carry out the tasks involved, e.g. Coastal Zone Authorities. [In the Netherlands specialized Water Authorities (Waterschappen in Dutch) are responsible for the management of the coast; the government, Public Works Department (Rijkswaterstaat) is also involved.] It is beyond discussion that such Authorities can only adequately operate if provided with governmental support and legal backing.

3.2. SOLUTIONS WITH STRUCTURES TO PROBLEMS AS MENTIONED

In this subsection some specific solutions with structures will be given to basic problems as briefly discussed in Section 2. The functionality of the different solutions with respect to the various problems will be the main point of interest. In all cases it is assumed that stable structures are designed. Appropriate structural design topics of the structures itself are not discussed. A recent overview of these topics can be found in 'River, Coastal and Shoreline Protection' (1995) and in the 'Shore Protection Manual'; CERC (1984).

Structural Erosion of Coasts

Some aspects of the case of structural erosion of coasts have already been discussed as a specific example in Section 3.1. Structures may be used as an alternative to resolve the structural erosion problem.

Series of groynes (or as an alternative: rows of wooden or steel piles) are in principle able to reduce existing sediment transport rates. The tuning problem (length and length/mutual spacing ratio) is, however, a difficult problem to be resolved. No general applicable design rules are available. Only some rules of thumb do exist. Groynes along the Dutch coast (first real groynes were built in 1776) were positioned by trial and error, so that gradually experience was gained.

In the example case of Figure 5 the existing sediment transport rates through the cross-section through point B had to be reduced by an (in this case) estimated factor of 0.5 in order to fulfil the requirements (i.e. achieving curve *b* in Figure 5; $S_{B\,new} \approx 0.5S_{B\,old}$). Notice that for arbitrary cross-sections between A and B basically different (larger) ratio's are required. Let point C be midway between A and B; then according to the example as shown in Figure 5: $S_{C\,new} \approx 0.7S_{C\,old}$.

Another point of practical concern is the absolute magnitude of the net yearly longshore sediment transports involved. In Figure 5 the basic erosion problem for stretch A–B is the loss of volume V m^3/year out of that stretch. By monitoring the coast for some time such a volume can be determined. The *difference* V of S_B and S_A is clearly determined in this way; the absolute values of the net yearly sediment transports of either S_B or S_A are in fact not automatically known. Since it is very difficult to calculate these sediment transport rates, errors can easily be made in the proper quantification of S_B and S_A. If in Figure 5 both S_A and S_B increase with ΔS, the difference V remains the same. In order to achieve also in this case (with increased transport rates), a constant sediment transport in section A–B, calls, however, for quite different reduction factors compared with the factors as mentioned previously.

If the erosion problem of stretch A-B has been resolved with groynes, an extra erosion with volume V in the area just at the right-hand side of point B has to be taken into account (lee-side erosion). Series of groynes do not effectively reduce the rate of dune erosion during a severe storm surge.

With the help of series of *emerging shore parallel offshore breakwaters* in section A–B also the structural erosion problem of stretch A–B can be resolved in principle. Such breakwaters are indeed able to interfere in the longshore sediment transports (are able to reduce these transports). Due to the (partial) shadow effects of the breakwaters, the general wave conditions landward of the series breakwaters are reduced, yielding smaller sediment transports. However, the proper tuning of a series of offshore breakwaters is a difficult task. In the example of Figure 5 the sediment transport through the cross-section at point B must be reduced from S_B to S_A. So near section B the remaining longshore transport should not be 0. Either in the area landward of an offshore breakwater near B or in the area seaward of that breakwater, still an on-going sediment transport is required. In the first case this calls for the forming of a so-called salient; in the latter case a tombolo may occur. Also with a proper series of offshore breakwaters lee-side erosion at the right-hand side of B is unavoidable. A series of emerging shore parallel breakwaters will also affect the rate of erosion of the dunes during storm surges. Whether that would (also) be an aim of using a series of breakwaters has to be clearly stated in the design phase of the project.

A series of *shore parallel submerged breakwaters* at some distance from the shore will undoubtedly reduce the wave heights landward of the submerged breakwaters (of course depending on the wave climate, the location of the breakwaters, the gap width between the breakwater segments and the crest height of the submerged breakwater segments relative to the still water level). By reducing the wave heights a reduction of the sediment transports can be expected. This holds as well for cases with only wave driven longshore sediment transports as in cases where a combination of wave driven currents and tidal currents occur. Although in the latter case the tidal currents landward of the series submerged breakwaters are, at a first approximation, unaffected, the resulting sediment transport will be highly reduced by the reduction of the wave heights in the area landward of the submerged breakwaters.

Shore parallel offshore breakwaters have not been applied along the Dutch coast. The tuning problem is felt to be too complicated. Besides the (unavoidable) lee-side erosion of a well-tuned system is not accepted in the Netherlands. The whole Dutch coast has to be preserved.

If an *'infinite' long submerged breakwater* is considered, no special additional problems are expected. In practice, however, 'infinite' long submerged breakwaters are not a realistic option. If, like in Figure 5 in stretch A–B, only locally provisions are required, at least two end sections of a submerged breakwater occur. Associated with the (partial) breaking of the waves above the crest of the breakwater, a mass transport of water over the breakwater takes place.

While in a pure two-dimensional case (cf. in a wave flume in a hydraulic laboratory) this mass transport is compensated for in the same cross-section, in a three-dimensional case the landward directed mass transport is often collected behind the submerged breakwater in an ever increasing shore parallel current. Along the end sections of the breakwater this current escapes in a seaward direction like a rip current. Examples are known from field and from laboratory tests (Van der Biezen et al., 1996) where severe erosion is generated landward of the submerged breakwaters.

As already was explained in Section 3.1 no proper solution is achieved in a structural erosion problem with the application of shore parallel seawalls or revetments along the front slope of the dunes or mainland.

Figures 7 and 8 show a remarkable example in the Netherlands of the 'growth' of a heavy sea dike over a few hundred years along a structural eroding part of the Dutch coast (Hondsbossche and Pettemer Sea Defence).

Fig. 7. The Netherlands with Hondsbossche and Pettemer Sea Defence.

Fig. 8. Picture of Sea Defence with adjacent coasts.

Along the west coast of the Netherlands over approximately 120 km nice beaches and, almost everywhere, vast dune formations exist. Over a stretch of about 5 km between Petten and Camperduin, dunes were almost missing because of a tidal inlet formerly existed in that area. To create a continuous coastline, already in the 15th century a sand dike had been put up in that region. In 1506 the first attempts to build 'groynes' of mainly wood and some stones were made. The groynes had to protect the sand dike. Storms in the past caused a lot of damage to the protection scheme. Many attempts to alter and to reinforce the groyne system were reported.

In the period 1839–1864 the toe of the sand dike was protected by connecting the landward root-ends of the groynes. Later on, after many adaptations, the present sea dike has gradually developed. The crest height of the Sea Defence is at the moment 14 m above MSL. Along the 5 km long Sea Defence some 30 stone groynes have been constructed. At present the Hondsbossche and Pettemer Sea Defence acts as a stronghold in the sandy coast (see Figure 8).

The adjacent sandy coasts were eroded by about hundred m during the last century. Both ends of the Sea Defence show at present a curve to smoothen the transition between the 'hard' and 'soft' parts of the coast. Along the Sea Defence only during Low Water some sand can be observed just in front of the toe of the dike. No normal beaches are present anymore.

The present policy of the Dutch government is to maintain the Dutch coastlines at the average position of 1990 (Ministry of Transport and Public Works, 1990). So nowadays the further retreat of the coasts at the north and south side of the Sea Defence is stopped. Beach nourishments are carried out in these areas on a regular time basis.

In summary: at large expenses in the past (and at present for maintenance) a 'strange' element in the sandy coastal system was developed. Although the Sea Defence has its own charm, the recreation potentials of this part of the coast are far less than along the adjacent sandy coasts.

Beach and Dune Erosion during Severe Storm Surges

If during design conditions the rate of erosion due to a severe storm surge is felt to be too large, the application of structures might be helpful to reduce the rate of erosion. Series of groynes do not 'work' to reduce the associated offshore directed sediment transports. Series of emerging breakwaters or submerged breakwaters are in principle able to reduce the wave heights landward from these structures and consequently may have a reducing effect on the rate of dune erosion. However, often an increase of the still water level is associated with a storm (the surge); consequently the effectiveness of these types of structures in reducing the wave heights is limited. Besides, the application of these structures has large (probably unwanted) implications for alongshore processes.

If a basically stable (stable: with respect to structural erosion) situation is considered, the use of seawalls or revetments may be useful to restrict the rate of erosion during a severe storm surge. With these structures the loss of material from the dunes or the mainland is physically prevented. The desired reduction of the retreat of the dunes or the mainland can be achieved. Since *behind* the seawalls or revetments no material can be eroded to fulfil the offshore directed transport capacities, one has to expect large erosion just *in front* of the structures. A deep scour hole can be expected; in the design of the structures this poten-

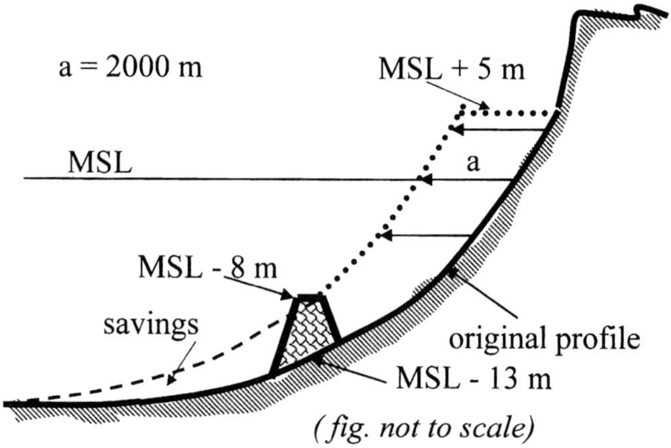

Fig. 9. Land reclamation with perched beach solution.

tial scouring has to be properly taken into account. Steetzel (1993) proposes a method to calculate the scour depth in front of seawalls and revetments.

Protection of Newly Reclaimed Areas
Assume a coast which has to be extended by a considerable distance (say in the order of 2 km in cross-shore direction; in longshore direction say over 20 km). In the present situation no serious erosion problems are assumed to occur. The area is extensively used as (beach) recreation area. One of the requirements is that after extension of the coast, again recreation beaches are available. To protect the newly reclaimed area simply by a dike or revetment is consequently an unacceptable option (see Section 2.3).

The more or less zero option for land reclamation would be an entire shift of the cross-shore profile of 2 km in seaward direction. This not only holds for the waterline, but in principle also for all other depth contours to a water depth where natural adaptations of the profile are hardly to be expected. In some cases this depth can be estimated at, say, 17 m below MSL. In order to achieve a 2 km shift of the coastline, per running m alongshore a volume of 2000 m $\times$ 22 m = 44,000 m^3/m sand is required. (The factor 22 m in the calculation is found by assuming a lower limit of MSL $-$ 17 m and an upper limit above MSL of 5 m.) With a longshore extension over 20 km, the total volume of sediment supply surpasses a volume of 800 million m^3. This is a really huge project!

The new coast has, with the zero option, a ‘foundation depth’ which is identical to the depth at the old coast. A large part of the volume that was calculated, is needed to make the new foundation. In order to restrict the volume of sediment needed for reclamation in the zero option, one could consider a longshore parallel submerged breakwater as an alternative to ‘support’ the upper part of the cross-shore profile. See Figure 9 for an example of an submerged breakwater; a perched beach is achieved; the toe of the horizontally shifted profile can be omitted.

In this example (a 2000 m seaward shift of the depth contours and a volume of 22 m^3 of sand needed to gain one m^2 of new land) 44,000 m^3 sand is required per m alongshore.

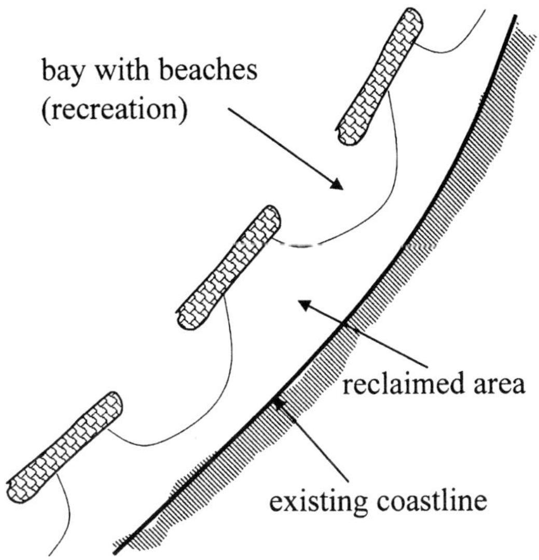

Fig. 10. Land reclamation with emerging breakwaters.

A submerged breakwater, supporting the upper part of the profile above MLS − 8 m, which is positioned in the original profile at MSL − 13 m, yields a reduction with approximately 13,000 m^3 sand per m (approximately 30% reduction of the required volumes of sand). Such a submerged breakwater is 5 m high; in this application 1 m^3 of stone 'saves' approximately 130 m^3 of sand.

If recreation beaches are required along the seaward boundary of the reclamation project, also a series of emerging shore parallel breakwaters might be applied (see Figure 10). Behind the gaps between the segments of the breakwaters typical (stable) bay-forms might be developed. The bays can be used for recreational purposes.

Stabilization of Dynamic Tidal Inlets

A lot of experience has been gained (especially in the United States) in the past with resolving the problem of stabilization of dynamic tidal inlets. In most cases long groynes or jetties at both sides of the inlet have been made. Since in most cases the natural sediment transport over the ebb tidal delta is fully inhibited, a proper design of the total solution calls for additional provisions in order to artificially restore the sediment transport across the tidal inlet. A sand by-pass system should therefore be an unavoidable part of a total design. To estimate in advance the required capacity of a sand by-pass system is a difficult task. Often that part of the total design is more or less postponed until the moment that nature shows by accumulation at one side and by erosion at the other, what are the real quantities involved. By postponing any measures (wait and see), sometimes no money is left at a later stage to build a proper sand by-pass system.

4. Recent Research in The Netherlands and Europe

Although the Netherlands has only a rather small sandy coast (approximately 300 km), that coast has offered many troubles to coastal managers over the past. The stretches of coast which suffer from structural erosion, are often 'repaired' with the help of artificial nourishments, which is a quite useful and successful method to resolve structural erosion problems.

Although artificial nourishments are very useful to resolve erosion problems, other methods are also possible. Recent research in the Netherlands was devoted to the applicability of such other methods in the struggle against erosion.

Series of experiments have been carried out in a wave flume and a wave basin of the Fluid Mechanics Laboratory of Delft University of Technology to study the effects of submerged shore parallel offshore breakwaters (see Van de Graaff et al., 1996; Van der Biezen et al., 1996). This research program in the laboratory was carried out within the framework of the Human Capital and Mobility program of the European Union.

Besides the experimental research a nationwide program has been recently carried out on the advantages of a combined application of beach nourishments and shore parallel structures. Mathematical models and a practical Design Support Model have been developed and used (See CUR reports on Beach Nourishments and Shore Parallel Structures, 1997a, b and c).

It has become clear that it is still very difficult to quantify the morphodynamic impact of detached offshore breakwaters. In a recent research project (see Bos et al., 1996), first attempts were made to calculate this morphodynamic impact. The results of calculations of the current pattern, induced e.g. by waves which approach perpendicular to the coast with a single offshore breakwater, show reliable features. The calculations were made with DELFT HYDRAULICS' compound model system DELFT2D/3D MOR. Also the calculated sediment transport patterns look reliable.

To model the impact of structures in general (port breakwaters, series of groynes, series of emerging and submerged breakwaters) on the morphology of the coasts where these structures are applied, is also a main topic in the Marine Science and Technology (MaST) project Surf and Swash Zone Mechanics (SASME) as partly sponsored by the European Union. Within a few years from now a fair increase of our modelling capabilities is to be expected.

References

Biezen, S.C van der, J. van de Graaff and J. de Later (1996) 3D Model Tests of the Influence of Submerged Breakwaters on a Beach Profile exposed to Regular Waves. *Proceedings Conference on Physics of Estuaries and Coasts*; The Hague, the Netherlands.

Bos, K.J., J.A. Roelvink and M.W. Dingemans (1996) Modelling the impact of detached breakwaters on the coast. *Proceedings 25th Int. Conf. on Coastal Eng.*; Vol. 2; Chapt. 157; Orlando; USA.

CERC (Coastal Engineering Research Center) (1984) *Shore Protection Manual*. US Army Corps of Engineers.

Coastal Engineering (1991) *Special Issue on Artificial Beach Nourishments*. Ed.: J. van de Graaff, H.D. Niemeyer and J. van Overeem. Eight papers dealing with several aspects of the use of artificial nourishments. Vol. 16, 1–163; Elsevier Science Publishers B.V., Amsterdam.

CUR (1997a) Beach Nourishments and Shore Parallel Structures. Phase I: Introduction and inventory. Center for Civil Engineering Research, Codes and Specifications, Gouda, the Netherlands.

CUR (1997b) Beach Nourishments and Shore Parallel Structures. Phase II: Modelling of sand transport. Center for Civil Engineering Research, Codes and Specifications, Gouda, the Netherlands.

CUR (1997c) Beach Nourishments and Shore Parallel Structures. Phase III: Evaluation by a design support model. Center for Civil Engineering Research, Codes and Specifications, Gouda, the Netherlands.

Graaff, J. van de (1986) Probabilistic design of dunes: An example from the Netherlands. *Coastal Engineering*, 9, 479–500.

Graaff, J. van de and E.W. Bijker (1988) Seawalls and shoreline protection. *Proc. 21st Int. Conf. on Coastal Eng.*, Malaga, Spain.

Graaff, J. van de, E.W.M. Claessen, M.D. Groenewoud and S.C. van der Biezen (1996) Effect of submerged breakwater on profile development. *Proc. 25th Int. Conf. on Coastal Eng.*, Volume 2; Chapter 188; Orlando; USA.

Larson, M. and N.C. Kraus (1989) SBEACH: Numerical Model for Simulating Storm-induced Beach Change. Report 1: Empirical Foundation and Model Development. Technical Report CERC-89-9. Vicksburg, Mississippi.

Larson, M, N.C. Kraus and M.R. Byrnes (1990) SBEACH: Numerical Model for Simulating Storm-induced Beach Change. Report 2: Numerical Formulation and Model Tests. Technical Report CERC-89-9. Vicksburg, Mississippi.

Ministry of Transport and Public Works (1990) A new coastal defence policy for the Netherlands. Rijkswaterstaat; The Hague.

NOURTEC, Innovative Nourishment Techniques Evaluation, Final Report (1997) Coordinating Institution Rijkswaterstaat – National Institute for Coastal Marine Management, the Hague, the Netherlands.

River, Coastal and Shoreline Protection (1995) *Erosion Control Using Riprap and Armourstone*. Ed.: C.R. Thorne, S.R. Abt, F.B.J. Barends, S.T. Maynord and K.W. Pilarczyk. John Wiley & Sons, Chichester.

Steetzel, H.J. (1993) Cross-shore Transport during Storm Surges. Ph.D. Thesis Delft University of Technology, Delft, the Netherlands. Also: DELFT HYDRAULICS Communication No. 476 and Report No. C1-93.05 of the Technical Advisory Committee on Water Defences (TAW) in the Netherlands.

THE DELTA PROJECT

G.J. SCHIERECK
Ministerie van Verkeer en Waterstaat
Afd. Q
P.O. Box 20901, 2500 EX Den Haag
The Netherlands

1. History

The history of protection against the sea in the Netherlands is a long one and a good example of leap frogging between man made solutions and man made problems. It started 2000 years ago with small scale interventions in the water management of the low lying soils of the Rhine–Meuse–Scheldt Delta, see Figure 1. Culverts for drainage, made of a hollowed trunk with a hinged cover, were placed in a dike, in order to prevent flood tides entering agricultural land, see Figure 2.

Although simple to our standards now, this certainly was a high-tech solution in those days. The results were higher yields to feed the inhabitants of the area, but also subsidence of the land due to the dewatering of the soil. This made necessary more dikes and more de-watering because of the lower level of the soil. Artificial means for drainage, like windmills and later steam engines, were used to cope with the subsidence, at the same time increasing this subsidence, see Figure 3. So, the story of the Netherlands as a country below sea level is caused by man himself. In this process of ages, parts of the Netherlands frequently flooded, after which the dikes were raised and strengthened again (see also Van der Ven, 1994). On February 1st 1953, the southwestern part of the Netherlands flooded due to a very high storm surge for which the dikes were too low, see Figure 4. Almost 2000 people drowned and a lot of damage occurred. Already before 1953, plans were made for a better protection of the area, but this disaster was necessary to trigger the decision to execute the Delta project.

2. The Project

2.1. INTRODUCTION

For the protection against flooding, two mainstreams of solutions were considered: raising all the dikes or damming the estuaries. Although the latter was more expensive, the choice for damming was made relatively easy. Damming had several advantages, like the possibility

J. Chen et al. (eds.), Engineered Coasts, 249–268.
© 2002 *Kluwer Academic Publishers. Printed in the Netherlands.*

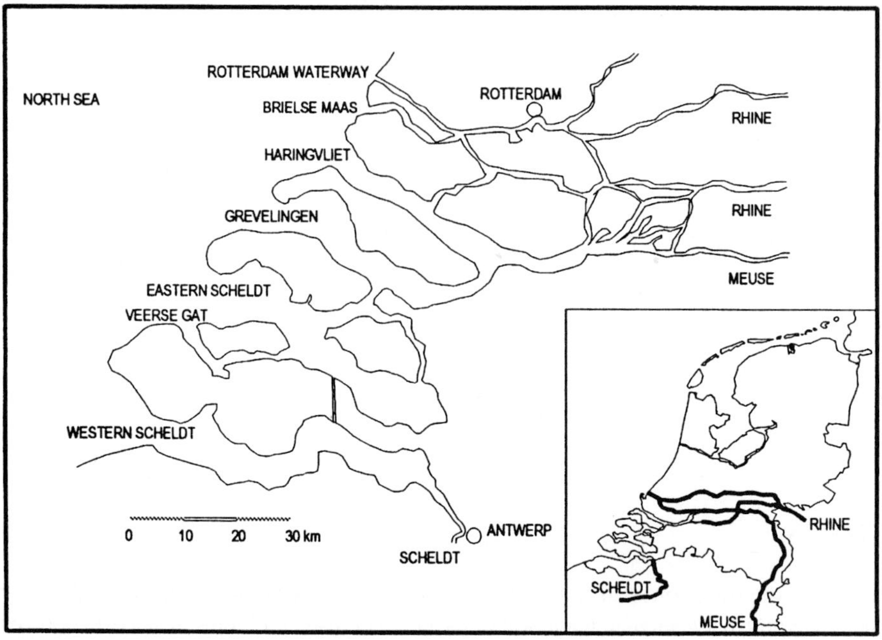

Fig. 1. The Rhine–Meuse–Scheldt delta in the Netherlands.

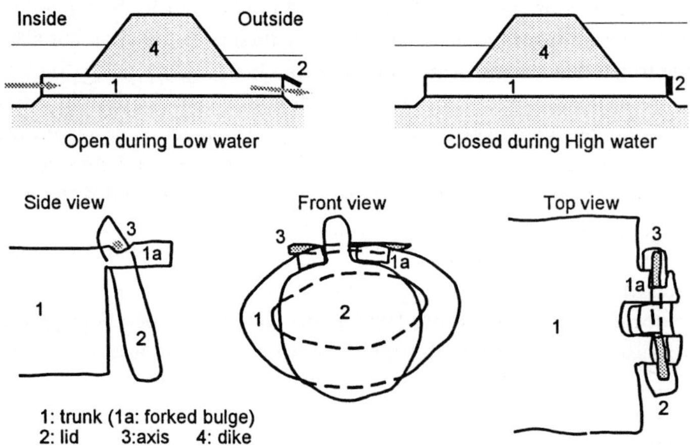

Fig. 2. Hollow tree culvert.

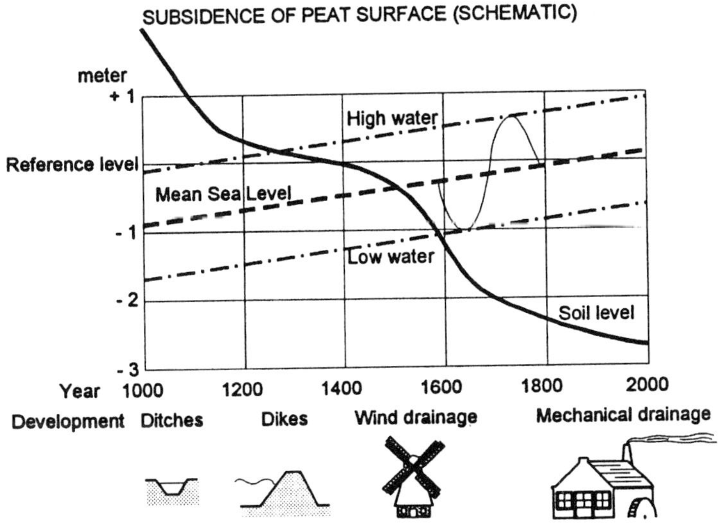

Fig. 3. Subsidence peat soil and sea level rise.

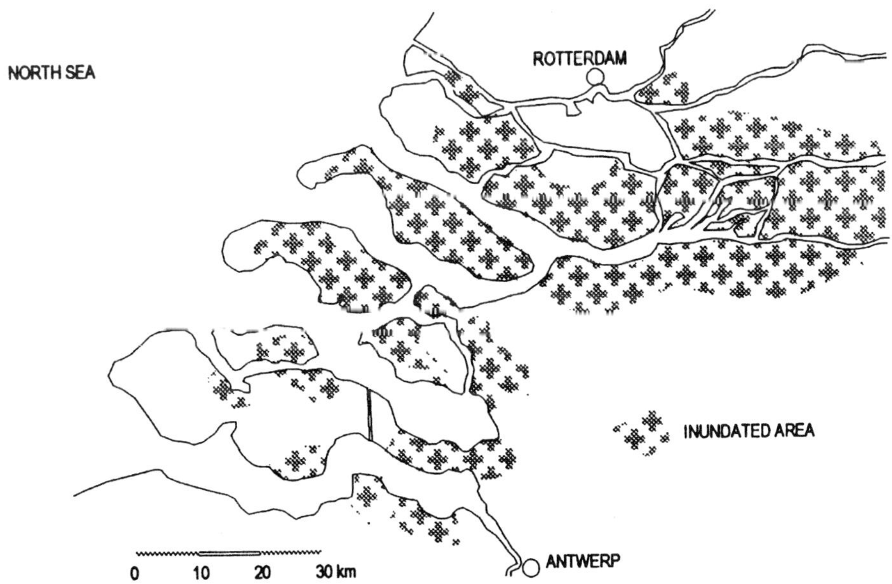

Fig. 4. Flooded areas in 1953.

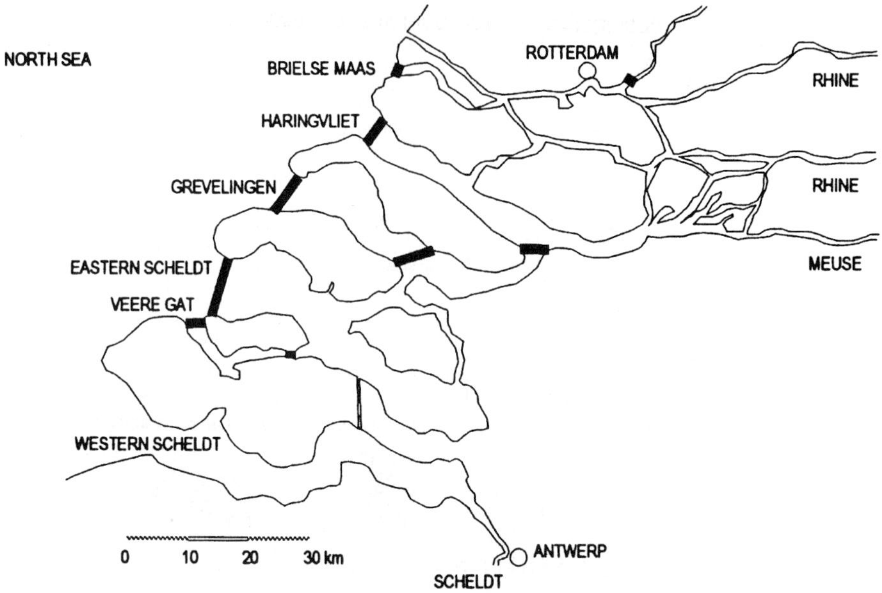

Fig. 5. The Delta project.

to create fresh water basins in order to improve the situation for agriculture and to create roads between hitherto isolated islands. Moreover, the general feeling was that this solution was more definite for the flooding problem. These additional advantages were decisive. The final proposal, as approved by government and parliament in 1958, consisted of the closure of the Haringvliet, the Grevelingen, the Eastern Scheldt and the Veerse Gat, see Figure 5. The Brielse Maas, south of the Rotterdam Waterway had already been closed in 1950. For the execution of these closures, several auxiliary dams were planned, which will be clarified in Section 3.3. The closures of the Western Scheldt and the Rotterdam Waterway were not envisaged in the original plan, because of the waterways to Antwerp and Rotterdam. Along these estuaries the dikes had to be raised. Also part of the project was a storm surge barrier in the Hollandse Ussel near Rotterdam. This was designed and constructed directly after February 1953, because the central part of Holland had a narrow escape then and the need for this barrier was undisputed. Major topics in the project were safety and water management. Although decisive for the future of the area, physical planning played a rather passive role. The environment did not play a role in the decision, but it did in the course of time and the original plans had to be changed, resulting in a storm surge barrier in the Eastern Scheldt instead of a massive dam. Also in the Rotterdam Waterway, finally a storm surge barrier was made, because of new data on the storm surge levels and the problems that occurred with the raising of the dikes. All these aspects will be dealt with in the coming sections.

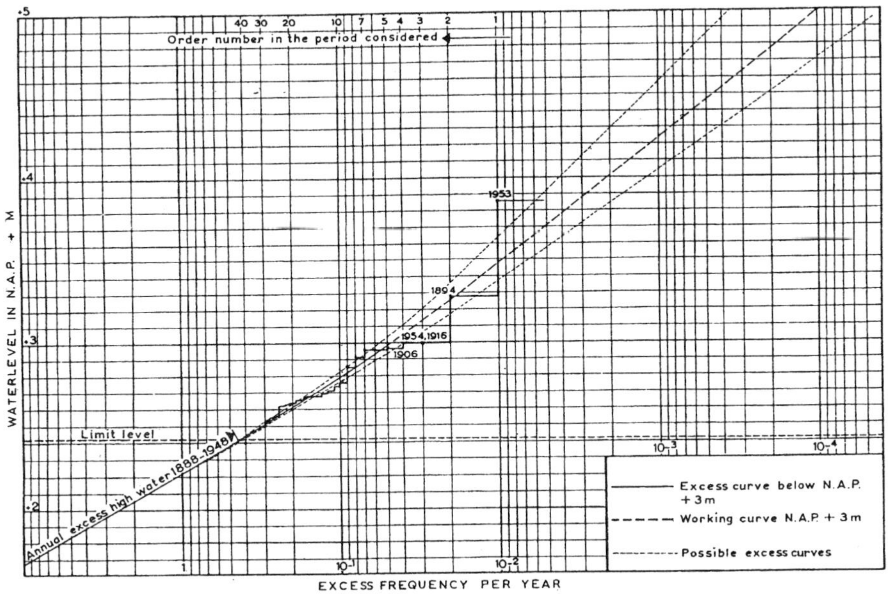

Fig. 6. Statistics of storm surge levels at mouth of Rotterdam waterway.

2.2. SAFETY

Before 1953, the height of the dikes was based on the highest known waterlevel at a certain location plus some margin. Every time the dikes appeared too low again during a high(er) level, this level automatically became the new highest known waterlevel. Already in 1939, statistical techniques were used to deal with this problem in a different way (Wemelsfelder, 1939). The idea appeared that there was no practical limit to the waterlevel, only the frequency of occurring becomes lower when the level increases. This knowledge should be used to design sea defences with a certain chance of flooding related to the involved risk. Therefore, the data on storm surge levels had to be extrapolated to values with a very low frequency, see Figure 6. The Delta committee, appointed by the government to counsel on the problem of flooding, finally chose a design level with a chance of exceeding of 0.0001/year (Delta committee, 1962). This approach was a breakthrough in the thinking on the design of sea defences and is still valid today although the time has come to adapt it again to modem risk analysis techniques.

2.3. WATER MANAGEMENT

An estuary is, by definition, a meeting place for fresh water from rivers and salt water from the sea. The supply of fresh water has always been a problem in the low lying areas of the Netherlands, related to the subsidence of the soil and saline seepage. This problem was aggravated by the extension of the port of Rotterdam. In an estuary there is a balance between river flow and salt water intrusion, see Figure 7.

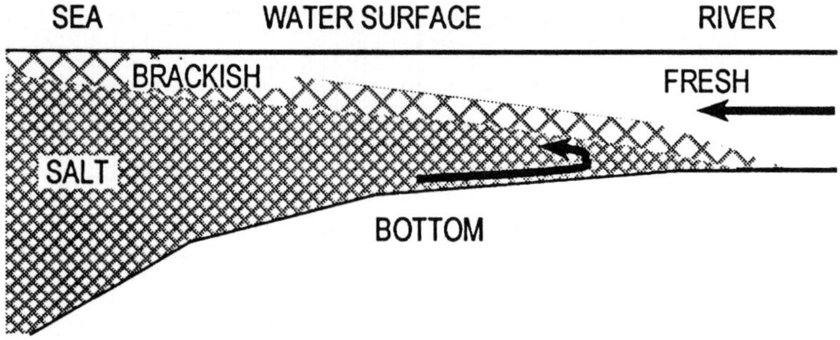

Fig. 7. Salt intrusion in estuary.

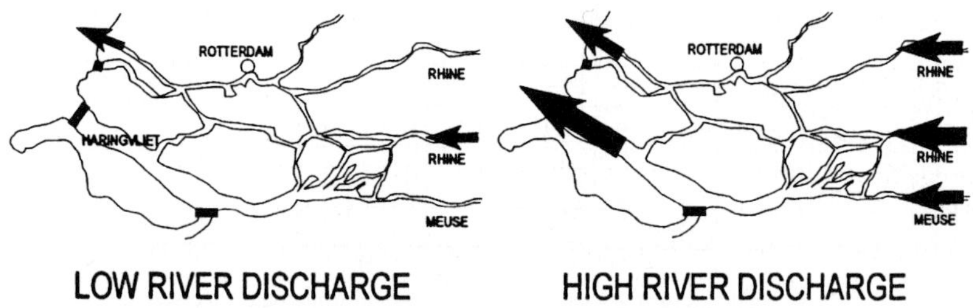

Fig. 8. Use of Haringvliet sluices during dry and wet periods.

Due to dredging, to accommodate larger ships, the depth of the Rotterdam Waterway increased considerably in the last century. Therefore, the power of the river water to push back the salt water relatively decreased and saline water intruded further inland. Also the construction of new harbour basins along the waterway contributed to the salt intrusion due to the increase of the tidal prism. The result was increasing problems in fresh water supply. The Delta project was used to seize the opportunity to improve the freshwater management in the southwestern part of the country as a whole. The main outlet of the river Rhine to the sea is the Haringvliet estuary. This function had to be maintained, in order to avoid high levels upstream during high river discharges. So, the dam in the Haringvliet also includes a large outlet sluice. This is opened completely when the Rhine discharge is larger than 6000 m^3/s. When the discharge is lower than 1500 m^3/s, the sluices are completely closed and practically the whole discharge of Rhine and Meuse is forced to flow via the Rotterdam Waterway, see Figure 8. There it serves to push back the sea water as much as possible. For Rhine discharges between 1500 and 6000 m^3/s, the sluice is partly opened. The Haringvliet and the other estuaries were seen as important freshwater reservoirs. Salt intrusion in these basins was excluded by the closure dams.

2.4. PHYSICAL PLANNING

Before 1953 the development of the southwestern part of the Netherlands was handicapped by three physical factors:

- The area was cut by east-west running estuaries, separating the industrialized areas of Rotterdam and Antwerp;
- The island structure of the area made it isolated and fragmented;
- The saltwater tidal area gave restrictions for the agricultural sector in dry periods.

Due to all this, the area as a whole was underdeveloped and there was a surplus in the departure from the area leading to ageing (Plancommissie Zuidwest, 1954). The Delta project was seen as an excellent opportunity to deal with these shortcomings. Before execution of the Delta project, the main sources of income were agriculture, fishery (shellfish and crustaceans) and some industry along the Western Scheldt. The area had large potentials for recreation and horticulture, but due to the poor infrastructure, these potentials could not be materialized. The dams in the framework of the project could not solve this problem completely. The closure dams were located at the western border of the area, giving a good connection between Rotterdam and the recreational areas along the coast. Secondary dams (see Section 3.3) together with a bridge across the Haringvliet, made a good road connection between Rotterdam and Antwerp. The secondary dam through the Grevelingen was very useful to open up the central part of the area, but across the Eastern Scheldt there was still a missing link. There too, a bridge was built to complete the network. So, although no part of the Deltaworks, the bridges across the Haringvliet and the Eastern Scheldt completed the infrastructure to enclose the whole southwestern part of the country, see Figure 9 (from Plancommissie Zuidwest, 1955). Looking back into history it is remarkable that physical planning completely followed hydraulic engineering. Obviously, safety and water management dominated the project.

Another planning aspect was the Belgian wish for a better shipping connection between Antwerp and the Rhine, see Figure 10. This was incorporated in the course of the project. For this canal it was agreed that there would be no tidal movement. This could be done easily because the Eastern Scheldt estuary would turn into a freshwater lake, see Figure 5.

2.5. ALTERATIONS IN THE PROJECT

In retrospect, for a project with the scale of the Delta project and with an implementation period of several decades in an open society with growing population and economy and with the cultural changes that came with the 1960s and 1970s, it would be naive to think that execution would go on without any changes. Minor and major alterations of the plan have been carried out in the course of time. Only the major alterations will be discussed here.

Eastern Scheldt

The environment did not play an important role in the decision on the project. In the 1950s, in the aftermath of the second world war, the focus was on reconstruction and economic growth. Safety and (fresh)water management were the main items to justify the project, while the new roads on the dams were seen as a good opportunity to give an impact to the economy of a hitherto somewhat underdeveloped area. The only foreseeable disadvantage of the project was the disappearance of shellfish cultures, mainly in the Eastern Scheldt. This was taken for granted and alternative locations were sought after. At the end of the 1960s, however, a new awareness arose and the intrinsic value of estuaries became an important

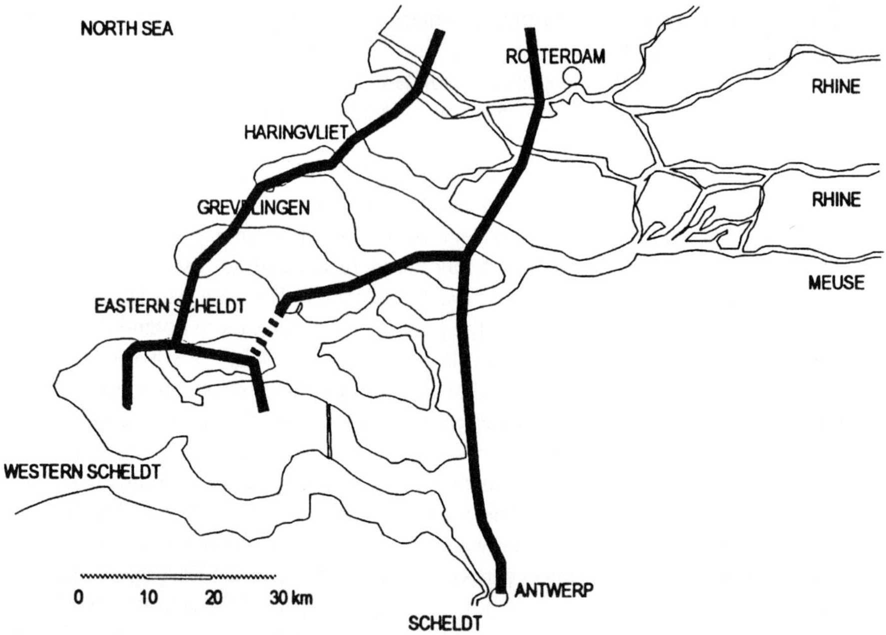

Fig. 9. Road connections in the Delta.

issue. Combined efforts of conservationists and shellfishers to influence the public opinion and policy makers finally resulted in a change of plans for the final closure of the Eastern Scheldt. Instead of a closure dam, a storm surge barrier was built. This barrier is normally open, letting the tides in and out. This barrier is combined with two secondary dams in the estuary, in order to separate the tidal estuary from the Scheldt–Rhine canal. Behind these secondary dams, a freshwater compartment is realised in order to make a tidal free shipping route from Antwerp to the Rhine (as was agreed between Belgium and the Netherlands) and to provide the adjacent areas with fresh water for agricultural purposes, see Figure 10. Moreover, these dams helped in maintaining the tidal difference in the estuary, because the area to be filled and emptied by the tides becomes smaller. For these alterations, a more integrated study was done than for the original Delta project, in which environmental and physical planning aspects were involved. Also, policy analysis techniques were used for the first time in the Netherlands for a hydraulic engineering project of this scale, see Rand (1977).

Grevelingen

Another adaptation of the Deltaproject concerns the Grevelingen. Originally meant as a freshwater reservoir, it was finally decided to keep this lake salt. From a study into the pros and cons of a salt or fresh water lake, it appeared that a salt water lake had more values, also economical. Biodiversity in the lake is large compared to the possibilities of a freshwater lake (which has to be filled with water from the Rhine) and there are possibilities for shellfish

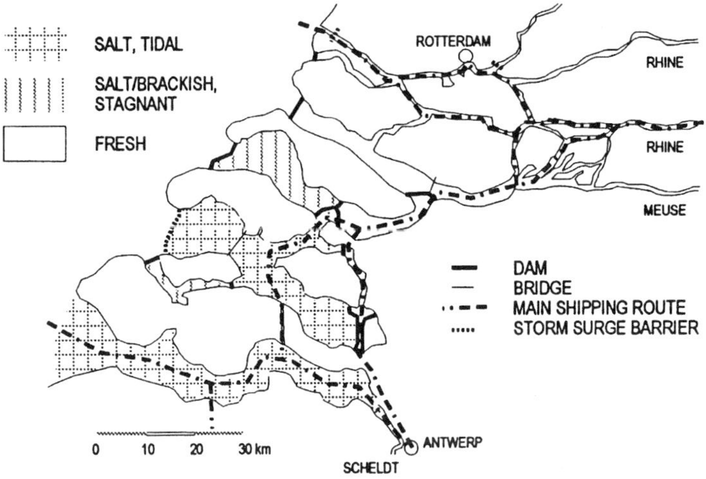

Fig. 10. Final outcome of Delta project (1985).

cultures. Fresh water for irrigation can be extracted from the lake behind the secondary dams of the Eastern Scheldt by constructing a pipeline (which has not been done yet). The lake is kept salt by a small inlet sluice in the closure dam.

Rotterdam Waterway
In the original Delta project, closure of the Rotterdam Waterway was not envisaged, because of the hindrance this would give to navigation. During the strengthening of the dikes along the waterway, a lot of difficulties were encountered, especially in urban areas because of conflicting interests between town planning and safety against flooding. When it appeared from new data that the design waterlevels had to be chosen higher to maintain the required safety level, it was decided to look for another solution. The final choice became a storm surge barrier with minimal hindrance to shipping, see Figure 11. When open, the waterway is hardly narrowed, while the closing frequency will be once every ten years on an average.

3. Technical Aspects

3.1. INTRODUCTION

The construction of closure dams in tidal areas is such a specific technique that some attention should be paid here to the technical aspects. Before World War II, there was experience with tidal closures on a small scale using traditional skills. By sinking willow matresses covered with stones, the depth of a tidal channel was decreased gradually. Because of the equipment used and the labour intensiveness of the method, only relatively small channels could be closed. Another method was to build a simple bridge from which stones were dumped. The closure of the Zuyderzee in the north of the country in the 1930s was done

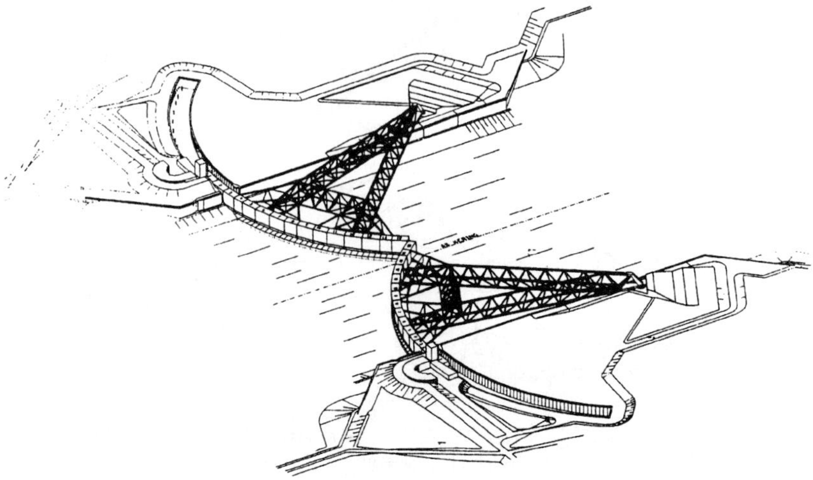

Fig. 11. Storm surge barrier in the Rotterdam Waterway.

with large cranes and lumps of boulder clay which was available in the area (coming from moraines in the ice age). Oddly enough, World War II itself meant a breakthrough in closing techniques. The British had bombed the dikes of the island of Walcheren in 1944 in order to chase away the German occupation forces. After that, the gaps had to be closed again. The country, being devastated, missed the equipment to perform the closure works while there was much pressure of time to save the island. This led to many improvisations and innovations. One of the major improvements was the use of large caissons (45 m long) designed as breakwater elements for artificial harbours at the allied landing stages in Normandy. These large rectangular elements did not fit very well in the channels and a strong current under and at the sides of the caisson gave much erosion ('piping'), jeopardizing the stability of the caisson. To stop that, torpedo nets, another remnant of the war, were used. The Dutch writer Den Doolaard has described this operation in his novel "Roll back the sea" (Den Doolaard, 1946).

3.2. TIDAL FLOW IN CLOSURE GAPS

When a tidal channel is being closed, the gap in the mouth will act as a hydraulic resistance. Due to this resistance, the total flow through the gap decreases when the gap becomes smaller. This will lead to less tidal motion inside the estuary and this will increase the difference in waterlevel between the sea and the estuary which, in turn, will increase the velocity in the gap. Figure 12 shows the tidal curves inside and outside an estuary for different closing stages.

Hence, the smaller the gap, the smaller the total flow, but the larger the flow velocity. The maximum velocity can be approximated by using Torricelli's law: $v = \sqrt{2}\,gz$, where z is the waterlevel difference, which is equal to the tidal amplitude when the gap is almost closed. With tidal amplitudes of 1 to 1.5 m in the Delta area, this leads to velocities of around 5 m/s. These high velocities will erode the original bottom, unless it is protected which can be done

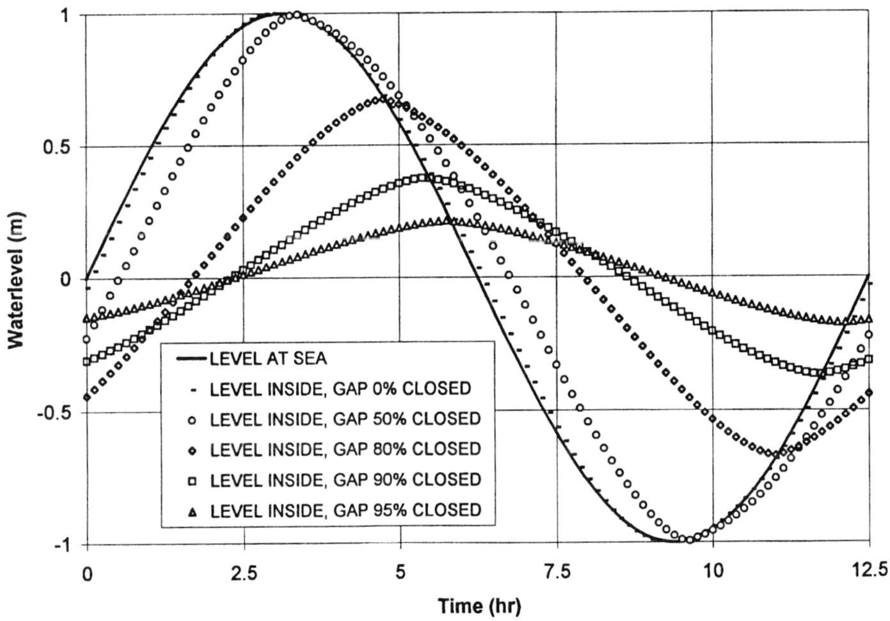

Fig. 12. Changes of tidal levels inside estuary due to closing operation.

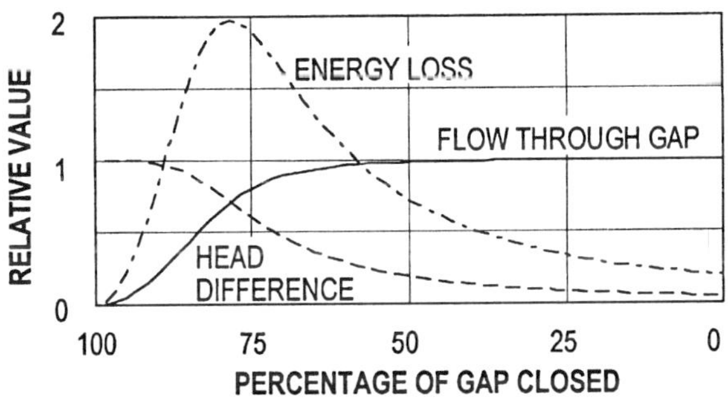

Fig. 13. Energy loss during closure operation.

with matresses of all kinds. But also then, at both sides of this protection, erosion will take place. This erosion is related to the energy loss of the water movement. Figure 13 shows the total flow through a gap and the waterlevel difference across the gap. The product of these two is a measure for the energy loss. The figure shows a maximum when the gap is closed for 70–80%. This is the critical phase of most closure operations. When erosion becomes too large, the stability of the closure dam as a whole can be jeopardized.

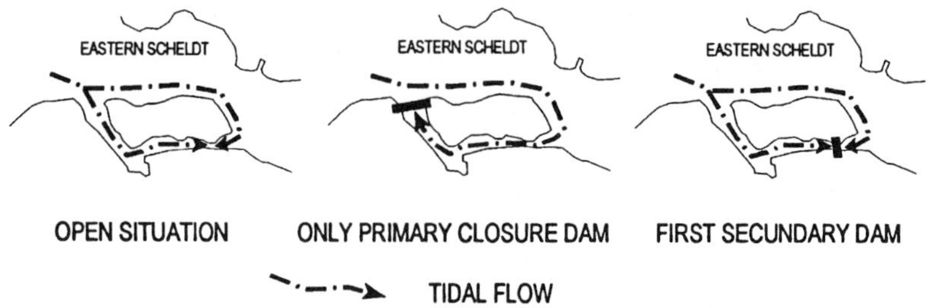

Fig. 14. Function of secondary dams.

3.3. TIDAL FLOW IN DELTA AREA

The consideration in Section 3.2 is limited to the situation when the mouth of an estuary is closed and there are no other connections with the sea. In the delta, the situation is more complex. For instance, when the Eastern Scheldt is closed before any other dam is made, the tide can still come from many sides. This, in turn, will lead to a considerable change in velocities in the area. At locations where there was no velocity at all, because of meeting of the tides there, this will give much erosion which can endanger e.g. bank protections. Therefore, auxiliary dams are needed to prevent that. Figure 14 clarifies this process.

3.4. CLOSURE METHODS

There are many methods and many materials with which a tidal estuary can be closed. The choice depends mainly on costs (availability of material and equipment) and reliability. The scale and complexity of the Delta project was such, that at the start it was not yet known how to close the Eastern Scheldt. Therefore, the execution started with relatively small closures and every closure was an exercise for the next one. Here, three methods will be discussed briefly.

Sand Closure

Sand is available in large quantities along the Dutch coast. So, the first option would be to close a gap by dredging sand and pumping it into the dam to be made. When a gap is closed gradually, the velocity will increase to about 5 m/s. Since the loss of sand is proportional with the velocity to a power 4 to 6, the sand losses increase tremendously. A practical limit for the maximum velocity in sand closures is about 2.5 m/s. This means that the mouth of an estuary practically never can be closed with sand completely. Only parts of the closure dam can be built with sand.

Rocks

Rocks are available in sizes large enough to withstand the velocities at the end of a closure operation. When no rocks are available, concrete cubes can be a good substitution. These elements can be dumped by means of barges, trucks, cableways or even helicopters. Most closures in the Delta project with this material were done with cableways, see Figure 15A.

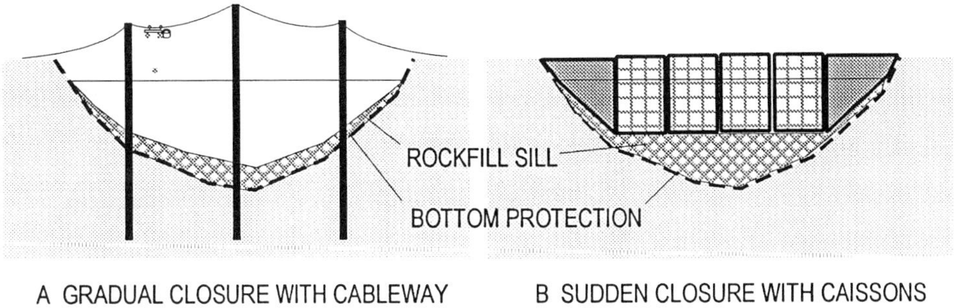

Fig. 15. Closure methods for final gaps of an estuary.

This had also the advantage that the dam was raised vertically, giving less attack on the bottom than when the gap is closed horizontally by stones being dumped from two sides.

Caisson

After the experiences with Walcheren, caissons were improved by making them open. With shutters they were made floatable and with steel gates they were closed once all the caissons were in position. By doing so, a relatively large opening is maintained and shut within minutes, see Figure 15B. This means that the critical phase in Figure 13 can be avoided as much as possible.

3.5. THE EASTERN SCHELDT STORM SURGE BARRIER

The opposition against the closure of the Eastern Scheldt made it necessary to come with new techniques. The traditional method to build a barrier with moveable gates was to make a construction pit in which the structure is built 'in the dry'. This would either mean a temporary complete closure of the estuary or to make alternately a pit in one of the three tidal channels. The first alternative was deemed unacceptable for the environment (the closure, being the crux of the opposition, would last several years) while the second was unacceptable for the safety which would be reached only after many years. Therefore, the barrier had to be built on the spot without interrupting the tidal movement. The experience with caisson closures was the starting point of the new designs. The final solution was reached with large concrete piers which were floated into position with a special barge. These piers were connected with concrete beams on top of which the steel gates were placed. The bottom needed a long and heavy protection against the day to day high velocities through the barrier and for the situation when one of the gates would not be closed during a storm, see Figure 16. This construction method leads to a situation in which the critical closure stage of Figure 13 will last forever. In the early design stages this lead to much objections among the generation of hydraulic engineers who had their experience from the first caisson closing of the dike breaches in Walcheren, see Section 3.1. The danger of scour and piping every day for many decades was an unprecedented engineering challenge.

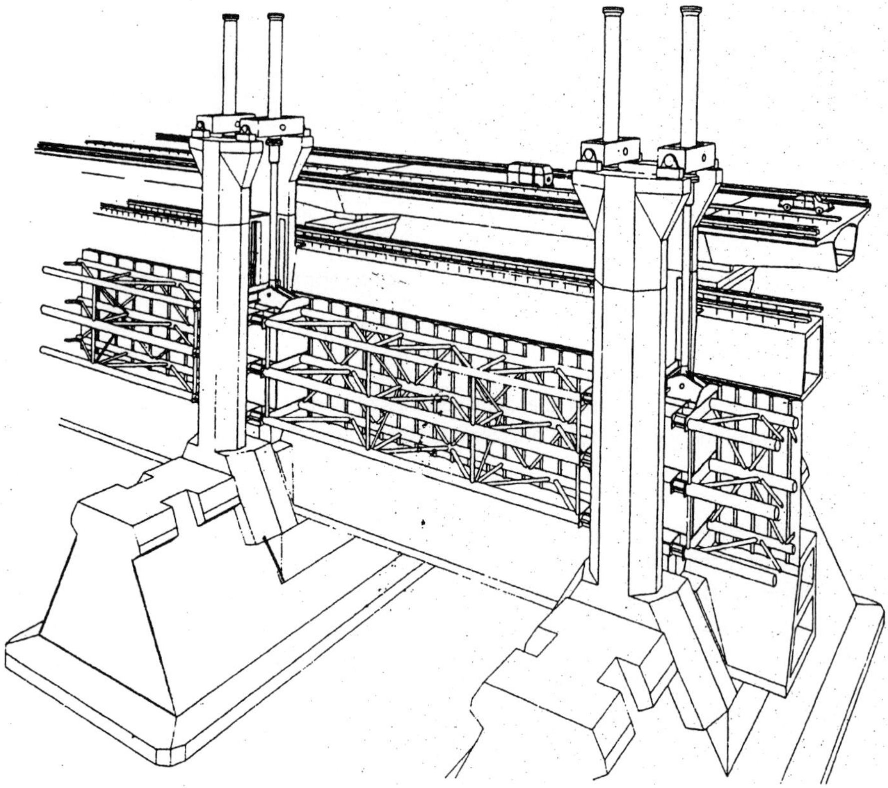

Fig. 16. Storm surge barrier Eastern Scheldt.

4. Environmental Impacts

4.1. INTRODUCTION

One of the major changes in society during the implementation of the project was a growing awareness of the importance of environmental aspects. In 1971 this led to the establishment of the Environmental Research Division within the Delta Department, being responsible for the whole project. This division appeared to be invaluable for an integrated approach of the further planning and execution. As a matter of fact, the change in thinking about large coastal engineering projects in the Netherlands was greatly influenced by the experience with the Eastern Scheldt estuary. Only some environmental aspects can be mentioned here briefly.

4.2. PHYSICAL ASPECTS

Decrease of tidal movement, desalinisation, changing river velocities are some catchwords to summarize the physical changes in the delta. These changes again induced other processes.

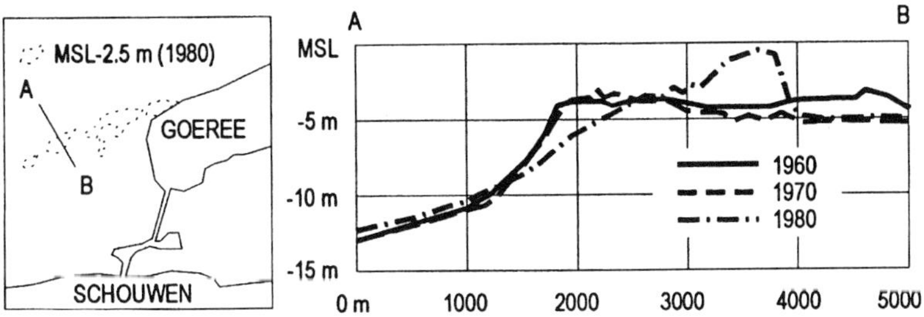

Fig. 17. Morphological development along Delta coast.

Bank Erosion

In a tidal environment there is a balance between erosion due to wave action and sedimentation due to tidal flow. When the tide disappears or diminishes, the erosion will win. This is what happened with the banks and tidal flats in the Haringvliet and the Grevelingen. Extensive protection measures are necessary now to prevent massive erosion.

Fore Delta

One of the effects of the project was the emerging of large tidal flats in front of the closure dams, see Figure 17. The Delta committee already predicted sedimentation of the tidal channels in the mouths of the estuaries, but the extension of the new tidal flats was somewhat unexpected.

Sedimentation Haringvliet

In the eastern part of the Haringvliet, the river flow velocities have decreased. In fact the Haringvlietdam acts as a barrage and the estuary becomes a reservoir. This sedimentation as such is not a problem, but the quality of the (Rhine) sediment turns it into an environmental problem.

4.3. WATER QUALITY ASPECTS

General

The main source of fresh water in the delta is the Rhine. The water quality of this river was bad up to the 1970s. This was one of the main reasons for the opposition against closing the Eastern Scheldt which should have been filled with Rhine water. After implementation of the internationally agreed measures, the water quality gradually improved. Figure 18 shows the contents of six heavy metals at various locations in the delta compared to the Rhine (= 100). At least until the quality really has improved, the management of the fresh water lake, in which the Scheldt–Rhine connection is located, is done with minimal flushing.

Haringvliet

The sediments in the Haringvliet after closure contains many pollutants. Figure 19 shows some indication of the sediment quality (1 is relatively clean, 4 is heavily polluted). Although not directly threatening the water quality itself, life that depends on the bottom (like

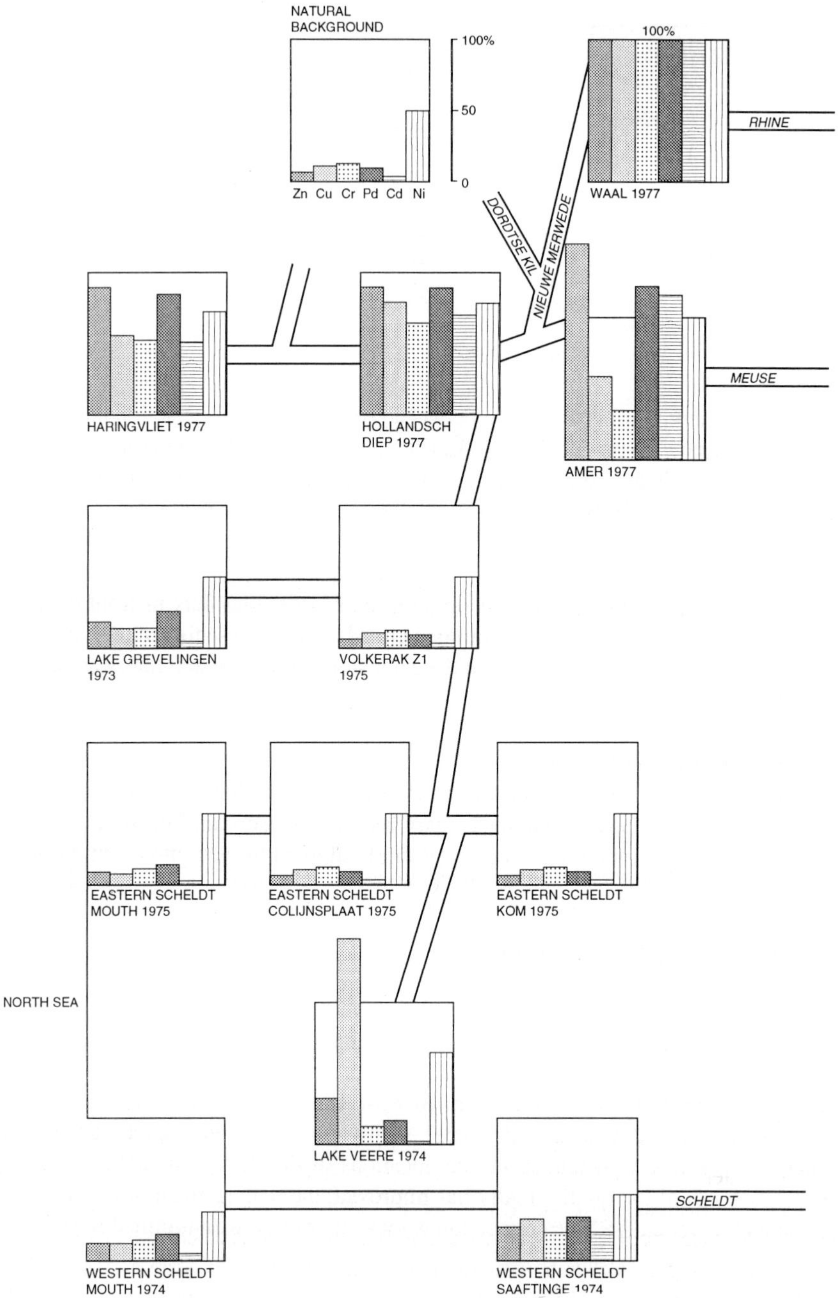

Fig. 18. Contents of heavy metals in Delta waters (from Saeijs, 1982).

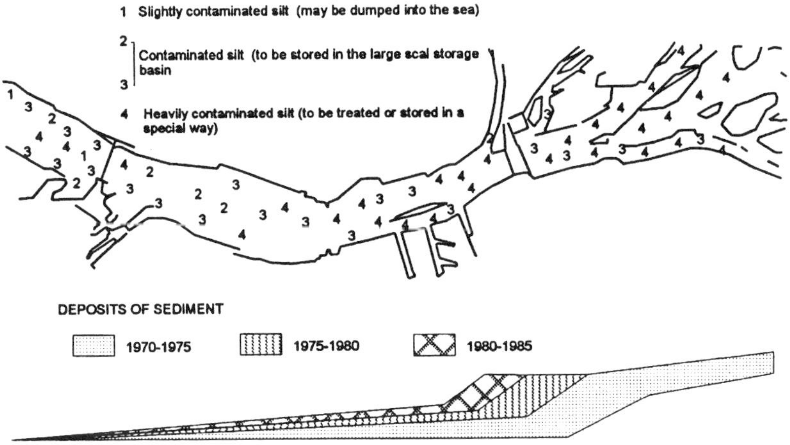

Fig. 19. Pollution of sediments in Haringvliet area.

invertebrates being at the base of the food chain) appears to be influenced. For a healthy ecosystem some treatment of the bottom will probably be necessary.

4.4. ECOLOGICAL ASPECTS

Here again only some examples can be shown. For more details, see Saeijs, (1982).

Birds

Just as an example of the influence of closure of a tidal basin, Figure 20 shows the population of two bird species, oystercatchers and grebes, in the Grevelingen. There is no river inflow into the Grevelingen, so here only the influence of disappearance of the tides plays a role. Oystercatchers feed themselves with invertebrates on the tidal flats, the majority of which died after closure of the estuary. Grebes feed on fish species that flourished after the tides disappeared, while also the water became more clear (less sediment transport).

Biodiversity

Figure 21 shows the well known relation by Remane, showing a minimum in biodiversity for brackish water. Although many other factors play a role, e.g. pollution and nutrification, it gives an idea for changes in a delta, where the relation between river and sea is altered drastically. The ideas behind this curve were among the reasons to build a small sluice, to refresh the Grevelingen lake with sea water, in order to reach chloride contents of more than 16‰. Of course, in the estuaries that became fresh water reservoirs, the species have changed completely. The general picture for the Eastern Scheldt is a decrease in seagrass as a result of erosion of tidal flats and decrease of freshwater supply, a shift in the number of birds per species e.g. as a result of reduced tides and increased recreational activities.

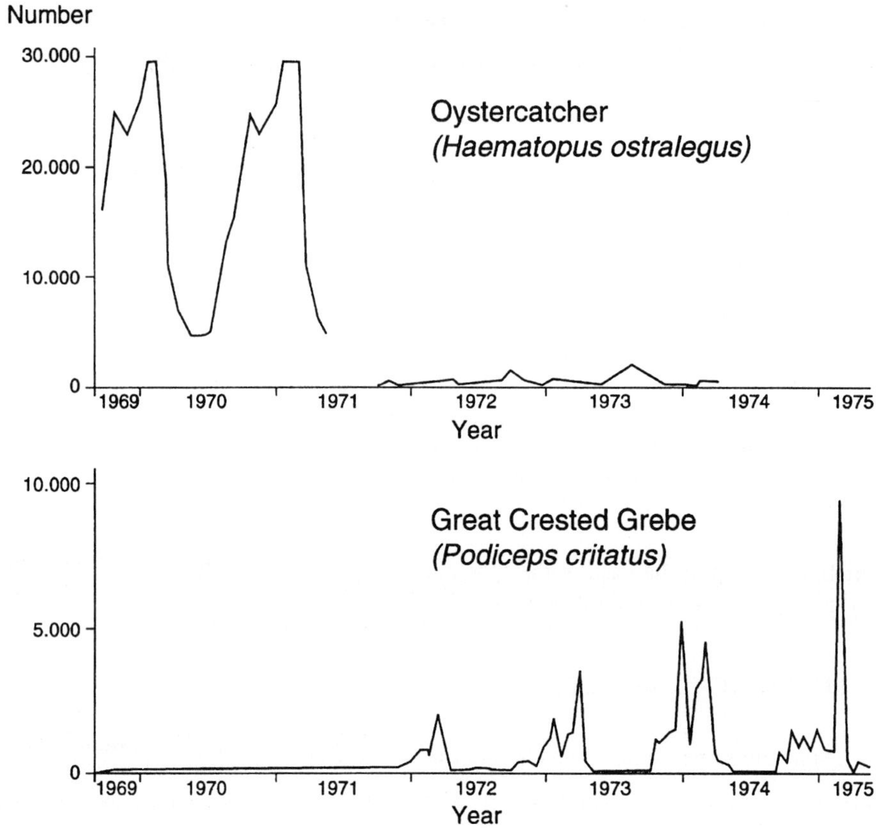

Fig. 20. Response to closure of two bird species in the Grevelingen (from Saeijs, 1982).

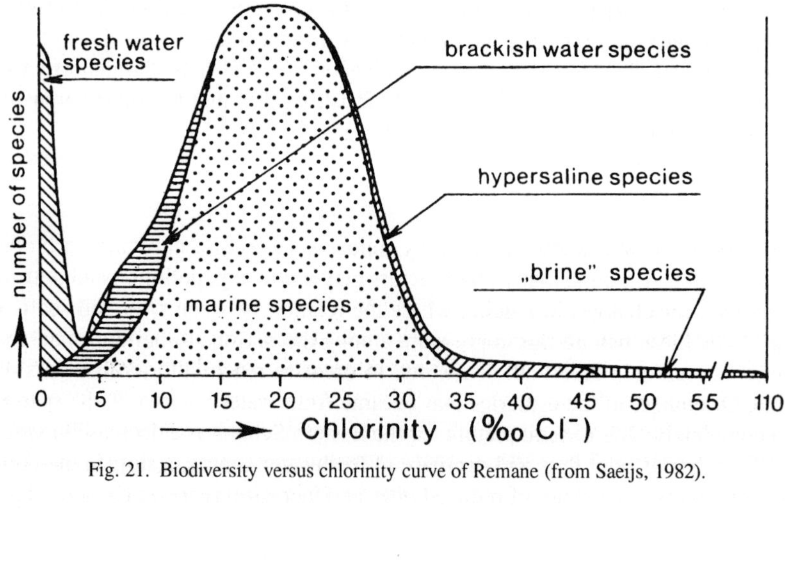

Fig. 21. Biodiversity versus chlorinity curve of Remane (from Saeijs, 1982).

5. Evaluation

5.1. INTRODUCTION

The delta area is dynamic in many ways. Therefore, an evaluation is always coupled to the moment at which it is written. This relativization becomes already clear when the original plan is compared to how it finally was executed. Moreover, an evaluation is complicated by the fact that one will never know how the area would have developed without the project. Again, only some aspects can be mentioned here.

5.2. SAFETY

After completion of the Deltaproject, no problems with safety against flooding have occurred. But it should be said that also no storm surge as in 1953 has frequented the area since then.

5.3. WATER MANAGEMENT

Since the construction of the Haringvliet sluices in 1970, hardly any problem with the supply of fresh water has occurred, not even in the very dry summer of 1976. This was also favoured by the fact that for e.g. drinking water, sources more upstream were used.

5.4. SOCIAL ECONOMIC

The expectations at the start of the project on recreational potential has come out beyond any prediction. Especially coastal recreation and aquatic sports have grown tremendously and have become a very important source of income in the area. Also many people live in the area commuting to their jobs in the Rotterdam area. The expectations on horticulture have not yet become reality. Although the connections with the large cities in Holland have improved, it appeared that the coherence of the existing horticultural infrastructure around Rotterdam was too strong for large-scale dispersion. Recently however, due to lack of space in that area, horticultural activities come more and more under pressure. It is not out of the question that the prediction will become true after all. The better road infrastructure did cause new business though: many transportation companies have settled in the area. The overall social economic effect is that the area is no longer underdeveloped compared to other regions in the country.

5.5. ENVIRONMENT

Until now, no insurmountable problems with the environment have come to light. The already mentioned polluted sedimentation in the Haringvliet still has to come to a solution. Erosion of banks and tidal flats in the former estuaries keeps asking attention and maintenance. Only in the very dry summer of 1976 botulism occurred in the river branches where the velocities had decreased because of the closures. The salt water environments of the Eastern Scheldt and the Grevelingen show some changes, but not yet in a very clear pattern. An extensive monitoring program is in operation to follow the developments.

5.6. GENERAL

Although large-scale projects like this can have impacts that go on for ages, until now the total balance seems very positive. This is certainly true for the economy. The environment has seen the disappearance of several tidal estuaries, but until now, the consequences seem manageable. One of the lessons with a project like this is that with such a long implementation time, policy makers, engineers, ecologists etc. have to remain flexible in order to react properly on changes in the system and in society. This does not change after completion of the project. One of the disadvantages of the project is a diminished interaction between sea and river. Migrating fishes, like salmon, possibly have difficulties in passing the Haringvliet sluices. This is part of a bigger problem. The water quality in the Rhine has long been too bad for salmon and also the artificial bank-protections along the river have deprived them of spawning grounds. So, for an overall solution to this problem, more measures have to be taken. In fact, discussions are going on at the moment to use the Haringvliet sluices more as a storm surge barrier than as an outlet sluice and to make the estuary more brackish again, restoring somewhat the old salinity gradient. The same is true for the Eastern Scheldt. Now there is a strict separation between salt and fresh water, but it could be advantageous to create again some fresh water flow to the estuary, bringing more nutrients which can be beneficial to biodiversity and fishery. All this means that a project like this is not finished as soon as the contractors have gone home. Monitoring and adjustment are essential. It might also mean that as soon as man starts changing his environment, he is forced to go on with it forever.

References

Delta committee, Final report, part 1, State printing and publishing office, The Hague, 1962.

Den Doolaard, A., *Roll Back the Sea: A Novel* (translated from the Dutch novel *Het verjaagde water*), Heinemann, 1949.

Plancommissie Zuidwest, First Interim Report: Southwest Netherlands, Terugblik en toekomstbeeld (Review and prospects, in Dutch), Staatsdrukkerij, 1954.

Plancommissie Zuidwest, Second Interim Report: Planologische consequenties van de plaats der dammen in het Deltaplan (Planning consequences of the location of the closure dams in the Deltaproject, in Dutch), Staatsdrukkerij, 1955.

Rand Corporation, *Protecting an Estuary from Floods – A Policy Analysis of the Oosterschelde*, Santa Monica, 1977.

Saeijs, H.L.F., *Changing Estuaries*, Rijkswaterstaat communications, Government publishing office, The Hague, 1982.

Van der Ven, G.P., *Man-Made Lowlands; History of Water Management and Land Reclamation in the Netherlands*, Matrijs, Utrecht, 1996.

Wemelsfelder, Wetmatigheden in het optreden van stormvloeden (Patterns in the occurrence of stormsurges, in Dutch), *De Ingenieur*, No. 9, pp. 31–35, March 3 1939.

THE ROTTERDAM HARBOUR: THE CONNECTION WITH THE NORTH SEA AND EUROPOORT

G.J. SCHIERECK
Ministerie van Verkeer en Waterstaat
Afd. Q, P.O. Box 20901, 2500 EX Den Haag
The Netherlands

D. EISMA
Netherlands Institute for Sea Research, Texel;
Faculty of Earth Sciences, Utrecht University
Utrecht, The Netherlands
Present address: Paulineweg 17,
1865 AD Bergen aan Zee, The Netherlands

1. The Connection with the North Sea

Rotterdam, located on the most northern branch of the Rhine–Meuse deltaic system, existed already as a fishing village in the 11th century A.D. and as early as 1100 A.D. was protected by (low) dykes. Flooding was a regular threat: in 1164 the entire village was destroyed during a flood and again in 1421. The nucleus of the present town existed from the 15th century on and gradually developed into a sea port. The first, small, harbour had been made in 1350 (van Dam, 1990) and the connection with the North Sea was along the main river branch that passed Rotterdam towards the west and opened in a funnel-shaped mouth towards the North Sea. The shifting sand banks and narrow channels were not a large problem as long as the ships were small. They were more vulnerable to stormy weather: the harbour, some 35 km inland, provided a better shelter than would have been provided by a harbour on the coast. At those times such harbours did not exist on the North Sea coast: fishing ships were drawn high on the beach.

At the end of the 14th century the small Rotterdam harbour traded with England, Portugal and the Mediterranean in competition with neighbouring towns like Dordrecht, Vlaardingen and Brielle, the latter two using the same inlet. But when ships increased in size, this inlet became a problem: by 1800 alternatives routes were used to reach the sea through a much larger inlet further to the south, the Haringvliet (Figure 1). The use of this inlet increased the distance from the sea to 95 km. Therefore, to shorten this, a canal was cut in 1827–1829 through the island of Voorne (Figure 1), which brought the travelling distance down again to about 40 km. The canal was soon too small for the larger sea-going ships that came into

J. Chen et al. (eds.), Engineered Coasts, 269–278.
© 2002 *Kluwer Academic Publishers. Printed in the Netherlands.*

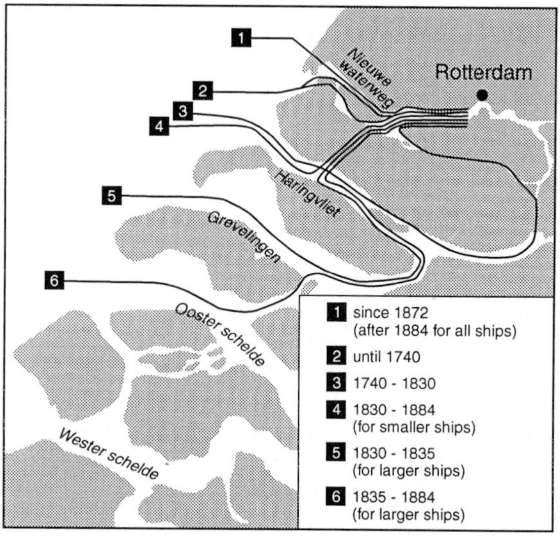

Fig. 1a. Accessibility of Rotterdam for merchant ships 1740–1884 (after Van de Ven ed., 1994).

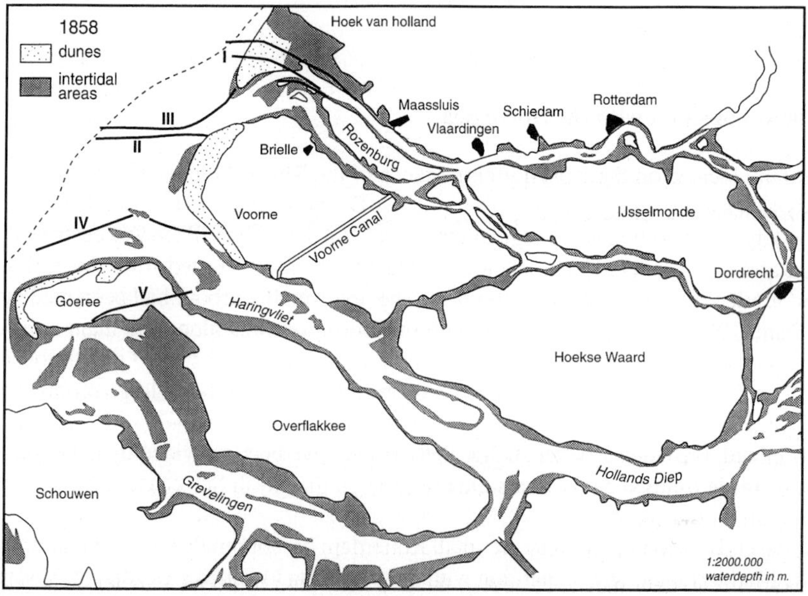

Fig. 1b. The Rhine–Meuse estuary before construction of the Waterweg with proposed situation of the artificial channel and other proposed improvements. I: Waterweg Plan-Caland 1856; II: Plan Caland 1856 (dams); Plan Greve 1857 (dams); IV: Plan Greve 1857 (dams); V: Plan Conrad 1836 (canal).

use, as it had only a depth of 5.50 meter and only ships with a length less than 71 meter and not wider than 14 meter were allowed to pass. This resulted in a very unfavourable position of the port of Rotterdam compared to Antwerp and Hamburg which had a much better connection with the North Sea. Also the plans for the Suez Canal and for a regular service to North America made it imperative for Rotterdam to have a better seaway. Around 1850 the situation in the inlet west of Rotterdam, the Meuse inlet, was such that at low tide the waterdepth in the channels was not more than 2.5 meter so that only around high tide and when weather conditions were favourable, ships could pass that had a draught of not more than 3.5 meter. In 1855 it was decided to improve this and the then recent improvements at the mouths of the Seine, Rhone and Clyde were studied. At the same time a high water level in the canal through Voorne was considered (which would give problems with the water level in the polders alongside the canal), as well as improvements in the Haringvliet inlet, which at that time was the main route. A proposal by P. Caland was accepted to cut a new channel through the broad spit (with up to 9 meter high dunes) that separated the Meuse inlet and its main inward channel from the North Sea (Figure 1; de Groot and Marinkelle, 1917; Ringers, 1953). A similar plan had been proposed already in 1737 by N. Kruik (Cruqius) but was never executed. The Caland proposal was for an open channel (i.e. without sluices) that would be kept open by the tides. It also involved the closure of part of the channel (named De Scheur) that along Rotterdam led to the Meuse inlet: this would prevent the new channel to be silted up on the landward side. Dams of stone and wood would extend into the North Sea to protect the mouth and prevent siltation on the seaward side. It is of interest that the size of the largest ships that existed at that time, was not used as a measure for the dimensions of the channel, as they were considered to remain exceptional, which is ironic in view of the later developments in ship size. The depth was therefore to be 7 meter below high tide level. The existing river channel – of the Meuse and the eastern part of the Scheur – was to be widened as far as upstream of Rotterdam because the tides were expected to become stronger and penetrate further inland.

In 1863 it was decided by the government of that time, that these works were to be carried out and financed by the state, as the costs would be too high for the towns (Rotterdam, Delfshaven, Schiedam, Vlaardingen, Maassluis) that would primarily benefit. Those who made this decision, in particular Thorbecke, a major statesman and then cabinet minister, were well aware of its international importance: the new channel would make the Dutch coast accessible to international shipping and "would provide a funnel, that would connect the world with the northern half of Europe and with all the countries behind it: the larger the funnel, the more would go through it". The dams were constructed first, in 1864–1866 and at the end of that period extended 1000 meters into the North Sea. They were made by lowering mattresses of willow matting to the seafloor loaded with stone ballasting. As many were lowered as needed to reach high tide level; then the dam was covered by a basalt pavement and wooden partitionings. The northern dam reached 2.20 above Dutch Ordnance Level (approximately mean sea level), the southern dam to about 1 meter. The latter was lower because the flood comes from the south and during periods of high water level would turn more easily into the shipping channel and maintain a greater waterdepth. During the following years the dams were enlarged so that in 1876 they had a length of 2300 meter, with the northern pier (as the dams were now called) reaching 3.85 meter above N.A.P.

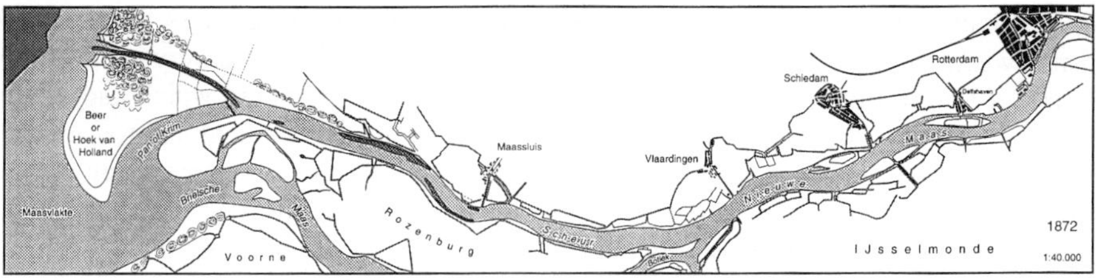

Fig. 2. The Waterweg at about 1987. Dark areas > 0.65 m below low tide level. The distance from Rotterdam to the sea by way of the Waterweg is approx. 35 km.

and the southern pier 1.80 meter. They had a width of 9 to 10 meter and extended down to 7 meter waterdepth below high tide level. The construction of the piers went surprisingly well, in spite of the winterstorms from the north and northwest that raised sealevel by 1.5 meter. This was largely due to the use of heavy stones, which hardly moved anymore after they had been placed.

The channel was excavated in 1866 to 1868 and the closure of the Scheur was completed in 1873 so that from that time on one single channel led from Rotterdam to the sea (Figure 2). The tides were supposed to keep the channel open and to excavate it further down to the required depth of 7 meter below low tide but this did not happen: about half the sand that was eroded from the channel by the tides between 1868 and 1880 was deposited at the mouth between the piers and in front of the piers on the seafloor. Because of this the waterdepth only remained between 2.50 and 3.30 meter below low tide level so that the channel was mainly used by fishing ships. This miscalculation was to be expected in a time when no physical or numerical modeling existed and decisions depended more on intuition than on exact tidal measurements and calculations. The problem was finally solved by dredging, which was stimulated by the improvements in the dredging technology of that time, and itself contributed to the development of such improvements. In 1902 a depth of 8.00 to 8.70 meter below low tide level could be maintained and large merchant vessels could pass through the channel. The dredging, however, that was to continue to this day and still keeps dredgers working the year round, indicates that the channel made in 1866–1868 was in fact a mistake and not to be repeated. A present construction of such works would not only have the advantage of more measurements and physical and numerical modelling, but would also have taken more advantage of the natural conditions of waterflow and sediment transport during and after construction. It should also be realized, however, that after the present Delta Works in southern Holland were completed (Chapter 12 of this book) sedimentation occurred in areas where it was not expected; at the tip of the now former islands erosion was expected (and dams were designed to prevent this) where now a broad sedimentation area exists.

2. Improvement and Enlargement: 1876–1995

The new channel, called the Waterweg (Waterway) and the increase in trade, which led to an increase in merchant shipping, resulted in the construction of a succession of new

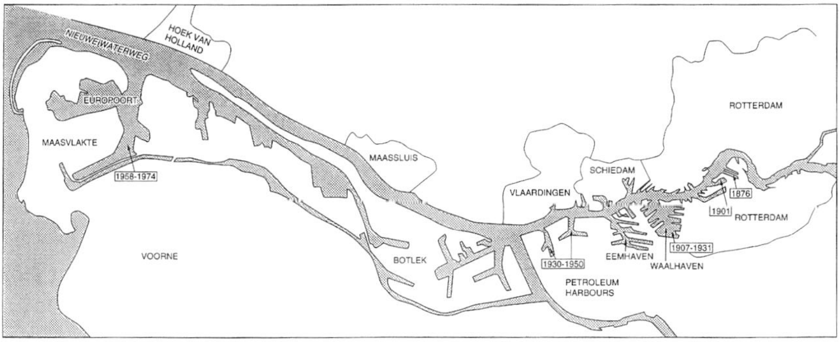

Fig. 3. Harbour development between Rotterdam and the sea after 1876.

harbours in Rotterdam, first on the south bank, starting in 1876, and from 1890 on also on the north bank extending as far as Schiedam (van Dam, 1990). In 1872 there had been only the small harbours in Rotterdam and the other towns on the north bank: Delfshaven, Schiedam, Vlaardingen and Maassluis with on the south bank only the small inner harbour in Rotterdam. By 1901 a series of smaller harbours had been made in Rotterdam as well as larger ones on the south bank including the railway harbour, the Rhine harbour and the Meuse harbour with small docks at Charlois. Up to the first World War no further harbours were made on the south bank but harbour construction continued on the North Bank, while on the south bank work started in 1907 on the construction of the large Waalhaven, which was completed in 1931. Meanwhile also Schiedam and Vlaardingen had built new, comparatively small harbours. The harbour expansion on the south bank was to the west, which brought the newer harbours nearer to the sea. This continued on the south bank (where expansion was not hindered by the presence of other towns) by the construction of the Eemhaven and the Petroleumharbours at Pernis. During the Second World war many harbours were destroyed by the Germans, who occupied Holland from 1940 to 1945, but shortly after the war, together with the reconstruction of what had been destroyed, the expansion continued with the large Botlek harbours (started in 1954) and Europoort, which was begun in 1958. The first part of Europoort could be used in 1960, which reflects the speed of the developments: harbour space was already fully booked before the construction was finished. The last expansion was into the sea where offshore a shallow area had been formed, the Maasvlakte. Besides the number of ships that came to Rotterdam and the adjacent ports, also their size increased, from 17,000 DWT (with a draught of 9 meter) to 300,000 DWT (with a draught of 22 meter. Channel depth therefore had to increase too but it also stressed the importance of port construction nearer to the sea, which would reduce the distance the largest ships had to travel inland. Thus plans were made for the construction of Europoort harbour, west of the Botlek harbours, besides for enlargeing the channel.

For the construction of Europoort the (now former) islands of Roozenburg and De Beer were used together with a newly reclaimed area, the Maasvlakte, that extended into the coastal sea. The harbour area was limited to a narrow, 2 to 3 km wide strip along the Waterweg channel, in order to protect the agricultural area further south from penetration of harbours and industry (Figure 3).Based on earlier ideas that a larger seaward extension of the entrance piers would result in a greater depth of the channel, a design was made, based on physical modelling, of two extended piers at a larger distance from each other, which would give a wider as well as a deeper entrance. Its construction, however, proved problematical as during its construction the safe passage of ships through the entrance and channel could not be guaranteed.

A solution was found by starting the construction of the southern pier far to the south on the coast and to proceed from there with a wide seaward curve towards the Waterweg entrance. In this way the existing entrance would be protected already in an early stage so that on the leeward side the construction of the north pier could be started. Another problem remained, however, which was the question how far the new entrance would have to extend seaward to reach sufficient waterdepth without a necessity of continuing excessive dredging. A solution to this question was postponed several times because of the rapid development of tanker size. At first the entrance had to be navigable for tankers up to 80,000 tonnes, but soon 100,000 ton tankers were used and larger ones were under construction. But almost at the same time the new large dredgers that came into service at the Waterweg entrance proved a great success so that maintaining a sufficient waterdepth had become less of a problem. This also made the construction of a new entrance less urgent, so it was ready only in 1974 (instead of 1970). Meanwhile the advent of supertankers had made another requirement more important: the need for safe manoeuvering of these tankers and the need to reduce speed gradually so that tugs could make fast, for which a distance of about 8 km was needed. As this had to be resolved before the tankers had actually been built, an answer was found through physical modelling of both tankers and harbour entrance. A lucky circumstance was that tracer studies had shown that at the harbour entrance and its vicinity along the coast no significant sand transport occurred parallel to the coast either towards the north or to the south. This made the design of the harbour entrance considerably less complex. A speed reduction channel was created on the south side of the Waterweg channel separated from the main shipping channel by a dam.

The new southern pier, which was to have a length of more than 8 km, was made of sand over the first 4 km, which was possible because of the increased capacity of the new large dredgers, so that much sand could be supplied within a short time. The remainder of the south pier and the entire northern pier were made of stone and concrete blocks. For the transport of large amounts of blocks special ships were developed: the design of the ships, and the size, shape and weight of the blocks were related to the expected wave characteristics and the possibilities of handling the blocks. The weight of concrete blocks was not sufficient so they were made heavier through the admixture of basalt. Cube-shaped blocks were the most convenient to handle. The curved shape of the southern pier allowed the shallow area within to be reclaimed for future use as a harbour area.

3. Traffic Control

Traditionally a pilot service has taken care of bringing the ships into port when the ships became larger, but the increase in harbour size (with at present about 300 ships daily entering Europoort) made navigation on the Waterweg more complex. When only pilots were operating, this became increasingly hazardous, even where transverse traffic (ferries, local transport) was reduced to a minimum of necessary services. A better traffic control was required, which led to changes in the way the pilot services operated (there were several of them) and to a system of seven radar towers along the Waterweg between the entrance and Rotterdam manned with traffic controllers who could give directions to ships how to proceed. This was an improvement for the larger seagoing ships that entered or left the Waterweg, but did not function well for the (usually small) boats that carried goods from the Waterweg ports further inland, as well as for the ships that passed the harbour entrance at sea. A near disaster in 1966 with a merchant ship that ran aground on the north pier was needed to have this problem solved by enforcing the necessary government regulations.

4. The 'Eurogeul' Channel

As the North Sea floor off the harbour entrance is shallow (mostly less than 20 meter depth below low springtide level) and for a large part covered with sanddunes that may reach a height of 15 meter, further engineering was needed to assure safe approach of the larger ships (oil tankers and ore carriers). Thus around 1971 work started to dredge a channel from deeper water (more than 25 meter deep) further offshore to the harbour entrance, and removing the top of some large sanddunes. A problem was to estimate the sand transport and deposition which might lead to deposition and shallowing of the channel. This would be the basis for estimating the amount of dredging that had to be done. The sand dunes are at least partly mobile and their tops are lowered through erosion during the winter storms, while they are being build up again during the summer up to about their original height. Sand may thus be deposited in the channel but on the whole the sanddunes proved to be rather stable with no marked displacement either in the flood or the eb direction. Sand deposition remains limited and in practice the need for dredging is small compared to the dredging needed in the Europoort entrance and the harbour basins further inland.

Large ships now coming through the Straits of Dover follow a deep-water channel with separation of north-going and south-going traffic. At about 52° North the Eurochannel branches off towards the east which is marked by a buoy and a large anchorage area (Figure 4). The Eurochannel is meant for the large deep-draught ships (oil tankers and ore carriers) that go to the Europoort harbour; it has a depth of 25.4 meter at the western end which decreases to 23.75 meter off the harbour mouth. Smaller ships that go eastward follow a route south of the Eurochannel, those going westward a route north of it. Traffic control is by radar that reaches as far as the deep water. The Eurochannel has a length of 45.65 km and a width of 600 meter. Off the harbour mouth it turns eastward and continues as the Meuse Channel, which leads directly into the Europoort harbour mouth. The floor of the Eurochannel consists of sandwaves over the first 22.5 km from the western end and then becomes almost flat. There are two turning points in the channel: one about halfway and one near to the junction with the Meuse Channel. Alongside the channel on both sides is a 300 meter wide zone where the waterdepth decreases to about 22 meter. The Meuse Channel has a length of 11.35 km and a depth of 23.75 meter; its width decreases from 600 meter at the junction with the Eurochannel to 500 meter in the harbour channel. For the larger ships approaching or leaving the harbour the time of safe entry is determined

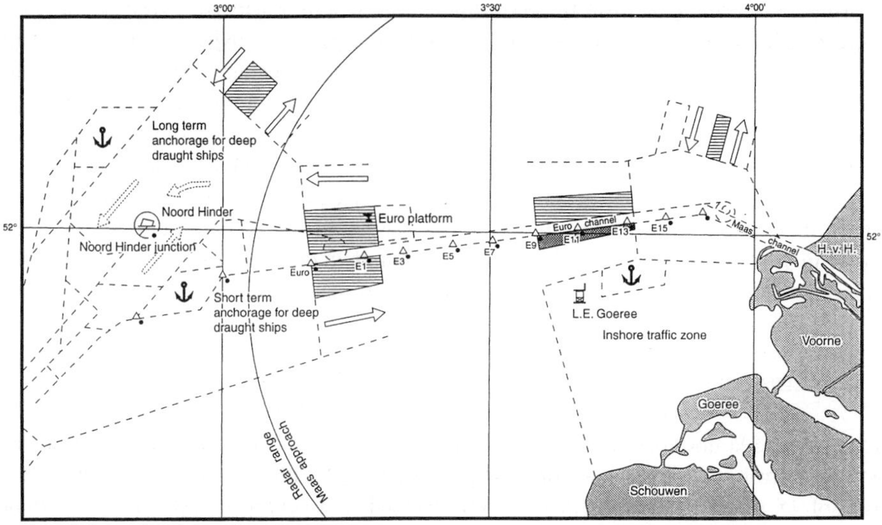

Fig. 4. The Eurochannel in the Southern North Sea. Arrows: obligatory traffic direction (drawn lines) and optional (safe) traffic direction (broken lines) After Rijkswaterstaat et al. (1995).

following a probabilistic regulation based on the navigational characteristics of the ship, the tides (waterdepth) and wave conditions. To be able to do this, the waterdepth and wave climate are measured at permanent stations with additional measuring campaigns. Errors in the estimation of the safe entry period occur (mainly) because of variations in wave height, effects of the bottom sand waves and a water density effect near the harbour entrance. These uncertainties are estimated thorough probability estimates.

5. Opposition to Harbour Developments

In the 19th century when the Waterweg was made, opposition to its realization concerned mainly feasibility and costs, and was not strong because the development of the port of Rotterdam, and later of the other smaller ports along the Waterweg, was considered to be in the national interest. This gradually changed in the 20th century with the advent of pollution problems and the threatening destruction of natural areas, that were of importance because of their vegetation, wildlife, beauty or combinations of them. This was enhanced by the growth of the cities and the increasing population density, in particular in the western part of the Netherlands. In the Waterweg area this opposition appeared for the first time when for the construction of the new harbour entrance and Europoort the nature reserve on the island of De Beer had to be sacrificed. Although this nature reserve, like so many others in The Netherlands, had originally been the result of canalization and dike construction, it had been left in peace so that sand dunes, beaches and dune valleys could be formed in a natural way and birds settled in large numbers. At first only planological arguments were used to leave an open space off the coast further south where also a large nature reserve was (and still is) located. Gradually also the ecological values of the nature reserves at De Beer and further south were recognized. A compromise was finally reached: the island of De Beer was to be used for the construction of the new harbour entrance and Europoort down to a demarcation line about 3 km from the coast further south which would protect the nature reserve there. This compromise was the basis for the much more far-reaching compromise that was reached when the Delta Works were in progress (see Chapter 12).

6. Pollution Problems

Pollution, not only of the water but also of bottom muds began to be a serious problem in the Waterweg and the harbours after about 1960. Pollution from the cities and industries along the Rhine, in Germany and France as well as in the Netherlands, began to be noticed around 1920. Not much was done to reduce or prevent it because of the economic crisis and the Second World War, but after that the growth of industry, cities, agriculture and traffic led to serious pollution of the bottom muds, in particular in the inwardly located harbours along the Waterweg. This led to problems with the disposal of dredge spoils. The muds in the inner harbours, which receive most of the pollution (with trace metals as well as organic waste, chemical waste and oil), could not be applied anymore on land, where they had been used in parks, sports grounds, children's playgrounds etc. These muds had also become too polluted for disposal in the coastal sea, as the muds dumped there are at least partly transported to the coastal areas further north and deposited there. Therefore, for these very polluted muds a large storage basin was constructed in the Europoort area, down to a depth of 29 meter below Dutch Ordnance Level (approximately average sea level) and surrounded by dikes up to 23 above this level. The total capacity is 90×10^6 m^3 (of dry mud). Although only part of the dredged mud has to be stored here, the total amount of mud that is yearly dredged is in the order of 12×10^6 m^3 (dry mud), so that in the near future a new storage basin will have to be made if pollution is not reduced. A possible area for such a basin is seaward of the present storage area, where part of the nearshore seafloor has become very shallow. Partly because the pollution is also an international problem, pollution is not expected to be soon reduced significantly.

7. Salt Problems and Hydrology Related to the Delta Works

The enlargement and deepening of the channel in the Waterweg allowed deeper inland penetration of saline water. Also storm surges would have a larger effect further inland. The salt problems, affecting agriculture, freshwater sources and freshwater inland areas, as well as agriculture and horticulture, and the recent construction of a stormbarrier in the Waterweg are discussed in Chapter 12, as both aspects are strongly interrelated with the Delta Works. Also the seaward flow of freshwater through the Waterweg is at present regulated through the Delta Works: at low river discharge the Haringvliet sluices remain closed so that more water can flow through the Waterweg, which helps in maintaining sufficient waterdepth. Also for this aspect the reader is referred to Chapter 12.

References

van Dam, T., 1990. *De Rotterdamse haven 650 jaar*. De Bataafse Leeuw, Amsterdam, 160 pp.

de Groot, A.T. en A.B. Marinkelle, 1917. 1866–1916, De Waterweg langs Rotterdam naar Zee. Ministerie van Waterstaat, Den Haag, 124 pp.

Ringers, J.A., 1953. *Caland en de betekenis van zijn werk voor Rotterdam*. Ad Donker Uitgever, Rotterdam, 60 pp.

Ferguson, H.A., 1991. *Benedenrivieren in de jaren zestig*. Ministerie van Verkeer en Waterstaat, Dir.-Gen. Rijkswaterstaat, Den Haag, 60 pp.

Rijkswaterstaat Directie Noordzee, Gemeentelijk Havenbedrijf Rotterdam and Regionaal Loodsencorporatie Rotterdam-Rijnmond, 1995. Informatie voor de vaart met geulgebonden schepen naar de haven van Rotterdam, 61 pp.

Van de Ven, G.P. ed., 1994. *Man-Made Lowlands*. 2nd edn. Kon.Bibliotheek, Den Haag, Stichting Matrijs. 294 pp.

THE NETHERLANDS: THE ZUYDER ZEE PROJECT

MARCEL J.F. STIVE and RONALD E. WATERMAN
Netherlands Centre for Coastal Research
Delft University of Technology
Faculty of Civil Engineering
P.O. Box 5048, 2600 GA Delft, The Netherlands

1. Introduction

The closing off and the partial reclamation of the Zuyder Zee has resulted in the gain of 166,000 hectares of new land (Figure 1). This new land appears to lend itself for agriculture, urban development, recreation and nature conservation.

As early as in 1667, the first idea for reclamation was published by Hendric Stevin. However, the idea was conceived prematurely: it could not be realised technologically. In the 19th century, the idea was given serious consideration, new plans were constantly made, but never carried out. Not until 1891 a plan, drawn up by Cornelis Lely, was published which turned out to be feasible. It was a plan of great simplicity: the closure of the Zuyder Zee in the 'neck' with a 30 km long dam. The enclosed part, fed with fresh water from the river IJssel and by rainfall, would subsequently turn fresh. Within this area, from then on to be called Ijsselmeer (Lake Yssel), five polders were to be constructed for agricultural use, with a joint surface of 220,000 hectares. The remaining part was to be a fresh water basin, of great importance for a country having to cope with salinization problems. As a result of the closing off, the coastline of the Netherlands was reduced by nearly 300 km, which made it much easier to defend.

A plan of great simplicity, but a large-scale and expensive operation, and the realisation would span a very long period of time. As so often in history, coincidence determined that in 1918 the Parliament decided to pass the act to close off and partially reclaim the Zuyder Zee. The first important factor was that Lely, the designer of the plan, developed into a formidable statesman and as Minister of Public Works and Water Management repeatedly put the act forward in Parliament. The second factor was that during the First World war the Dutch people became painfully aware of the extent of dependence on foreign countries for their food supply: the wish to produce more home-grown agricultural products was very strong. The third and probably decisive factor was the 1916 storm surge that caused destruction in the area surrounding the Zuyder Zee.

The name Waddenzee (Wadden Sea) is of a relatively recent date. In the second half of the 19th century it was used for the first time and became accepted on 22 September 1932,

J. Chen et al. (eds.), Engineered Coasts, 279–290.
© 2002 *Kluwer Academic Publishers. Printed in the Netherlands.*

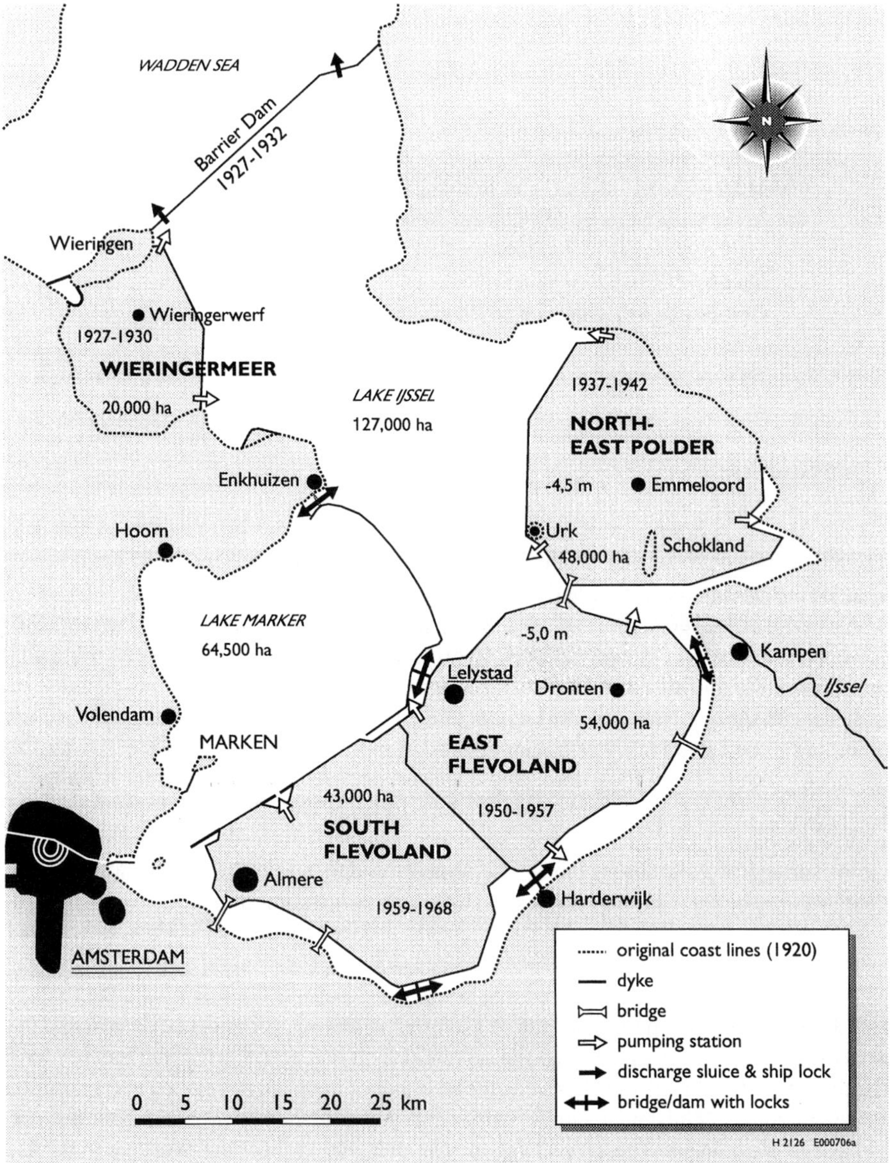

Fig. 1. Situation of the Zuyder Zee Werken (South Sea Project) in The Netherlands.

when the name of IJsselmeer was adopted as well. Before that day, the present Waddenzee and IJsselmeer were both called Zuyder Zee (South Sea).

2. The Birth of the Zuyder Zee

From a geological point of view, the Zuyder Zee and the Waddenzee are very young. The area mainly consisted of peat and clay lands at the beginning of our calendar. Only in the south, there was an inland fresh water lake called the Almere. This situation was the result of the marine infilling of the topographic depression between the Northern Texel High and the Rijn-Maas Delta, which was left at the beginning of the current interglacial, Holocene period (Beets et al., 1992).

During a significant decrease in the rate of sea-level rise around 5000 year BP, the tidal entrances along the Holland coast choked due to an abundance of sand, thus creating the relative low-lying central area. Around 1200 AD, the North Sea encroached the area once again but now from the North, and the Zuyder Zee came into being. The situation created around 1250 AD did not change until the beginning of the 20th century. The Zuyder Zee and the Waddenzee then formed a marginal sea together with a surface of about 670,000 hectares on the south-eastern side of the North Sea. A row of dune islands with deep sea inlets in between formed the divide between the North Sea and the marginal sea. Tides and storm surges penetrated via the sea inlets into the Waddenzee and from there into the Zuyder Zee.

Tides, topography and soil properties varied considerably in the various parts of the marginal sea. Near the sea inlets, the average tidal amplitude was 1.6 m. Going south, this decreased to a few centimetres in the far south. During storm surges, high water levels occurred that reached heights of 3.2 to 3.3 m + NAP in the southern Basin (1825 and 1855).

3. Plans to Close Off and Reclaim the Zuyder Zee

The estuaries of the Waddenzee, Zuyder Zee and the Delta have been called the clutches with which the sea held the Netherlands in its grip and which kept on trying to encroach further and further inland. At an early stage, plans were made to start an offensive on these fronts. Evidence of this is the, as far as we know, first proposal to close off and partly reclaim the estuaries of the Waddenzee, Zuyder Zee and the Delta, which was presented in 1667. Its originator was Hendric Stevin, hydraulic engineer and son of the well-known mathematician and builder of fortifications Simon Stevin. The objective was to: 'get rid of the fury and the poison of the North sea from the United Netherlands [...] together with all brackish waters from Seelant, Hollant and West-Frieslant.' With the 'fury' the storm surges are meant, with the 'poison' the salt water.

Stevin proposed to dam up all sea inlets between the Wadden islands of Noord-Holland to Ameland and build a dike from Ameland to the Frisian coast. Drainage sluices would have to be built into the dams, so surplus water could be discharged at low tide. Thanks to the surplus of fresh water into the lake, a gradual desalinisation would be the result. The larger part of the lake could be reclaimed.

Of course, the proposals of 1667 could not be realised then. Incentives, social organisation and a high technological level were essential. It took more than hundred years before these conditions were starting to be met. After the French centralist model the Bataafsche Republic founded in 1798 the Rijks Waterstaat, which promoted the education of army and civilian engineers. In 1850, the first polder of the Haarlemmermeer, a two hundred year old plan of Leeghwater, was reclaimed. At that time engineers and civilians started to make plans for reclaiming the Zuyder Zee. Between 1848 and 1886 some 126 plans had been published. The political field investigated, held meetings, even designed laws, but could not reach a decision. Men of name and fame decided it was time for private action and founded the Zuyder Zee Association in 1886. They claimed for subsidies, gave out shares with little success, and finally privately funded the necessary studies. In six years time they were able to produce eight Technical reports, describing the technical feasibility of the closing off and reclamation of the Zuyder Zee.

The leading force behind these Technical reports was Cornelis Lely (1854–1929), who followed civil engineering studies at the Polytechnic School of Delft. Although his initial career was not very successful, his belief in a future career did not die. His ambitions came close to fulfilment when he became an engineer at the Zuyder Zee Association. In the third Technical report he formulated his plans. The five next reports were trivial exercises and before the last one was completed, Lely became Minister of Public works. It took however 27 years, six governments, a storm surge disaster and a World War until on 14 June 1918 his plan was endorsed by law. A State Committee to guide the project, and a special execution agency Zuyder Zeewerken was founded. Three years before the actual closure of the Zuyder Zee by the Afsluitdijk, Lely died. The reclamation of the polders took four decades, and resulted in the largest land reclamation in the world.

4. Hydraulic Works

4.1. ENCLOSURE DIKE/BARRIER DAM

Prior to the construction of the Enclosure Dike, comprehensive research had to be conducted. One important field of research that had to be developed, concerned that of tidal hydraulics in inlets, because it was realised that closing off the Zuyder Zee would have large repercussions on the tidal motion in the Waddenzee. Specifically, the level of the future storm surges that had to be resisted by the Enclosure Dike and the coast of Holland and Friesland was of course very important. While some, mainly qualitative, physical understanding of tidal motion had been developed, the government decided that this problem deserved thorough research.

This research was to be conducted by the State Committee, established in 1918, and Lorentz was asked to be the chairman. Hendrik A. Lorentz (1853–1928) was a theoretical-physicist, appointed professor in Leiden in 1878. In 1902 he received the Nobel price for physics together with Zeeman.

Lorentz, bright indeed, had no specific experience in the field of hydraulic engineering. J.Th. Thijsse (1893–1984), a young engineer of the Directorate of the Zuyder Zeewerken, was appointed second secretary. The story goes (Bijker, 1994) that Thijsse accompanied Lorentz on a trip on the Waddenzee to at least experience the sea once. During the greater part of the trip Lorentz sat down, quietly overlooking the water, and mumbling at times: 'It has to be possible to calculate this.' And indeed he proved it to be possible. By linearizing the quadratic terms in the tidal equations he developed an amazingly simple and robust method to predict the changes in the tidal movements due to the construction of the Enclosure Dike. With this method it appeared to be especially possible to predict the changes in the current velocities. One of the calculated results, an increase in the velocities in the inlets to the now smaller Waddenzee was first regarded with some suspicion, also by the members of the Committee. But when the Barrier Dam was closed off in 1932, the predictions proved to be very good. This mathematical model was the largest model at that time (Staats Commissie Lorentz, 1926 and Thijsse, 1922 and 1933). In this model the equations were analytically solved, so it cannot be compared with the numerical models of today.

Also with respect to the design of the structures and the execution of the works the existing experience proved to be insufficient. The final closing gaps were much larger and had much greater discharges than before. This was equally valid for the design of the discharge sluices. Therefore Thijsse was sent to participate in model investigations in Karlsruhe, where a hydraulics laboratory had already been established at that time. This may well be regarded as the direct introduction of science in hydraulics engineering in the Netherlands. Soon after this, in 1927, the Delft Hydraulics Laboratory was established and Thijsse was put in charge.

Returning to the closing off of the Zuyder Zee. Although the velocities in the closing gaps of the Zuyder Zee dam were higher than they had been in any of the executed projects until then, they could be closed with basically the same techniques as used before. Only the scale changed. Large cranes were used to dump large quantities of qualitatively good clay in the final gap, of which the bottom was protected against scouring by fascine matresses with a heavy stone cover.

The larger part of the enclosing dam was to consist of sand, which was found in the neighbourhood in large quantities. It could be dredged with a bucket dredger or suction dredger, to be transported to the dike body in barges or hoppers and to be deposited at the designated spot. Hoppers are vessels of which the bottom can be opened, so the sand can be unloaded. In this way, the dike body could be raised to a level, that depended on the draught of the hopper and the size of the bottom flaps. Above that level, the sand was supplied from the barges. This method had two drawbacks. First, the sand body is permeable, so ground water currents may occur. Furthermore, it was impossible to keep the sand in place unprotected because of the rates of flow that were to be expected during and after the dike construction. In order to seal the dikes effectively, a so-called clay casing or claylining was incorporated on the dikes. To offer the necessary water tightness, fascine mattresses had always been used. People departed from the idea that osier could provide adequate protection. In view of the cost of osier, people had been searching for alternative solutions for a number of years.

The clay needed for the lining of the dike body could not be borrowed by suction dredgers, but had to be dredged by buckets. It was feared however that the dredged clay was too weak. That is why drying grounds had been provided where the clay was to be left. During the first work for the Enclosure Dike, the construction of submerged bottom dams in the gullies of the Amsteldiep, it was decided to experiment with a clay dam. When drillings were carried out to trace any amounts of clay that could be won, they came across a spot at

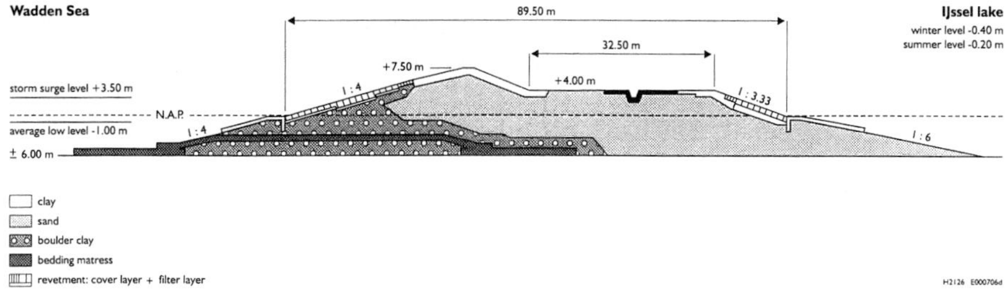

Fig. 2. Cross-section of Enclosure Dike/Barrier Dam.

the east bank of the Amsteldiep where, according to fishermen, 'darned tenacious clay' could be found. This turned out to be boulder clay which was very suitable for dike construction. When processed in large lumps, good boulder clay can withstand velocities up to 4 m/s without eroding large quantities of the material. So it could very well be incorporated into the dike bodies and used to seal the closing gaps. On the basis of these priorities, boulder clay has played a mayor role in closing off the Zuyder Zee.

The layers of clay to support the dike body could present problems with regard to settlement and possibly dike slidings as well. For instance, the western section of the dike in the Amsteldiep had to be built on weak layers of clay, 4 to 15 m thick. Since the problems had not been sufficiently been anticipated here, it had cost quite large amounts of earth and money before a reliable dike body had been built. Later on, the dikes were constructed on top of improved soil that had been created by dredging away the weak clay layers for the greater part and filling up the resultant trench with sand.

The osier need for the fascine mattresses that had to protect the bottom from scouring, were particularly abundant in the Zuid-Holland osier beds. The fascine mattresses were rectangular and consisted of two frames, encasing one or more layers of spread loose twigs. The frames were made up of fascines, which are tied bundles of osier.

The rip-rap, use to reinforce the mattresses, and the facing stones, which served to protect the slopes of the dike against the waves, mainly came from quarries in the Eiffel and the Belgian Ardennes.

The final cross-section of the Enclosure Dike was decided upon in 1925 on the basis of several studies. For the determination of the crest level the criterion was based on the storm surges that had raged in the period 1825–1926. If the Zuyder Zee would have been closed off in that period, a highest level of 3.50 m + NAP (Dutch Ordnance Level ≈ Mean Sea Level) would have occurred at Den Oever and about 3.75 m + NAP at the Frisian coast according to the prediction. The storm surge level was raised by the uprush of waves. On the basis of the tests, it was concluded that a slope of 1:4 was most favourable. After having made various calculations, a crest level varying from 6.7–7.4 m + NAP, depending upon the location, was arrived at. The profile and the structure of the dike body are depicted in Figure 2.

4.2. THE DIKES AROUND THE IJSSELMEER

The fact that there was no longer any tidal influence made it much easier to build the dikes around the polders in the IJsselmeer than had been the case for the Enclosure Dike. Determining the crest level of the dikes was also a much easier job than determining the crest level of the Enclosure Dike. The old dikes along the Zuyder Zee had to stand up to the forces of the sea. Therefore, the height varied from 4 m to well over 6 m + NAP. Lely calculated necessary crest levels of 2.5 m + NAP for the Wieringermeer Polder, 2.7 m + NAP for the later South Flevo Polder and 3.0 m + NAP for the Northeast Polder and the later East Flevo Polder. Initially, these heights were more or less adopted. When the parliament dealt with Zuyder Zee act, however, the motion Bongaerts–De Muralt was passed. In this motion, it was stated that one should not rely on the Barrier Dam if it involved the safety of hundreds of thousands of people who lived and worked some meters below sea level. The motion to which the minister did not object, required that 'dikes of all impolderings will have such dimensions, that they will be able to withstand the highest storm surges if the enclosing dam has been burst for a longer period of time over a considerable length or if the sluices have collapsed.' That is why, originally, polder dikes were not calculated to be much lower than the original Zuyder Zee dikes. Additionally, new calculations were made for situations that might occur in the IJsselmeer. This led to crest levels that were more or less similar to the crest levels calculated on the basis of the motion Bongaerts–De Muralt.

4.3. WATER MANAGEMENT SYSTEM

To transform the original Zuyder Zee into the fresh water IJsselmeer the 32.5 km Barrier Dam or Enclosure Dike was constructed, linking the provinces of North- Holland and Friesland and equiped with 25 gravity-flow discharge sluices and 2 sets of locks for shipping, complete with a four lane motorway and bicycle path. The salt water of the former South Sea was gradually replaced by fresh water due to the action of river outflow of the river IJssel (a branch river of the Rijn), precipitation and gravity drainage towards the Wadden Sea through the opening of the discharge sluices during the ebb-tides into the Wadden Sea. The Zuyder Zee project was coupled with the creation of a sequence of four very large polders reclaiming 1650 km^2 land from water:

The original plan was to create five polders of which by now only four have been realised, reclaiming 1650 kmš land from water (Figure 3).

Period of creation	Name of Polder	Area in ha	Pumping stations number	power (MWatt)	Maintenance pumping water (Mm/yr)	costs (US$/yr)
1927-1930	Wieringermeer	20,000	2	3.3	160	200,000
1937-1942	North East	48,000	3	6.1	400	500,000
1950-1957	East Flevo	54,000		5.9		
1959-1968	South Flevo	43,000	1	3.5	800	900,000

The creation of these polders started, as usual, with the construction of the perimeter-dykes and the pumping stations, followed by pumping out the encircled water. These initial pumped-out water quantities were 700 million m^3 for the Wieringermeer, 1,500 million m^3 for the North East Polder, 1,600 million m^3 East Flevo, and 1,400 million m^3 water for the South Flevo Polder (Duin and Kaste, 1990). Subsequently the sowing of reed (using

Fig. 3. Zuyder Zee Werken/South Sea Project. a – Barrier Dam; b – Wieringermeer Polder; c – Northeast Polder; d – East Flevo Polder; e – South Flevo Polder; f – Dike Enkhuizen–Lelystad.

aeroplanes) was to enhance the evapo-transpiration, thereafter reed burning-off, followed by the creation of canals, ditches, furrows, drainage pipes. Ploughing was the next phase and planting crops improving the initial soil structure (rape-seed, winter wheat, barley) transformed the former sea-bed in highly productive farm land as a base for living, working, infrastructure, recreation & tourism and nature development.

Maximum depth of the new land is about 6.5 meter below Mean Sea Level. The situation of the polders below sea level gives rise to seepage of water. This fresh water has a considerable salt content depending on the proximity of the sea. The pumping stations of the Wieringermeer annually remove about 750,000 ton salt, while the combined East and South Flevo Polders remove about half of this quantity. One of the consequences of draining is subsidence of the reclaimed land due to compacting of clay and peat layers. Subsidence rate depends on the location of the polder and soil type (clay, light clay, loam, sand, peat). Average subsidence rate experienced in The Netherlands during the last millennium is in the order of 4 meters, with a maximum value of 1 meter per century. Another price to pay is the daily pumping of excess rain- and seepage water, resulting in sizeable maintenance pumping water quantities, which are in the same order of magnitude of the initial pumped-out water quantities. The total costs of the annual maintenance pumping amounts to 2 million US$, keeping the ground water at preferred levels in these four polders (1650 km^2).

It is interesting to note that in the North East Polder two former islands are incorporated: Urk (an important fishing village) and Schokland. The design of this polder and the Enclosure Dike/Barrier Dam was such that the functioning of the fishing port of Urk was guaranteed and even augmented. Of great interest is not only the creation of the new land but also the urban and regional planning and the use with regard to the new territories.

soil use in %	Wieringermeer Polder	North East Polder	East Flevoland	South Flevoland
agriculture	87	87	75	50
woodland & nature	3	5	11	18
cities	1	1	8	25
dikes, roads, water	9	7	6	7

A remarkable feature is the shift in functional uses of land area in time from 1930 to 1970: a large increase in the area of urbanisation and woodland & nature in the newest polder (South Flevoland), while the agricultural grounds are decreasing. This shift is related to the general urbanisation trends, the large increase of productivity of the horti- and agriculture per hectare during the last decades and the increasing appreciation for nature conservation, development and open-air recreation.

The newest polders are relatively close to the Rim City Holland. Here we have the fastest growing city in The Netherlands: Almere, before long reaching 250,000 inhabitants (in 1997 100,00 inhabitants), and the capital of the Province of Flevoland: Lelystad or Lelycity, named after the originator of the Zuyder Zee Project.

The remaining fresh water bodies are: IJssel Lake 127,000 ha, Marker Lake 64,500 ha, the border lakes: IJssel Lake 5,000 ha, Gooi and Eem Lake 4,500 ha. These are very important as fresh water reservoirs, for inland navigation and for recreation. Especially the border lakes fulfil an important role in the regulation and control of the ground water levels in the adjacent hinterland. The general principle of a polder system is depicted in Figure 4.

5. Impact of the Zuyder Zee Enclosure Dike on the Waddenzee

Through the work of the State Committee Lorentz, the impact of the construction of the Enclosure Dike, the Afsluitdijk, on the tidal velocities in the Waddenzee inlets was anticipated. However, the consequences in terms of geomorphology of the Waddenzee tidal basins could not be predicted. Of the tidal basins the basinal area of the Texel Inlet has been affected most, and it is yet not well-known whether this basin has reached a new dynamic equilibrium state.

The Texel Inlet basinal area has been reduced considerably, which implies that the new basin created a sand and silt deficit. At the same time, however, the increased tidal velocities in the inlet are indicating an increase in tidal prism. The consequences of these changes are complex. The increased tidal prism can be expected to imply a sand deficit for the ebb-tidal delta as well. Together with the sediment deficit of the basin this implied an increase in the sand demand from the coasts adjacent to the Texel Inlet. Reconstruction of the observations has shown (Stive et al., 1990) that the coasts of Texel and the north of North-Holland have been supplying the sediment required, but not in the amounts that the Texel Inlet basin required. As a consequence the ebb-tidal delta has played the role of sediment source in this complex interacting system, which has been leading to a structural decay of the ebb-tidal volume. This process may be expected to be ending at some stage, but our capabilities to predict this are rather limited and only slowly developing (Stive et al., 1996).

6. Discussion and Conclusion

In the 20th century, the history of water management and land reclamation in the Netherlands is dominated by two large-scale projects: the Zuyder Zee project and the Delta project. The Delta Project was an answer to the 1953 storm surge disaster in the Southwest of The Netherlands. It consisted of the complete or partial closure of the various estuaries by a series of dams and storm surge barriers, drastically shortening the coastline (see also Chapter 12 by Schiereck). Both projects have drastically changed the topography and the environment of the Netherlands over a surface of around 1,000,000 hectares: more than a quarter of the country. The decisions of 1918 and 1958 to carry out these projects were made almost unanimously; the decisive factor was formed by the attacks of the traditional enemy and friend of the Netherlands: the sea. Now, enough time has elapsed to conclude that both projects have been a success in resolving the main issues of protection against flooding and of management of water. On the other hand, now that these issues have been resolved, there is a growing awareness that the projects have resulted in a loss of natural values and of the naturals system's resilience. The coming decades are expected to be devoted to answering the societal and political wish towards restoring and where possible strengthening of the natural situation, which is expected to enhance the natural system's resilience.

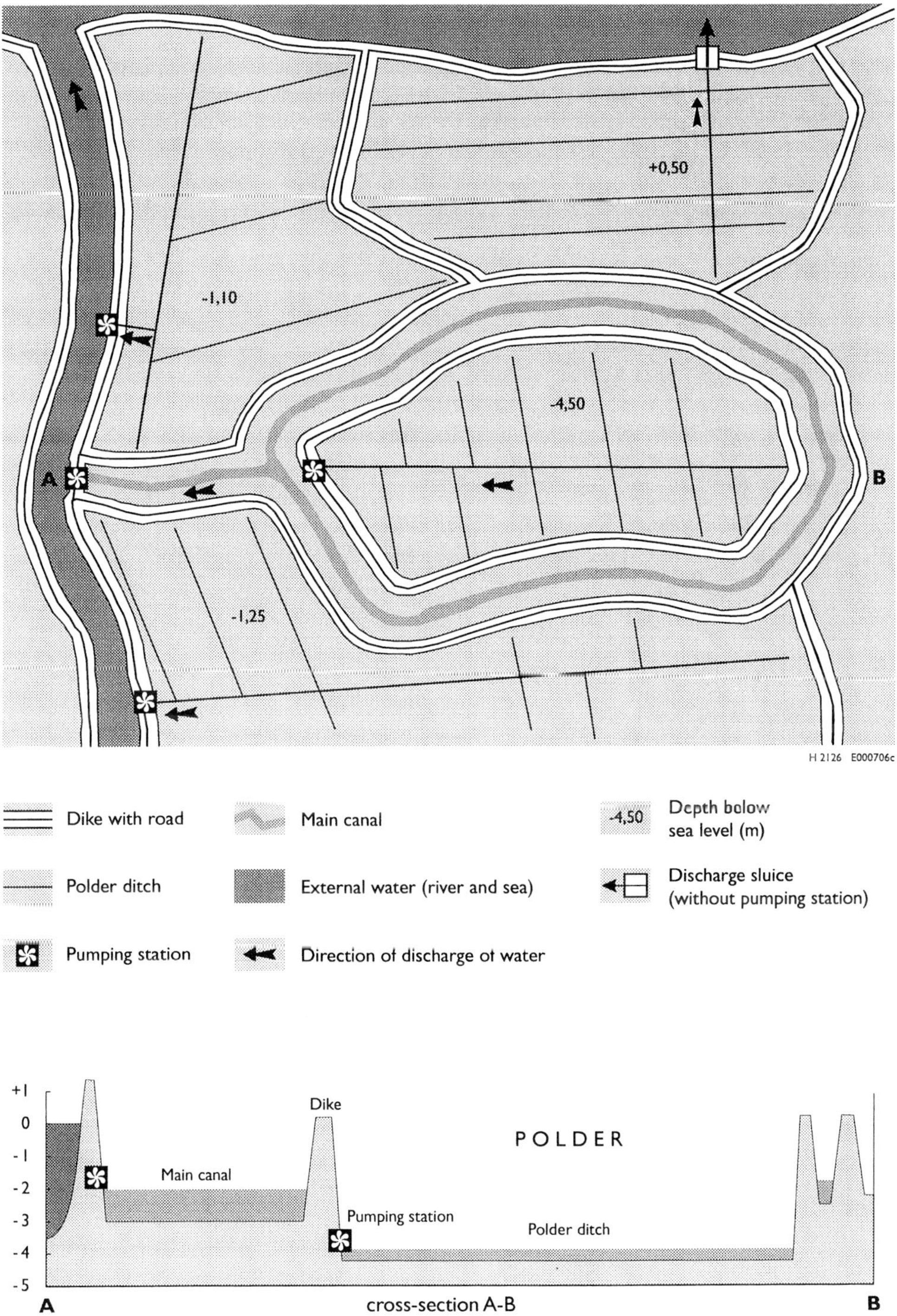

Fig. 4. General principle of polder systems.

Acknowledgements

Important parts of this text describing the history of the project and the construction of the enclosing dam are based on the comprehensive documentation on the Zuyder Zee works in the book "Man-made lowlands: a history of water management and land reclamation in the Netherlands", Ed: G.P. van de Ven, 1993. Other sources of importance were Bijker (1996) and Waterman (1997).

References

Beets, D.J., Van der Valk, L., and Stive, M.J.F., 1992. Holocene evolution of the coast of Holland, *Marine Geology*, 103, 423–443

Bijker, E.W.1996. History and heritage in coastal engineering in the Netherlands. In: *History and Heritage of Coastal Engineering*, Ed. N.C. Kraus, ASCE, pp. 390–412.

Staatscommissie Lorentz, 1926. Report in Dutch, Staatsdrukkerij.

Stive, M.J.F., Roelvink, J.A. and De Vriend, H.J., 1990. Large-scale coastal evolution concept, *Proc. 22nd Int. Conf. on Coastal Eng.*, ASCE, New York.

Stive, M.J.F., Capobianco, M., Wang, Z.B., Ruol, P. and Buijsman, M., 1996. Morphodynamics of a tidal lagoon and the adjacent coast, *8th International Conference on Physics of Estuaries and Coastal Zones*, The Hague, The Netherlands, September 9–12, 1996

Thijsse, J.Th., 1922. *Berechnung von Gezeitenwellen mit betrachtlicher Reibung*. Vortrage aus dem Gebiet der Hydro- und Aerodynamik, Innsbruck. Springer, Berlin.

Thijsse, J.Th., 1933. L'influence de la fermeture du Zuyderzee sur le regime des marees le long de cotes Neerlandaises. *Bulletin de l'Association permanente des congres de navigation*, No. 15.

Waterman, R.E., 1997. Large-scale coastal land reclamations, *Proceedings of Cosu 97 – Coastal Ocean Space Utilisation*.

THE THAMES BARRIER

CHARLES J. VOS[†]
Delft University of Technology
Faculty of Civil Engineering
Delft, The Netherlands

1. Historical Background

The history of the British East coast is different fro the continental coast along the North Sea, like in Holland, where over 2000 years the people lived in alluvial lowlands, that were frequently flooded and had to protect themselves anyhow against the Sea. The British topography provided sufficient high territory for people to live and to protect themselves against the sea until the era of the industrial revolution, when a fast increasing population started to live and work on land that was seriously under threat of being flooded by the Southern North Sea. Apart from this demographically defined need for an artificial improved coast, the sea itself has caused requirements for coastal improvements as well.

Since the end of the last Ice Age, some 20 000 years ago, there has been a steady increase in the level of high water in the tidal Estuary of the Thames, being created 8900 years ago, in relation to the adjoining land.

Archeological evidence indicates that at the time of the Roman Occupation of Britain the Thames in the region of London Bridge was not tidal and that under normal flow conditions the water level was about present mean tide level.

Subsequent history records some major floods. A surge tide in 1099, flooding of the palace of Westminster in 1236 and in 1663 Samuel Pepys, Secretary of the Admiralty, who left a famous diary, reported: "There was last night the highest tide that ever was remembered in England to have been in the river, all Whitehall having been drowned." Casualties on large scale or major damage however, as reported from floods in the past on the continent are not recorded.

Exceptional high water levels have gradually increased at London Bridge from nearly 4.27 meters above Ordnance Datum Newlyn in 1791 to 5.41 meters in 1953, the most recently recorded flood disaster.

The diagram, shown in Figure 1, shows the gradual increase in high tide levels since 1791. Various factors are responsible for the increase in high water level. These include a general rise in mean sea level, an increase in tidal range and settlement of the South-East corner of Britain resulting from long-term geological movements following the recession of the ice cap which covered the country as far south as the Thames Valley.

J. Chen et al. (eds.), Engineered Coasts, 291–308.
© 2002 *Kluwer Academic Publishers. Printed in the Netherlands.*

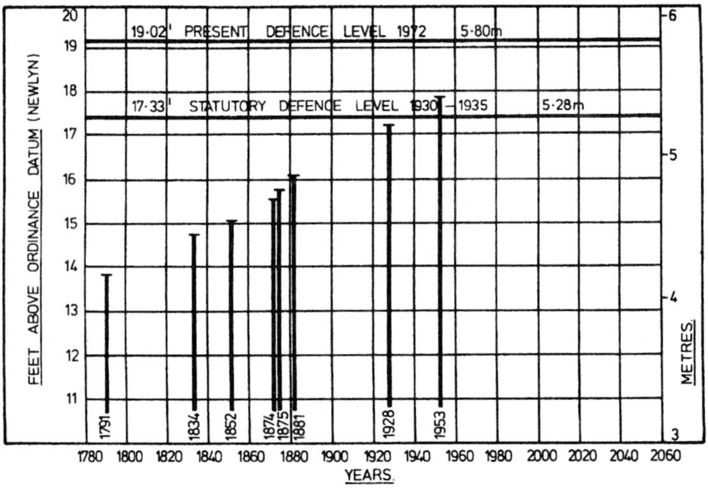

Fig. 1. Increasing high tides at London Bridge.

From time to time, high waters occur which greatly exceed the predicted tide levels. These variations are due to exceptional surges which occur when an area of low pressure moves in from the Atlantic Ocean and travels down the North Sea. Associated strong Northerly winds help to drive the raised water level into the narrowing southern end of the North Sea and the Thames. The increase of water levels caused by these meteorological conditions, by incidence with astronomical initiated surge tides, is a matter of pure chance. This seems to have taken place more or less in 1099, where the flood occurred one day after new moon.

Statutory powers to coordinate and implement protective works were first given to the Metropolitan Board of works in the 1879 Thames Flood Act. Extensive studies initiated by Government following the 1953 floods led to the 1972 Barrier Act which empowered the Greater London Council to construct the Thames Barrier.

In conclusion Figure 2 illustrates the area of Greater London before the construction of the Barrier and the associated downstream embankment raising along the Essex and Kent Coasts being at risk. The whole extent was in 1970 a densely built up industrial and residential area of 116 square kilometers containing hospitals, power stations, the underground railway system, power and telecommunication networks as well as a large multi-million population with its housing and social facilities.

2. Sites, Schemes and Final Site Selection

From 1956 onwards to 1961, under the authority of the ministry of housing and Local Government with the technical assistance of various consulting engineers, like Rendel Palmer & Tritton and Sir Bruce White Wolfe Barry & partners various schemes were produced in-

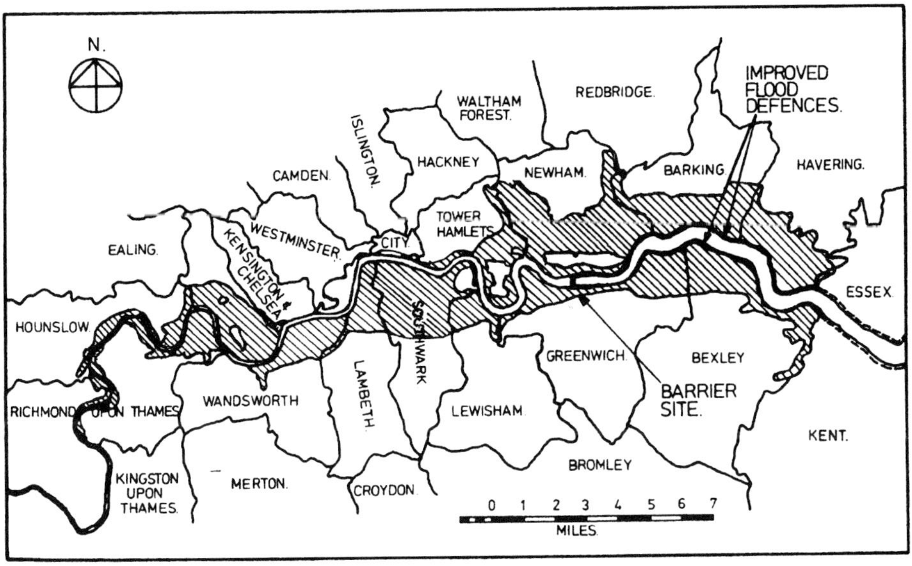

Fig. 2. Area affected by surge flood tide and the final Barrier position.

cluding one at Long Reach, 30 kilometers from London Bridge, providing two navigational openings 152 meter wide with an airdraft of 67 meters.

These schemes were considered by a Steering Group and were rejected by the navigational interests as not providing for safe two-way navigation for the number of ships then using the dock systems of the Port of London Authority and the numerous riverside berths. Considerable development work was done to deal with the objections and in the early 1960s, further schemes were prepared for a site near Crayfordness, just upstream of Long Reach, which provided a clear opening of some 427 meters to satisfy the navigational interests. The influence over the years of the navigational interests and shipping requirements resulted in a long gestation period for the final scheme, which in the end was not without its beneficial effect on requirements affecting the structure. When the studies started, all the London dock systems – St. Katherine's, East India and Millwall, Surrey and the Royals – were in use and there was substantial movement of many large vessels along the River.

Progressively trade has moved down the River to new facilities at Tilbury or away from the River with well known regrettable labor problems. Docks have closed and today they are virtually no longer used for handling of cargo.

As a consequence it was possible eventually to reduce the one navigational opening of 427 meter, once considered as a minimum for safe two-way traffic, to multiple 61 meter openings each for single-way navigation. Also the airdraft requirement became an anachronism, having fixed to suit large liner fixed masts, which was rendered as irrelevant in the adopted scheme. The huge impact of navigational considerations in the planning of a Barrier scheme was caused by an era where a dramatic increase in the size and consequentially, in the draft of ships and the ratio in between length of quay and storage and process area of

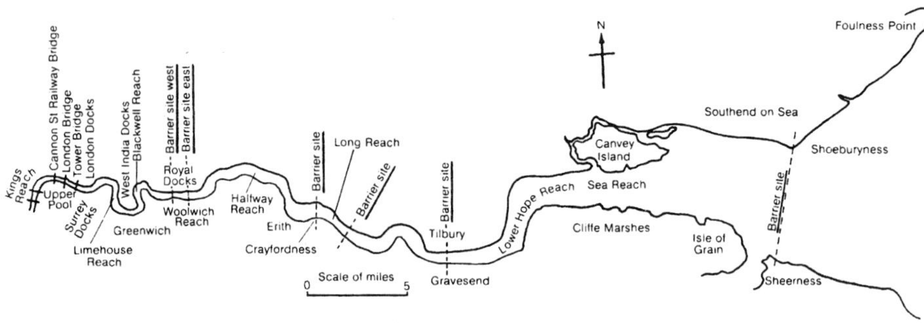

Fig. 3. Proposed Barrier sites.

the load were changing world-wide in a very fast way. The first development of container transport contributed as well to these planning problems. Following figures illustrate this trend.

When the Woolwich Reach site was finally selected, the Royal Docks were still in use, but as the site is upstream of the main lower entrance, vessels using that dock system did not necessarily have to pass the Barrier. In 1984, traffic through the barrier was of the order of 5800 sea-going ships per year as compared with 9400 in 1977. In addition, some 40 000 rivercraft, pleasure boats and river tankers, pass through the Barrier each year. The daily total is about 140.

In 1967 the Greater London Council (GLC) took over responsibility for the project. At that time the cost of possible damage to London was estimated at £3 billion and it was assessed that the capital would be out of action as a communication center for up to one year should severe flooding occur.

The GLC set up working parties of experts, including numerous public bodies, like: the Port of London Authority, Hydraulics Research Stations, the consulting firms Rendal, Palmer & Tritton and Sir Bruce White Wolfe Barry and Partners again, to continue the investigations. The investigations included hydraulic and siltation studies, mathematical model studies, hydraulic characteristics of various gate designs and many other detailed studies. All kind of solutions, ranging from raising the flood banks along the estuary, the construction of an outer estuary with an overspill of the incoming surge tide onto marshland and the construction of different types of barrier and barrage structures were considered. A barrier distinguishes itself from a barrage as being a structure normally kept open and not causing any substantial change in the river tidal regime, only being closed to hold up an incoming surge tide. Combinations of these solutions were of course part of the work being performed.

Considering factors of standard of protection provided, speed of construction, costs, environment, impact and effects on navigation and river regime, the choice of a barrier appeared to be the best solution. This conclusion started a second scheme of investigations. Some 40 further schemes for sites between Cannon Street and Gravesend, respectively 7 km upstream and 8 km downstream from the final site, were prepared and evaluated.

Woolwich became the preferred site and a scheme having 137-m opening, considered adequate for two way navigation, was prepared.

Table 1. Evaluation of main characteristics of different Barrier sites.

Site	Length flood defenses downstream	Minimum opening size m	Foundation type	Gate waterway HWOST m	Width bed HWOST m	Depth to cross section sq m	Area in
Cannon Str.	44	46	LC	1 2 3	213	11.0	1950
Blackwall	35	61	W & R	1 2 3 4	350	13.4	3900
Woolwich	30	61	C & TS	1 2 3 4	550	13.7	4830
Halfway	21	305	C	3 4	670	15.5	8180
Crayfordness	14	427	C	3 4	790	17.4	10130
Gravesend	0	427	C	3 4	853	20.4	12360

Gate Type: 1 Drop Gate Foundations, 2 Rising Sector, 3 Drum gate, 4 Retractable.
LC: London Clay, W & R: Woolwich and reading Beds, C: Chalk, TS: Thanet sand, HWOST, Typical Mean spring Tide (2.86 m high-water at Southend)

Table 2. Comparative costs of different solutions on different sites for a Thames flood protection scheme in £ million.

Site cost	Total	Structure ment	Develop- up stream	Banks down stream	Banks	Land sation	Compen- delay	Shipping
Bank Reconstr.	65				55		10	
Cannon	46.5	45		2	33	1	6	
Blackwall	47	150		2	24.5	1	4.5	
Woolwich	45	22.5		2	14.5	1	2.5	2.5
Halfway	50	29	20	8	7	1	1.5	1.5
Crayfordness	48	33	20	8	2	0.5	0.5	2.0
Gravesend	53	36	20	8	2	0.5	0.5	4.0

Table 1 illustrates the basic findings that had to be generated for further evaluation for the different sites.

After establishing these site characteristics a cost evaluation could be made, being summarized in Table 2. The result shows that a range of sites using such methods of evaluation keeps being in the same magnitude of costs. This meant that although the Woolwich site proved to be challenging, further investigations had to be carried through, to avoid risks. It had to be proven that a sound foundation was possible. Navigation required a long straight approach from either side in order to avoid risks of collisions. For reason of economy a minimum amount of navigation openings was required. For construction and positioning of permanent shore installations both sides of the river had to offer space and the site should be such that the environmental impact of the barrier would be acceptable. Last but not least construction costs and time of the project had to be a minimum.

Taking into account all these considerations, dropping the idea of having raised banks only, being a solution with the highest costs and the longest time for realization and introducing a Western and an Eastern location at Woolwich, a final selection could be made. Data used for this final site selection are summarized in Table 3.

Table 3. Summary of final considerations for the selection of a site for the Thames Barrier.

Site	Foundation	Approach	Navig. opening m	Land	Constr. time yrs	Total costs % highest	Environ-mental merits
Cannon Street	+	+	46	–	7	88	+
Blackwall	–	–	61	–	6	89	–
Woolwich West	++	++	61	++	5	85	++
Woolwich East	++	+	61	–	5	85	++
Halfway	–	+	305	–	7	94	++
Crayfordness	++	–	427	+	7	90	++
Gravesend	++	+	427	–	7	100	+

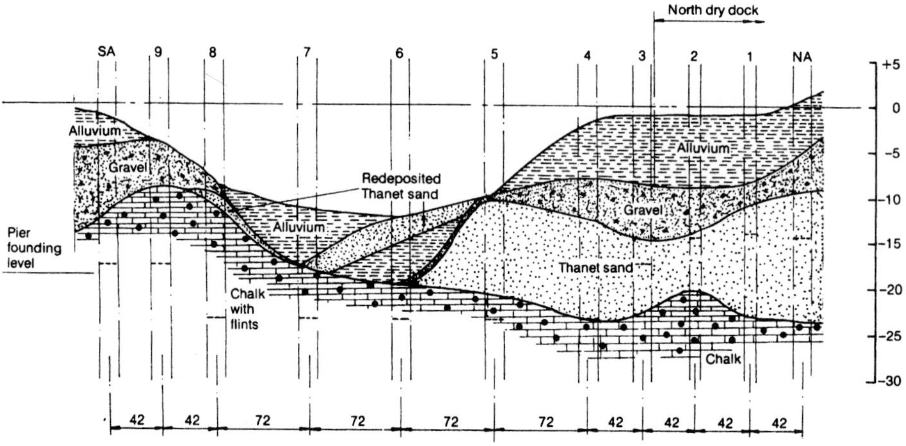

Fig. 4. Geological section across the Barrier axis.

3. Site Characteristics, Further Investigations and Their Influence on the Design

Having selected the site, a final geotechnical investigation had to be carried out, particularly in relation to the design of the foundations to the piers and the abutments. Figure 3 shows a geological cross section across the barrier axis, showing main characteristics of the soil. As can be seen from Figure 4, the northern end of the Barrier is founded in Thanet Sand, being fine and dense, with a relative high angle of friction. The southern part of the barrier is founded on Chalk. Investigations showed that there was a general absence of solution and solifluction features in the chalk The chalk at the southern end of the Site is frost shattered to a depth of a few meters below the surface, but where the Chalk is overlain by Thanet sand frost shattering was found to be absent. Several small faults were also located at the site during final geotechnical investigations.

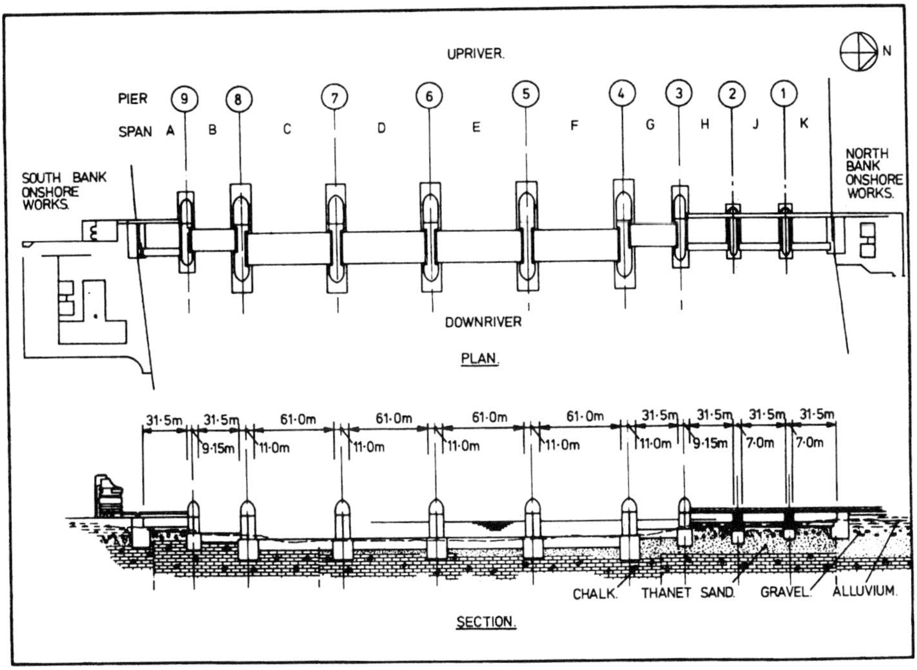

Fig. 5. General arrangement of the Barrier and its foundation.

The outcome of these geotechnical investigations were combined with possible solutions for the actual foundation. Deep sheetpiled cofferdams, the use of bored piles or diaphragm walls combined with shallower sheetpiled cofferdams and the use of open dredged caissons were considered. Other methods of foundation were immediately discarded because of their obvious impracticality or their disruption of the Thames bottom and the final foundation layers. The open caisson method, although proposed in one of the later tenders, raised many problems with draft during construction and placing, as well as structural problems for transferring forces from the complex shaped upper part of the structure to the foundation. A solution with a shallow construction pit combined with bored piles and diaphragm walls just did not prove to be economical. Consequentially a layout and cross section of the foundation of the piers as shown in Figure 5 was adopted.

Apart from the geotechnical investigations, leading to the design of the foundation, hydraulic studies, using physical and mathematical studies were required to establish the design load on the Barrier and the establishment of the riverbed protection necessary to prevent local scour at the Barrier. These studies were also required to substantiate the effect of the Barrier on the existing estuary regime.

Four large scale estuary surveys, two smaller scale local surveys of the reaches adjacent to the Barrier site, in-situ silt settling tests and the establishment and operation of stations at six locations for continuously monitoring the suspended solids content of the river were the main objects of field measurements.

Table 4. Design levels for the Thames Barrier.

	Design surge	Design reverse head	Extreme surge
Downriver water level	+6.9 m	−1.8 m	+7.2 m
Upriver water level	−1.5 m	+4.3 m	−2.7 m
Differential head	8.4 m	6.1 m	9.9 m

For physical model testing two different models were used. These were, the 'Thames Estuary Model' extending from Teddington to Southend on a horizontal scale of 1/600 and a vertical scale of 1/60 and the 'Silvertown Barrier model' which reproduced the tidal river between Teddington and Tilbury to horizontal and vertical scales of 1/300 and 1/60 respectively. Two different Barrier types with different type of doors and the position of the Barrier at four different sites were tested here.

Site investigations, physical model testing and additional mathematical models answered questions regarding tide control, flow distribution and sediment distribution as required for the assessment of the effects of the Barrier within the range of the design criteria. These were finally fixed as follows in the Parliamentary Act in 1972 that gave authority to the GLC to proceed with detailed design and construction.

- A 1000 year return flood relevant for conditions existing up to the year 2030.
- A surge tide level of 2 m above the 1953 level with a differential water level of 9.9. m across the gate.
- A reverse head water level differential of 6.1 m such as would occur if the gate had to be left closed during the low tide period of a long period surge.
- A navigational depth of 9.25 m at mean tide level to match the guaranteed level in the Woolwich reach.
- Minimum disturbance to the river bed including protection against failure of a gate to close.
- Maintenance of navigation at all times during construction.
- restriction of the waterway cross-section not to exceed 25% for the finished structure and 30% during construction.
- No restriction in time on the strategy of actual closing the Barrier anticipating a surge.

This resulted in the design levels at Woolwich Reach as represented in Table 4.

4. Some Important Components

4.1. THE RISING SECTOR GATE

The concept of the gate, illustrated in Figure 6, the so-called rising sector gate, was invented and detailed during the design period of the Barrier. The main features are that this door can lie in the bed of the river under normal conditions, be rotated into a vertical position at times

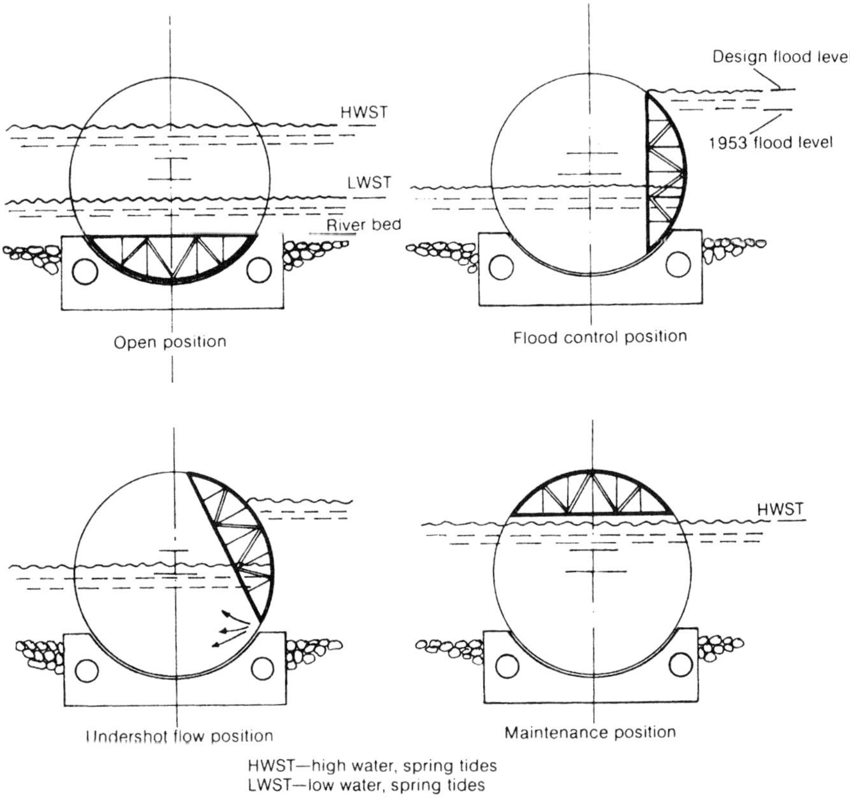

Fig. 6. Basic concept of rising sector gate operations.

of flood and be rotated completely above the water for maintenance operations. It may be obvious but not always well understood today, that intensive high technology was needed to develop this simple idea to the finished form.

The rising sector gate could not be used to span wider openings previously required before the decision of closing the Surrey Docks was taken in 1970, as the ratio of diameter to span would not fit the practical sizes required for the Barrier. The radius of the gates is 12.20 m and the height of the door when in vertical position is 20.12 m with a largest depth of 5.3 m. The opening span for the four main navigation openings of 61 m (200 ft) is the same as the main span of the Tower Bridge through which ocean-going ships had for many years safely passed into the Pool of London.

The openings on both sides of the main openings have rising sector gates as well, although the span is reduced to 31.5 m. The remaining openings, three at the South and one at the North are being closed by conventional falling radial gates with a span of 31.5 m each as well. The steel sector gates are attached against circular gate arms at both sides of the gate. These cellular steel structures have a radius of 12.2 m and a thickness of 1.5 m. They house the counterweight to the door of 450 tons of cast iron per arm. The arms are fixed by hinges

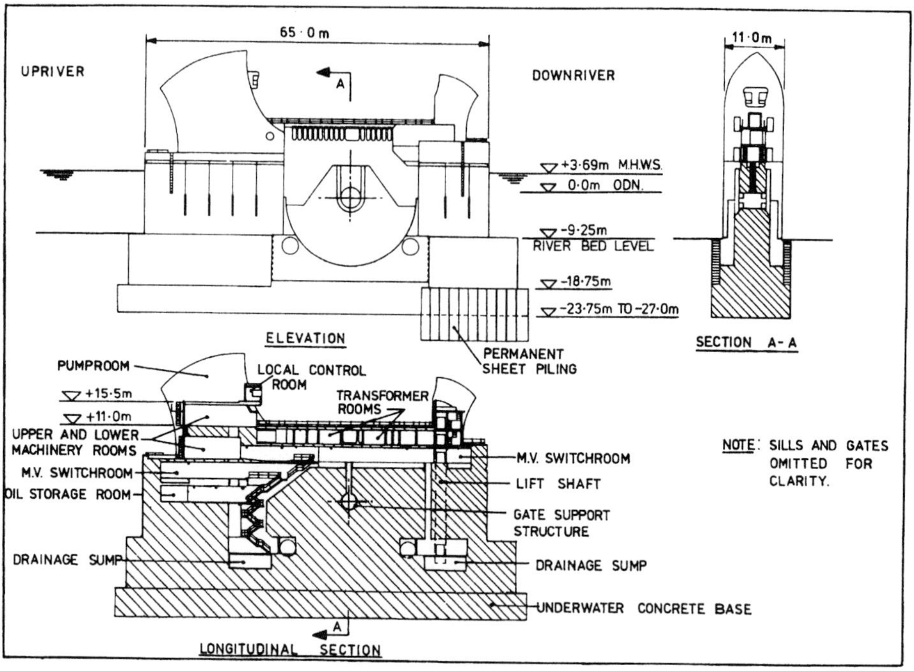

Fig. 7. Cross sections and vieuw on Thames Barrier Piers no. 4 thru. 8.

to the pier and a hinged system of beams and rods to two hydraulic jacks per gate-arm. These jacks can make a stroke 3.13 m and develop a force of 1526 tons.

The sector gates are protected, when in downward position, by concrete sills, with a scallop on top to house the door, spanning in between the piers.

An often heard criticism against doors that are resting on the bottom, regarding the vulnerability for damage and foreign objects and the difficulty of maintenance, is mainly overcome in this concept. Maintenance can easily be performed by rotating the door to the upward position. Foreign objects, like silt and small bottom transported objects can be flushed by closing all doors except one during high tide. Larger objects will hardly be capable of damaging the sector-doors, as they are protected by the reinforced concrete sills.

A real test of this philosophy took place on the 27th October 1997 at 6.50 am, when a Dutch Hopper dredger “MV Sand Kite”, in fog, with poor visibility collided against pier 5 and sank on gate F with 5000 tons of sand and gravel on board. The barriers Pier 5 bull-nose on the down stream side absorbed the impact and showed scratches in the concrete, just as Pier 4, but no serious damage was observed. The vessel was refloated by a salvage crew after discharging most of its load. About 2000 tons of sand and gravel remained however on the flat upturned face of Gate F. Flushing by closing the adjacent doors gave only poor results. Diver inspection further showed gravel wedged in between the concrete sill and the curved outer surface of Gate F. Suction dredging had to be carried out very careful as no damage to the scour protection was allowed. So called shovel dredging for a first clean up and a completion by hydro-jet dredging cleaned up even most of the material entered in between the gate and the sill. Doing so the gate could be raised for further inspection 10 days later on November 7th. As expected severe damage was observed to the paint surface only, which was repaired next spring. All with all a demonstration that the design philosophy was right.

4.2. THE PIERS

The piers and the abutment structures are generally of mass concrete construction. This is required in order to mobilize sufficient weight and bearing surface to resist the horizontal forces from the gates and to transfer these forces within the allowable limits to the chalk and the Thanet-sand foundation strata. For this reason the piers are constructed within a 15 m wide and 74 m long cofferdam in which a solid underwater concrete plug of 5 m thickness was placed. On top of this, a narrower reinforced concrete structure is constructed in the dry, to support the sills, the gate support structure, the gate operating machinery and to house the various operating facilities.

4.3. THE THAMES BARRIER BED PROTECTION

A most essential part of the design was of course the geometrical extend and detailing of the river bed protection against scouring from the river bed. Partial closure with the four large gates operating undershot and failure of any of the rising sector gates to close were the main conditions to be investigated, as these were supposed to raise the largest velocities at bottom level to initiate scour.

Two models, one of 1:25, modeling one pier flanked by two half sector gates and one of 1:50, representing the whole barrier including 1 km up and downstream of the Thames, were actually made and tested. The amount of scour of the bed without protection, the design of bed protection and scouring beyond the protected area were the main goals to be achieved. By far, the condition of one door failing to close was the governing design parameter for the bed protection, as peak velocities of 9 m/sec. were observed. After 30 tests with the models a design could be finalized. Figure 8 shows a final version of the protection which was adopted after some considerations for actual placing the material were implemented after consulting the specialized contractors to carry out the job.

5. Architecture

The architectural treatment of the piers of the barrier has been generally favorably received. To obtain this reaction was the intention from the GLC right from the beginning, as is clear from a statement in 1973. There it was stated, that: 'The Barrier will be one of Britain's finest engineering achievements as well as a tourist attraction. So in visual as well as in practical terms it will be something of which every Londoner can be proud'. This statement not only concerned the Barrier structure itself, but the vicinity of the Barrier, as well. It was chosen to give each pier a clear and separate identity so that the presence of the navigational channels will always be implied, even when they are not actually visible. This was also to ensure that the form of each pier will contribute to the overall visual effect of the Barrier as a whole.

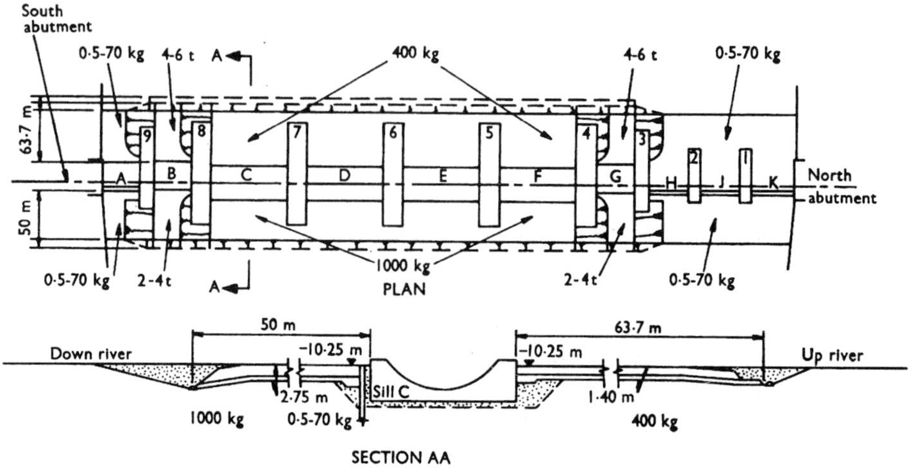

Fig. 8. Thames Barrier bed protection design layout.

Fig. 9. The Thames Barrier as a unit spanning 520 m of river.

The requirement of the superstructures to house the primary gate operating machinery needed a covered span of considerable volume extending to the full width of the piers. A double curved set of shells was found to be a good solution. This introduced a recognizable form, which could be incorporated in the shore installations, the Barrier obtained a memorable form. The changing patterns in the silhouette of the piers introduce a variety of interest when viewed by river users and pedestrians at shore. The large linking mechanism which operates the gate-arms required no protective enclosure and is shown in its dynamic engi-

Fig. 10. A Thames Barrier Pier as a recognizable unit, with a sector door in flood control position.

neering forms, which provides a dramatic contrast to the static forms of the superstructure, particularly when the gates are seen in motion during operation of the Barrier.

The shells are detailed as a timber structure clad in stainless steel sheeting with profiled aluminum gable ends, which offered the best prospects for reliability, durability and appearance.

6. Construction and Contracts

6.1. THE CONSTRUCTION PROCESS

Although the design almost dictated the way construction had to take place, certain changes were made before and during actual construction.

Most important for construction was of course the way safety of shipping through the works under construction was scheduled. The basic idea of having two 60 m openings at

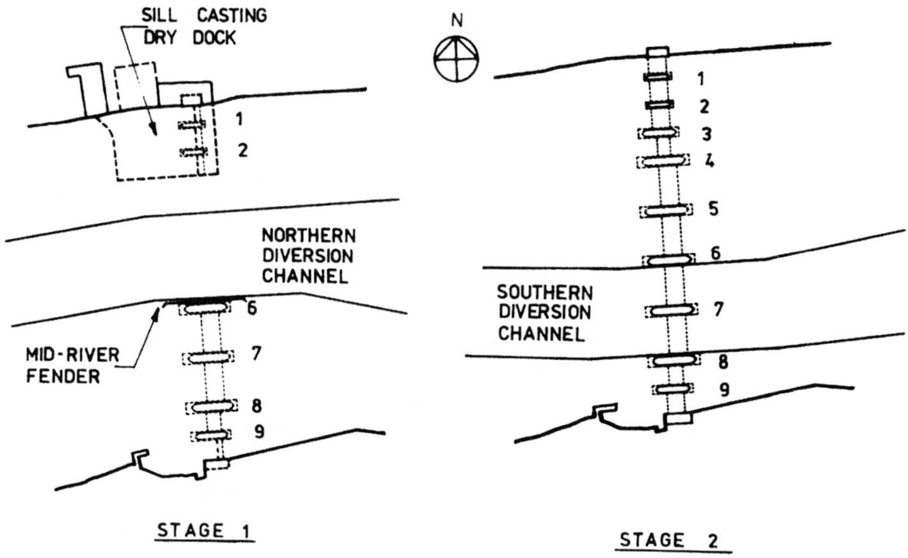

Fig. 11. Stage 1 and stage 2 of constructing the Barrier.

least was met by the idea of constructing and finishing the bank and the first four piers from the south.

After completion construction would start from the North. In practice, just because serious delays in the start of construction, a modified scheme was realized, allowing for less stringent requirements to the width of shipping lanes to be maintained. The result of that is shown in Figure 11.

After dredging the Northern channel, construction started with the heavy pile driving operation of the Southern piers, where the distance from the top of the pile to the bottom of the excavation was 30 m. These piers were then excavated and strengthened with 4 layers of struts from which 3 had to be placed under water with the assistance of divers.

The bottom was cleaned and a 5 m thick layer of under water concrete was cast. After the installation of a pressure relief system underneath this concrete prop the cofferdam was pumped dry and the remaining reinforced concrete work on the piers could be made. This procedure is presented in Figure 12. The sheetpiles being used were the largest ones available at the time in the UK. For the Northern piers, more heavy and stiff German Peiner piles were used, allowing the use of only one stiffening strut frame, which led to a considerable gain in construction time and productivity. After completion of the Southern Piers, the sills, being prefabricated in a construction pit at the North bank, in which the Northern abutment and the first two piers were constructed as well, were floated to their position between the piers. They were suspended from a hydraulic jacking system attached to the piers, immersed and lowered with the aid of the hydraulic jacks until they were positioned on their bearings at the piers. The gate arms and the steel sector doors were lifted in by heavy floating shear legs

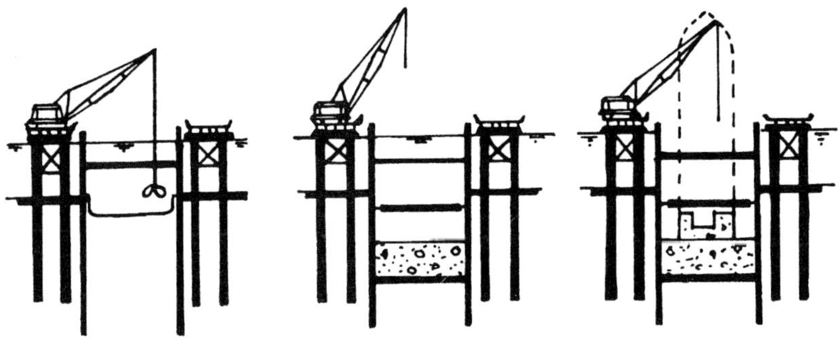

Fig. 12. Stages in cofferdam construction.

and fitted to the piers. In the mean time construction of the bed protection with a variety of dredging equipment was carried out.

The total construction involved 36 000 tons of steel, including 10 000 tons of reinforcement and 214 000 m^3 of concrete.

6.2. SCHEDULE, CONTRACTS AND COSTS

A tender for the civil engineering works was issued in June 1973 to 5 international joint ventures of civil engineering contractors. Two of them withdrew and only 3 heavy qualified tenders were submitted in November 1973. Negotiations took more time than foreseen, which caused the withdrawal of a 3rd contractor. No contractor would in fact take a risk for such a large job, lasting at least 5 years under normal civil engineering conditions of contract. At that time there was a period of high inflation and considerable industrial unrest. A contract with CTH, a joint venture from two British contractors, Costain and Tarmac and a Dutch contractor, HBM, was finally signed in July 1974 for starting the job on site in January 1995. Inflation was fully compensated and only a limited risk for industrial disputes was accepted by the contractor.

After two years the contract would be reviewed. All joint ventures tendering had a Dutch contractor in their team, as the heavy pile driving required, the placing of the sills like immersed tunnels, and the installation of bed protection were jobs, uncommon in the UK, but familiar in Dutch Civil Engineering practice.

Summer 1976, at the time of the review, progress was poor. After 18 months working the schedule was 6 months behind program, most of it due to low productivity. All parties involved found a new way out and work was organized slightly different in order to avoid industrial disputes as much as possible. Late 1978 a surge tide only 0.2 m lower than 1953 at London Bridge occurred and the contract was changed again in order to obtain a strong incentive for the Joint Venture to get the Barrier operational before the 1982–1983 flood season. Substantial changes in the construction of the Northern Piers proposed by the con-

Fig. 13. Construction in progress.

tractor were effectuated, which resulted in more speed and better productivity. This enabled a first operation of the Barrier in 1982, still far ahead of the official opening by H.M. Queen Elisabeth II in May 1984.

Gates and machinery were tendered in two separate contracts, but were let to one consortium in the end. They were fabricated at yards elsewhere in the UK and transported on barges to the site for erection by shear-legs.

Just as time, it is obvious that costs ran out of schedule as well. The original costs of the Barrier at October 1973 price level were £110.7 million. The completion costs of the Barrier were about £440 million, which is about 4 times as much! Of the increase of £329.3 million, it can be calculated from the price index figures used that about 70% has been due to inflation. Of the remaining increased cost, about 5% has been due to refinement of design, like more reinforcement in certain parts of the concrete, the use of stairs instead of ladders due to new safety legislation and increases in foundation depths after findings during pile-driving and excavation. About 10% can be attributed to construction difficulties not covered in the original tender, mainly related to unexpected problems because of soil conditions met. The remaining 15% can be attributed to low productivity. This covers the whole range from optimistic output figures in the original tender, to overt and covert industrial action and shortcomings in the construction methods and organization of the Southern piers.

7. Operating Now and Into the Future

Apart from the accident with the hopper dredger in 1997, no serious incidents took place. Most years the Barrier had to be closed only once or not at all, like in 1984, 1986, 1991 and 1997. This was quite different in 1990 and 1993 when the Barrier had to be closed 6 times and with 1995 with 5 and 1996 with four closures. These figures do not include closures for testing and training. From the information collected during closures it became evident that more data have to be recorded during different combinations of tide and flow in order to improve the strategy for closing of the Barrier, which is of course a function of River Thames flow and the expected tidal high water at Southend.

During the design of the Barrier, design levels were based on figures that could be expected by 2030. As the technical design live of the Barrier reaches far beyond that point, it became an urgent task to explore design criteria after 2030.

Sea level rise is the first one to be studied further. Figures of 6 mm per year until 2030 and 7 mm annual after that until 2100 are agreed estimates for man made global warming alone. Nevertheless a figure of 8 mm until 2030 and 10 mm thereafter is advised as more realistic as these figures take account for historic rise and the shape of the Thames estuary as well. This means that without any change to the defense-works the standard of defense will fall below the 1:1000 year event by 2030.

Another effect of the rising tide levels will be the higher frequency of closing the Barrier. Upstream defenses as raised during the 1970s as temporary protection will be sufficient until the middle of the 21st century. Downstream defenses will have to be designed and executed for a substantial higher level. This means an increase or a total replacement. The Thames Barrier itself will require a higher defense level. This may be achieved by increasing the height of the sills. A large program of modeling and investigations is planned in order to define the works to be carried out in the 21st century in vieuw of possible refined operational procedures being possible because of more sophisticated data recording and analysis. It is envisaged that the strategy resulting from the present report and the proposed investigations has to be reviewed at 5 or 10 year intervals.

8. Concluding Remarks

It is obvious that a first closure of the Thames Barrier in 1982 after a flood-disaster in 1953, which could easily have had a far more disastrous character, like in the Netherlands, is not the way people think it should have been. The government however has only limited power of doing something as it can only implement acts of parliament. Preventing of flooding is mainly a matter of local authorities. Only in cases where, like in 1953, a flood ran across the boundaries of many local authorities and the damage is high, the Central government can take initiatives. But to actually make a plan, design it, perform sufficient investigations within a world where demands, like for navigational interests are changing, is time consuming. On top of that the size of such a job is great and risks are high, because of extrapolation of technology and changing legislation during design and construction. It was a great effort to have Barrier be built. It was not because of an agreement on the details of a cost-benefit analysis, or because of a general agreement on probabilistic levels of safety to be achieved, but it was the wish to do something about the danger of a flood. 'DO IT' was the answer of an eminent mathematician, professor Bondi, bypassing the morass of statistics being created by all scientists involved. This part of the history is quite relevant. In the Netherlands action was taken by the Delta Plan. Shortly after 1953 the Dutch were already constructing their new artificial coastal defense works, being initiated by an organization, RWS, whose traditional task it is to keep the country dry and accessible. But in places like Petersburg and Venice, where problems of the same magnitude are present, no effective Barrier has yet been completed.

References

1. Cox, P.A. and Horner R.W. et al; *Thames Barrier Design*, Proceedings of the Conference held in London on 5 October, 1977. The Institute of Civil Engineers 1977. ISBN: 0-7277-0057-X.
2. Rendal Palmer & Tritton; "The Thames Barrier"; Company brochure 1985.
3. GLC (Greater London Council) Leaflet: Thames Barrier; 7168 1434 X.
4. Vos Ch. J.; "Flood prevention schemes from raising banks to ecological systems", 75th Anniversary of the Institute of Civil Engineers, Institute of Structural Engineers; London, July 1984.
5. CTH (Costain, Tarmac, HBM) "The Thames Barrier – Construction progress", leaflet from the Contracting Joint Venture CTH.
6. Gilbert S., Horner R.; *The Thames Barrier*, Thomas Telford NRA edition 1984, ISBN 0 7277 0249 1.
7. RKL-ARUP; "London Tidal Defenses, The development of a strategy for the 21st century", Environment Agency; London; August 1998.
8. Wilkes David J.; "MV SAND KITE collision with Thames Barrier", Information paper for Thames Regional Flood Defense committee; London, November 1997.

Subject Index

Author Index

(authors cited in the text, for other authors see the References at the end of each chapter)